EARTH SCIENCE

New Methods and Studies

EARTH SCIENCE
New Methods and Studies

Edited By
Roy H. Williams
Astronomer and Educator, Kopernik Observatory
and Science Center, Vestal, New York, U.S.A.

Apple Academic Press
TORONTO NEW YORK

Apple Academic Press Inc.
3333 Mistwell Crescent
Oakville, ON L6L 0A2
Canada

Apple Academic Press Inc.
1613 Beaver Dam Road, Suite # 104
Point Pleasant, NJ 08742
USA

Exclusive worldwide distribution by CRC Press, a Taylor & Francis Group

International Standard Book Number: 978-1-926692-57-9 (Hardback)

Library and Archives Canada Cataloguing in Publication

Earth science: new methods and studies/[edited by] Roy H. Williams.

Includes index.
ISBN 978-1-926692-57-9
1. Earth sciences–Textbooks. I. Williams, Roy H., 1966-

QE26.3.E27 2011 550 C2011-905381-0

Preface

Earth science is a broad field of study that encompasses many different disciplines. Fields of scientific study include meteorology, climatology, and geology. Included within geology itself are a large number of subfields, including petrology, stratigraphy, hydrology, oceanography, paleontology, structural geology, and planetary science. Furthermore, geoscientists utilize the broader hard sciences of chemistry and physics to advance their discoveries within all these disciplines. The integrated nature of many of these fields within the Earth sciences calls for the ability of scientists to use the latest technology.

Earth sciences careers can be found within university institutions and government agencies such as the United States Geological Survey (USGS) and the National Atmospheric and Oceanic Administration (NOAA). Geologists in the United States primarily work in industry. Most are working for the petroleum and natural gas industry. The goal of these geologists is to discover new oil and gas reserves by comparing the stratigraphic similarities of known oil and natural gas reserves. More recently, there has been a push for more offshore drilling. Searching for offshore oil reserves requires deep well cores that are examined for evidence of oil and natural gas. Surface geochemistry can also be used to find tell-tale signs of hydrocarbons leaking through the sea floor (called seeping), which can even hint at the type of source rock. Finally, with good seismic data, geologists can pinpoint the actual location of new oil reserves.

Hydrologists search for sources of fresh water and study the distribution of water within the water cycle. They attempt to mitigate the possibilities of floods and landslides, and they assess risk of drought. Applications of these forecasts may involve agriculture, forestry, and planning of urban development.

Another group of geologists in the United States are teachers in public schools and colleges. These teachers might not have all the rigorous geologic course work, but they must be trained to meet the needs of today's students. In high schools, they can teach a variety of classes such as Earth science, astronomy, and general science. At the university level, professors teach an assortment of introductory and advanced courses in geology. Teaching Earth science has been greatly assisted by recent technological advances, including the Internet, computer simulations, and smartboards.

The importance of the Earth sciences in predicting weather and climate, tracking pollution, drilling for petroleum, evaluation of soil, and monitoring ground water is paramount in modern society. Our everyday movement around the planet is assisted by accurate forecasting and real-time data to assist airliners, ships, and other forms of transportation around the globe. Issues such as global warming have caused world governments to begin dealing with carbon emissions. However, at the same time, geologists are searching for new reserves of oil and natural gas to quench the thirst of the developed world. Advances in satellite monitoring of the planet can help governments deal with natural disasters, human-caused disasters, and even a chemical or radiological terrorist attack.

Marine oil spills caused by either oil tankers or offshore drilling platforms can cause vast areas of the ocean to be contaminated. Accurate monitoring can bring responders to spill areas with the necessary equipment to protect vital coastlines and ocean life. Remote sensing instrumentation uses various wavelengths of spectra, such as microwave, thermal, near infrared, visible, and near ultraviolet, to view these spills. However, remote sensing from satellites does have its limitations due to cloud cover, and inclement weather can obscure remote sensing techniques, making aerial monitoring the only option.

Future trends will rely on the interconnectedness of the Internet, along with open source code, to bring data to various corners of the globe. The use of open source digital libraries connected via the Internet is called cyber infrastructure. Important tools include Linux, Apache, MySQL, and other open source coding platforms that will act as a catalyst in the spreading of information. Open source software such as Project for a Network Data Access Protocol (OPeNDAP) and Thematic Real-time Environmental Distributed Data Services are leading examples of open source software used by geoscientists around the world. Open source code is inexpensive and allows people to manipulate it to their needs. One prime example of open source protocol is a digital library sponsored by the National Science Foundation (NSF) called Unidata Internet Data Distribution System (IDD). There are over 170 colleges and universities participating in this online library. These cyber networks allow for the collection of geologic data planet-wide and to effectively distribute information to its users. Real-time geoscience data, such as seismic data, oceanic metrics, airborne pollution, and satellite images of various spectra wavelength, assist in a multitude of scientific research. In addition, improvements to early warning systems for natural disasters and other events could be tied into this cyberinfrastructure, allowing for a much faster reaction time from governments and the military.

— Roy H. Williams

List of Contributors

L. Alfonsi
National Institute of Geophysics and Volcanology, via di Vigna Murata 605, 00143 Rome, Italy.

L. Badiali
National Institute of Geophysics and Volcanology, via di Vigna Murata 605, 00143 Rome, Italy.

E. Baroux
National Institute of Geophysics and Volcanology, via di Vigna Murata 605, 00143 Rome, Italy.

B. Baum
Space Science and Engineering Center, 122 W Dayton Street, Madison, WI 53706, USA.

S. Berthier
Service d'Aéronomie du CNRS, Institut Pierre-Simon-Laplace, Université Pierre-et-Marie-Curie, 4, Place Jussieu—75252 Paris Cedex 05, France.
Sciences Laboratory of Climate and the Environment, joint laboratory CEA-CNRS, F-91191 Gif-sur-Yvette, France.

D. K. Bird
Department of Environment and Geography, Macquarie University, North Ryde, Sydney, NSW 2109, Australia.
Department of Geography and Tourism, Faculty of Life and Environmental Sciences, Askja, University of Iceland, 101 Reykjavík, Iceland.

A. Bono
National Institute of Geophysics and Volcanology, via di Vigna Murata 605, 00143 Rome, Italy.

P. Burrato
National Institute of Geophysics and Volcanology, via di Vigna Murata 605, 00143 Rome, Italy.

P. Casale
National Institute of Geophysics and Volcanology, via di Vigna Murata 605, 00143 Rome, Italy.

C. Castellano
National Institute of Geophysics and Volcanology, via di Vigna Murata 605, 00143 Rome, Italy.

P. Chazette
Sciences Laboratory of Climate and the Environment, joint laboratory CEA-CNRS, F-91191 Gif-sur-Yvette, France.

Hone-Jay Chu
Department of Bioenvironmental Systems Engineering, National Taiwan University, 1, Sec. 4, Roosevelt Rd., Da-an District, Taipei City 106, Taiwan.

M. G. Ciaccio
National Institute of Geophysics and Volcanology, via di Vigna Murata 605, 00143 Rome, Italy.

G. Cultrera
National Institute of Geophysics and Volcanology, via di Vigna Murata 605, 00143 Rome, Italy.

D. Danov
Solar-Terrestrial Influences Laboratory, Bulgarian Academy of Sciences, Sofia, Bulgaria.

D. Dominey-Howes
Natural Hazards Research Laboratory, School of Biological, Earth and Environmental Sciences, University of New South Wales, Sydney, NSW 2052, Australia.

F. Doumaz
National Institute of Geophysics and Volcanology, via di Vigna Murata 605, 00143 Rome, Italy.

F. Di Felice
National Institute of Geophysics and Volcanology, via di Vigna Murata 605, 00143 Rome, Italy.

Robert B. Finkelman
US Geological Survey, Mail Stop 956, Reston, VA 20192 USA and Ulli G. Limpitlaw, Earth Sciences, University of Northern Colorado, Greeley, CO 80639, USA.

A. Frepoli
National Institute of Geophysics and Volcanology, via di Vigna Murata 605, 00143 Rome, Italy.

F. T. Freund
NASA Ames Research Center, 242-4, Moffett Field, CA 94035, USA.
Carl Sagan Center, SETI Institute, 515 N. Whisman Road, Mountain View, CA 94043, USA.
Department of Physics, San Jose State University, San Jose, CA 95192-0106, USA.

G. Gisladottir
Department of Geography and Tourism, Faculty of Life and Environmental Sciences, Askja, University of Iceland, 101 Reykjavík, Iceland.

M. Gousheva
Space Research Institute, Bulgarian Academy of Sciences, Sofia, Bulgaria.

Kerem Halicioglu
Bogazici University, Kandilli Observatory and Earthquake Research Institute, Geodesy Department, Cengelkoy, 34680, Istanbul, Turkey.

Q. Hao
SOA Key Lab of Marine Ecosystems and Biogeochemistry, The P.R. China.
Second Institute of Oceanography (SIO), State Oceanic Administration (SOA), Hangzhou, Zhejiang, 310 012, The P.R. China.

P. Hristov
Space Research Institute, Bulgarian Academy of Sciences, Sofia, Bulgaria.

Douglas S. Jones
Florida Museum of Natural History, University of Florida, Gainesville, Florida, USA.

Georg E. Kindermann
International Institute for Applied Systems Analysis (IIASA), Laxenburg, Austria.
University of Natural Resources and Applied Life Sciences (BOKU), Vienna, Austria.

Michael Xavier Kirby
Center for Tropical Paleoecology and Archaeology, Smithsonian Tropical Research Institute, Balboa, Republic of Panama, Panama.

F. Le
SOA Key Lab of Marine Ecosystems and Biogeochemistry, The P.R. China.
Second Institute of Oceanography (SIO), State Oceanic Administration (SOA), Hangzhou, Zhejiang, 310 012, The P.R. China.

C. Lin
Second Institute of Oceanography (SIO), State Oceanic Administration (SOA), Hangzhou, Zhejiang, 310 012, The P.R. China.

Yu-Pin Lin
Department of Bioenvironmental Systems Engineering, National Taiwan University, 1, Sec. 4, Roosevelt Rd., Da-an District, Taipei City 106, Taiwan.

C. Liu
State Key Lab of Satellite Ocean Environment Dynamics, The P.R. China.
SOA Key Lab of Marine Ecosystems and Biogeochemistry, The P.R. China.
Second Institute of Oceanography (SIO), State Oceanic Administration (SOA), Hangzhou, Zhejiang, 310 012, The P.R. China.

Bruce J. MacFadden
Florida Museum of Natural History, University of Florida, Gainesville, Florida, USA.

P. Macrì
National Institute of Geophysics and Volcanology, via di Vigna Murata 605, 00143 Rome, Italy.

A. Marsili
National Institute of Geophysics and Volcanology, via di Vigna Murata 605, 00143 Rome, Italy.

M. Matova
Geological Institute, Bulgarian Academy of Sciences, Sofia, Bulgaria.

Ian McCallum
International Institute for Applied Systems Analysis (IIASA), Laxenburg, Austria.

X. Ning
State Key Lab of Satellite Ocean Environment Dynamics, The P.R. China.
SOA Key Lab of Marine Ecosystems and Biogeochemistry, The P.R. China.
Second Institute of Oceanography (SIO), State Oceanic Administration (SOA), Hangzhou, Zhejiang, 310 012, The P.R. China.

C. Nostro
National Institute of Geophysics and Volcanology, via di Vigna Murata 605, 00143 Rome, Italy.

Michael Obersteiner
International Institute for Applied Systems Analysis (IIASA), Laxenburg, Austria.
Institute for Advanced Studies (IHS), Vienna, Austria.

Haluk Ozener
Bogazici University, Kandilli Observatory and Earthquake Research Institute, Geodesy Department, Cengelkoy, 34680, Istanbul, Turkey.
Istanbul Technical University, Department of Geodesy and Photogrammetry Engineering, Surveying Technique Division, Maslak, 34469, Istanbul, Turkey.

J. Pelon
Service d'Aéronomie du CNRS, Institut Pierre-Simon-Laplace, Université Pierre-et-Marie-Curie, 4, Place Jussieu—75252 Paris Cedex 05, France.

A. Piscini
National Institute of Geophysics and Volcanology, via di Vigna Murata 605, 00143 Rome, Italy.

M. K. Ramamurthy
Unidata, University Corporation for Atmospheric Research, Boulder, Colorado, USA.

Ewald Rametsteiner
International Institute for Applied Systems Analysis (IIASA), Laxenburg, Austria.
University of Natural Resources and Applied Life Sciences (BOKU), Vienna, Austria.

P. Scarlato
National Institute of Geophysics and Volcanology, via di Vigna Murata 605, 00143 Rome, Italy.

Elif Sertel
Istanbul Technical University, Faculty of Civil Engineering, Department of Geodesy and Photogrammetry, Maslak, 34469, Istanbul, Turkey.

J. Shi
Second Institute of Oceanography (SIO), State Oceanic Administration (SOA), Hangzhou, Zhejiang, 310 012, The P.R. China.

S. Stramondo
National Institute of Geophysics and Volcanology, via di Vigna Murata 605, 00143 Rome, Italy.

A. Tertulliani
National Institute of Geophysics and Volcanology, via di Vigna Murata 605, 00143 Rome, Italy.

M. Vallocchia
National Institute of Geophysics and Volcanology, via di Vigna Murata 605, 00143 Rome, Italy.

Cheng-Long Wang
Department of Bioenvironmental Systems Engineering, National Taiwan University, 1, Sec. 4, Roosevelt Rd., Da-an District, Taipei City 106, Taiwan.

Yung-Chieh Wang
Department of Bioenvironmental Systems Engineering, National Taiwan University, 1, Sec. 4, Roosevelt Rd., Da-an District, Taipei City 106, Taiwan.

Hsiao-Hsuan Yu
Department of Bioenvironmental Systems Engineering, National Taiwan University, 1, Sec. 4, Roosevelt Rd., Da-an District, Taipei City 106, Taiwan.

A. Winkler
National Institute of Geophysics and Volcanology, via di Vigna Murata 605, 00143 Rome, Italy.

List of Abbreviations

ACP	Panama Canal Authority
AIRS	Atmospheric infrared sounder
ASTER	Advanced spaceborne thermal emission and reflection radiometer
AT	Air temperature
BB	Benthos biomass
BF	Bornova fault
BLUE	Best linear unbiased estimator
BOAI	Budapest Open Access Initiative
CALIOP	Cloud-Aerosol Lidar with Orthogonal Polarization
CALIPSO	Cloud-Aerosol Lidar and Infrared Pathfinder Satellite Observation
CC	Cephalopod catch
CTH	Cloud top height
DIN	Dissolved inorganic nitrogen
DLESE	Digital Library for Earth System Education
DO	Dissolved oxygen
DTC	Demersal trawl catch
E.L.	Emperador limestone
EAFS	East Anatolian Fault Zone
EC	Evacuation centers
EM	Electromagnetic
EOSDIS	Earth Observing System Data and Information System
EPA	Environmental Protection Agency
ERE	Environmental Research and Education
ESA	European Space Agency
ESS	Earth system science
FFT	Fast Fourier transform
FOD	First-order design
GALEON	Geo-interface for Atmosphere, Land, Earth, and Ocean netCDF
GCM	General circulation model
GCPs	Ground control points
GDP	Gross domestic product
GEOIDE	Group on Earth Observations Integrated Data Environment
GEOSS	Global Earth Observation System of Systems

GF	Guzelhisar fault
GHG	Greenhouse gas
GISs	Geographic information systems
GLAS	Geoscience Laser Altimeter System
GtC	Gigatonnes of carbon
GuF	Gumuldur fault
HELIX-Atlanta	Health and Environment Linked for Information Exchange in Atlanta
HIRS	High resolution infrared radiation sounder
HRV	High resolution visible
HRVIR	High resolution visible infrared
HWP	Wood products pool
IAVCEI	International Association of Volcanology and Chemistry of the Earth's Interior
ICESat	Ice, Cloud, and Land Elevation Satellite
ICP	Icelandic Civil Protection
IDDs	Iodine deficiency disorders
İF	Izmir fault
IFFT	Inverse fast Fourier transform
INGV	National Institute of Geophysics and Volcanology
Inv LAT	Invariable latitude
IOM	Icelandic Meteorological Office
ISCCP	International Satellite Cloud Climatology Project
IT	Information technology
IUGG	International Union of Geodesy and Geophysics
L.C.F.	Las cascadas formation
LHS	Latin hypercube sampling
LITE	Lidar in-space technology experiment
LT	Local time
m.y.	Million years
MODIS	MODerate-resolution imaging spectrometer
MS	Multiple scattering
NAFZ	North Anatolian Fault Zone
NASA	National Aeronautics and Space Administration
NDVI	Normalized difference vegetation index
netCDF	Network common data form
NOMADS	National Operational Model Archive and Distribution System
NPP	Net primary production
NRC	National Research Council

nSCS	Northern South China Sea
NSDL	National Science Digital Library
NSF	National Science Foundation
OGC	Open GIS consortium
OPeNDAP	Open-source Project for a Network Data Access Protocol
P.M.F.	Pedro Miguel formation
PA	Phytoplankton abundance
PBL	Planetary boundary layer
PDF	Probability density function
PM	Particulate matter
PP	Primary production
PPP	Purchasing power parity
PSCs	Polar stratospheric clouds
RASA	Regional aquifer-system analysis
RSS	Really simple syndication
SCS	South China Sea
SCSFRI	South China Sea Fisheries Research Institute
SF	Seferihisar fault
SGS	Sequential Gaussian simulation
SISs	Scientific information systems
SNR	Signal to noise ratio
SOA	State Oceanic Administration
SOD	Second-order design
SOQ	Service oriented architecture
SPd	Standardized population density
SPOT	Satellite Pour l'Observation de la Terre
SS	Sea surface
SST	Sea surface temperature
TCP/IP	Transmission Control Protocol/Internet Protocol
TDSs	Total dissolved solids
TEC	Total electron content
TF	Tuzla fault
THREDDS	Thematic realtime environmental distributed data services
TOD	Third-order design
USGS	United State Geological Survey
UT	Universal time
UTM	Universal Transverse Mercator
VOCs	Volatile organic compounds
WMO	World Meteorological Organization

XML	Extensible markup language
ZB	Zooplankton biomass
ZOD	Zero-order design

Contents

Chapter 1

Cyberinfrastructure and Data Services for Earth System Science Education and Research

M. K. Ramamurthy

INTRODUCTION

A revolution is underway in the role played by cyberinfrastructure and modern data services in the conduct of research and education. We live in an era of an unprecedented data volume from diverse sources, multidisciplinary analysis and synthesis, and active, learner-centered education emphasis. Complex environmental problems such as global change and water cycle transcend disciplinary and geographic boundaries, and their solution requires integrated earth system science (ESS) approaches. Contemporary education strategies recommend adopting an ESS approach for teaching the geosciences, employing pedagogical techniques such as enquiry-based learning. The resulting transformation in geoscience education and research creates new opportunities for advancement and poses many challenges. The success of the scientific enterprise depends heavily on the availability of a state-of-the-art, robust, and flexible cyberinfrastructure, and on the timely access to quality data, products, and tools to process, manage, analyze, integrate, publish, and visualize those data.

Concomittantly, rapid advances in computing, communication, and information technologies have revolutionized the provision and use of data, tools, and services. The profound consequences of Moore's Law and the explosive growth of the Internet are well known. On the other hand, how other technological trends have shaped the development of data services is less well understood. For example, the advent of digital libraries, web services, open standards, and protocols have been important factors in shaping a new generation of cyberinfrastructure for solving key scientific and educational problems.

This chapter presents a broad overview of these issues, along with a survey of key information technology (IT) trends, and discuses how those trends are enabling new approaches to applying data services for solving geoscientific problems.

Cyberinfrastructure and data services are transforming the conduct of research and education in the geosciences. For example, current day weather and coupled climate system prediction models and a new generation of remote-sensing systems like hyperspectral satellite instruments and rapid scan, phased-array radars are capable of generating massive amounts of data each day. Complex environmental problems such as global change and water cycle transcend disciplinary and geographic boundaries, and it is widely recognized that their solution requires integrated ESS approaches. Contemporary education strategies recommend adopting an ESS approach for teaching the geosciences, employing new pedagogical techniques such as enquiry-based learning and hands-on activities. The resulting transformation in today's education and research enterprise creates new opportunities for advancement, but also many new challenges. For example, the success of this enterprise depends heavily on the availability of a state-of-the-art, robust, flexible, and scalable cyberinfrastructure, and on the timely, open and easy access to quality data, products, and tools to process, manage, analyze, integrate, publish, and visualize those data.

An empirical observation that the number of transistors on a chip doubles every 18 months, in the information revolution are well known as philosophical retribution of Moore's Law. In education and research, rapid advances in computing, communication, and information technologies have also been revolutionized the provision and use of data, tools, and services. Similarly, the explosive growth in the use of the Internet in education and research, largely due to the beginning of the World Wide Web, is also well documented. Technological, social, and cultural trends have shaped the development of data services is somewhat less well understood. In shaping the use of a new generation of modern, end-to-end cyberinfrastructure for solving some of the most challenging scientific and educational problems, the initiation of digital libraries, web services, grid computing, open standards, protocols, and frameworks, open-source models for software, and community models have been important factors, both individually and collectively.

The purpose of this study is to present a broad overview of these and related issues largely from the author's perspective, along with a brief discussion of the how the above changes are enabling new approaches to applying data services for solving integrative, multidisciplinary geoscientific problems. To that end, the chapter focuses on documenting some of the changes in the conduct of geoscience and science education, highlighting the revolution in cyberinfrastructure and documenting how the resulting technological advances and approaches are leading to an evolution from once proprietary and centralized data systems to open, distributed, and standards-based data services that facilitate easier data integration and greater interoperability.

The layout of the chapter is as follows. A few key scientific and education drivers, IT trends that have shaped new approaches to providing data services in the geosciences. Specific issues related to data services, including ideal data service attributes, data categories, and data analysis and integration methods. Conclusions on how today's cyberinfrastructure and data services are reshaping the science and education landscape.

KEY DRIVERS

Data, information, and embedded knowledge are central to the advancement of science and education, as articulated in National Science Foundation (NSF) Geosciences Beyond 2000: Understanding and Predicting Earth's Environment and Habitability (NSF, 2000). The aforementioned report recognizes that progress in research and education in the geosciences will require "... a commitment to improve and extend facilities to collect and analyze data on local, regional, and global spatial scales and appropriate temporal scales," including real-time observing systems, and modern computational facilities to support rapid computation, massive data archiving and access, distribution, analysis, and management. Conversely, contemporary data services need to be firmly grounded in not only scientific and education drivers, and community needs, but are also greatly influenced by the myriad technological and sociological trends.

The following sections describe the key drivers and trends that have transformed data provision and access from centralized systems that were once based on proprietary architectures to modern, distributed data services. In the process, the new generation of data services has also reshaped their use in research and education in new and innovate ways.

Science Driver

Numerous national and international reports underscore the importance of interdisciplinary ERE. Among them are *Grand Challenges in Environmental Science* (NRC, 2001) and *Complex Environmental Systems: Synthesis for Earth, Life, and Society in the 21st Century* (NSF, 2003). National Research Council (NRC) report points to a growing recognition that "natural systems—ecosystems, oceans, drainage basins, including agricultural systems, the atmosphere, and so on—are not divided along disciplinary lines." Two of the grand challenges identified by the NRC, biogeochemical cycles, and climate variability, depend heavily on integration of data from several disciplines. Another excellent example is hydrologic forecasting, one of the four challenges prioritized as deserving immediate investment.

According to the former NSF director Rita Colwell (1998), "Interdisciplinary connections are absolutely fundamental. They are synapses in this new capability to look over and beyond the horizon. Interfaces of the sciences are where the excitement will be the most intense..." For example, studies on societal impact of and emergency management during hurricane-related fl998), involve integrating data from atmospheric sciences, oceanography, hydrology, geology, geography, and social sciences with databases in the social sciences.

The NSF decadal plan for Environmental Research and Education (ERE) also echoes the need for improving our understanding of the natural and human processes that govern quantity, quality, and availability of freshwater resources in the world. While recent advances in remote sensing, combined with a new generation of coupled models, are driving a new revolution in hydrometeorological predictions, future research, and education in this area will require finding and integrating observational and model data from the oceans, the atmosphere, the cryosphere, and the lithosphere, crossing the traditional disciplinary boundaries.

Similar multidisciplinary needs are emerging to solve certain disaster/crisis management problems. Two highly topical examples are fire-weather forecasting and environmental modeling for homeland security. In homeland security, for instance, there is a need to forecast the dispersal of hazardous radioactive, biological, and chemical materials that may be released (accidentally or deliberately by terrorism) into the atmosphere. For the latter scenario, detailed, four-dimensional information on transport and dispersion of hazardous materials through the atmosphere, and their deposition to the ground are needed at a resolution of individual community scales. Moreover, this information needs to be linked in real-time to databases of population, evacuation routes, medical facilities, and so on to predict the consequences of various release scenarios (e.g., number of people may be exposed to or will be injured by potentially dangerous concentrations of those materials).

In addition to identifying national priorities and computational grand challenges in the sciences, many of the NSF and NAS reports cited above have also documented infrastructure needs, including comprehensive data collection, management and archival systems, and new methods of data mining and knowledge extraction. For example, the NSF ERE Advisory Committee calls for building infrastructure and technical capacity with a new generation of cyberinfrastructure "to support local and global research and to disseminate information to a diverse set of users including environmental professionals, the public, and decision makers at all levels." Toward building the cyberinfrastructure, the ERE agenda foresees the need for a comprehensive suite of data services that will facilitate synthesis of datasets from diverse fields and sources, information in digital libraries, data networks, and web-based materials so that they can serve as essential tools for educators, students, scientists, policy-makers, and the general public. Similar needs for web-based real-time and archived data services, including digital library integration and fusion of scientific information systems (SISs) with geographic information systems (GISs), were expressed at the NSF-sponsored Workshop on Cyberinfrastructure for ERE (CIERE, 2003).

Growing numbers of universities are engaged in real-time modeling activities, and this number is expected to increase as advances in computing and communication technologies facilitate local atmospheric modeling. A new generation of models (e.g., the weather research and forecasting model, (Michalakes et al., 2001)) can predict weather on the sub 1 km scale, with the potential to address community-scale concerns. Providing initial and boundary condition data along with analysis and visualization tools for these efforts requires an extensive cyberinfrastructure.

The recent decades have also been marked by a revolution in our ability to survey, probe, map, and profile our global environment. For instance, a plethora of instruments and digital sensors mounted on geostationary and polar orbiting satellites scan vast areas of the Earth's surface round the clock. With their powerful ability to continuously and remotely monitor the global environment, observations from satellite platforms are increasingly replacing *in-situ* surface and upper air observations. Today, dozens of satellites are rapidly scanning and measuring the global environment and in the process generating an ever-expanding range of geoscience data to help us manage and solve some of the most vexing and complex multidisciplinary problems of the society.

This revolution in remote sensing technology and techniques and their many geoscience applications have had a profound impact on geoscience operations. At the same time, the complexity and explosive growth in the volume of remotely sensed has also transformed the provision and use of data from remote sensing platforms such as satellites, radars, and lidars.

Modern environmental studies rely on diverse datasets, requiring tools to find and use the data. The data discovery process has become an important dimension of the scientific method, complementing theory, experimentation, and simulation as the tools of the trade.

Education Driver

Challenges facing science education have been well articulated in a number of documents (e.g., *Shaping the Future* (AGU, 1997) and *Geoscience Education: A Recommended Strategy* (NSF, 1997)). They recommend adopting an ESS approach for teaching the geosciences, integrating research experiences into curricula, employing contemporary pedagogies, and making appropriate use of educational technologies. Science education should also be about teaching students the language of science and providing students with opportunities to engage in scientific inquiry and investigation (Lemke, 1990).

Shaping the Future also calls for an inquiry-based approach to science education. For example, hands-on, learner-centered education in meteorology depends on the availability of meteorological data and analysis and display tools of high quality. By supplying these data and tools, programs like Unidata have been instrumental in transforming learning in the atmospheric sciences. Digital libraries (exemplified by efforts like the National Science Digital Library (NSDL) and the Digital Library for Earth System Education (DLESE)) augment web-based learning resources with high-quality data resources that can be embedded in interactive educational materials. The Internet-based tools also open data access for faculty and students at small colleges where little system administration support is available for the installation of advanced data systems and applications. Engaging students with real-world data is a powerful tool not only for motivating students but also helping them learn both scientific content and principles and the processes of inquiry that are at the heart of science (Manduca, 2002). Earth science education is uniquely suited to drawing connections between the dynamic Earth system and important societal issues and making science relevant to students. Recent catastrophic events like the 2004 Indian Ocean tsunami, Hurricane Katrina, and the October 2005 earthquake in Northern Pakistan are three stark examples that drive home this point. These events also heavily underscore the importance of multidisciplinary integration and synthesis of data from the various Earth science disciplines. Working with such real-world events and actual data can place learning in a context that is both exciting and relevant. Another example is providing connections between classroom instruction and students' experience with their local environment (e.g., diurnal temperature changes and seasons), major weather events (e.g., tornadoes, hurricanes, blizzards), and climate events (e.g., global warming).

In essence, significant strides in advancing Earth science education can be made by incorporating new teaching techniques, active learning strategies, IT, and integrating real-world Earth and space science data into our curriculum. To accomplish these objectives, students will need to have opportunities for genuine inquiry and hands-on experience, so that the excitement of discovery is infused into all courses while students gain experience in the process of science. A critical component of successful scientific inquiry includes learning how to collect, process, analyze, and integrate data. Innovative data services that promote this perspective on student learning are needed and should be integrated into Earth science education at all levels.

The richness of students' exploration and experience depends, among other things, on the quality of the data available and the tools and technology they use. To that end, cyberinfrastructure provided by organizations like Unidata allows students to access the very databases and tools that are used by the scientific and operational communities, and provides an important pathway toward the pursuit of the long-sought goal of the NSF to integrate research and education. Distributed computing, data access, and collaboration are rapidly becoming the de facto means for learning and doing science, leading to the pervasive use of the World Wide Web in everyday life of a scientist, an educator, or a student. This reliance on the Internet and other technologies and applications is only expected to increase greatly in the future.

INFORMATION TECHNOLOGY TRENDS

Computers and information technologies are now playing a central role in this complex and ever-changing world in which we live and work, with the World Wide Web reshaping almost every aspect of our work, including education, research, and commerce. Computing, communications and IT trends of recent years have not only had a democratizing effect on daily life, but they have also changed the very nature of data services for education and research. For example, below is a partial list of key technologies and trends that have enabled a new generation of end-to-end data services in the scientific community:

- Internet
- Commodity microprocessors, enabled by Moore's Law
- World Wide Web
- Open source model
- Object-oriented programming
- Open standards, frameworks, and conventions
- Extensible markup language (XML)
- Web services
- Digital libraries
- Collaboratories
- Grid computing
- Data portals and federated, distributed servers
- GISs
- Ontologies and semantic web
- Data mining and knowledge discovery

This section highlights a few of the above key IT advances and trends that have revolutionized the provision and use of data in the geosciences. The highlights selected for further discussion are meant neither to provide a comprehensive overview of all key advances, nor are the specific examples cited sole exemplars of their myriad implementations. Taken together, however, the above technologies have enhanced the ability of data providers to better serve their communities, lower the costs for the users, and allowed a greater participation in the data activities in a new networked world.

The introduction of microprocessor based computer systems in the 1980s, combined with the increased connectivity of college campuses to the Internet, led to a transition from large scale, mainframe-based technologies to low-cost distributed systems, making it possible for widespread access to and use of scientific data. The wiring of universities for the Internet connectivity was a prerequisite for receiving data via, for example, the Unidata Internet Data Distribution System and the Local Data Manager, which use Transmission Control Protocol/Internet Protocol (TCP/IP) communication standards for data transport.

The advent of the World Wide Web (or simply the web) in the 1990s brought about a revolution in information services. It was directly responsible not only for the explosive growth of the Internet and increasing its numbers of users, it also accommodated the ability to provide interactive, remote services. In the process, the web radically transformed the sharing of data and information and resulted in greater use of communication infrastructures to create and store information and then to deliver it from providers to end users. The web also brought with it a massive proliferation of online educational materials, many of them based around extensive use of interactive services. Services and tools were created to help one communicate, search for information and data, and make information and data available on the Internet. In the process, library services evolved from local traditional collections to global resources provided on demand via the web, ushering in the era of digital libraries.

The 1995 NSF-sponsored Digital Libraries Workshop entitled "Interoperability, Scaling, and the Digital Library Research Agenda" defined digital libraries as: "An organized collection of multimedia data with information management methods that represent the data as information and knowledge."

Even though efficient retrieval of information is arguably the most important role of digital libraries, a potentially even more valuable contribution of digital libraries is their ability to preserve, catalog, and curate information, extend discourse, build communities that provide richer contexts for people to interact with information and each other, all toward the creation of new knowledge. According to Griffin (1998), the real value of digital libraries may ultimately prove to be their ability to "alter the way individuals, groups, organizations etc, behave, communicate, collaborate, and conduct business." In essence, digital libraries, much like other aspects of the web, are becoming powerful instruments of change in education and research.

The digital library era has also spawned a movement toward open access to scholarly literature. Suber (2003) defines open-access literature as one which is digital, online, free of charge, and free of most copyright and licensing restrictions, whereas the Budapest Open Access Initiative (BOAI), an international effort to make research articles in all fields freely available on the Internet, provides a slightly different definition: "literature with free availability on the public Internet, permitting any users to read, download, copy, distribute, print, search, or link to the full texts of these articles, crawl them for indexing, pass them as data to software, or use them for any other lawful purpose, without financial, legal, or technical barriers other than those inseparable from gaining access to the Internet itself. The only constraint on reproduction and distribution, and the only role for copyright in this domain, is to give authors control over the integrity of their work and the right to be properly acknowledged and cited." The core belief of BOAI is that the removal of access barriers to literature will have a democratizing effect and will result in "accelerated research, enriched education, and make scholarly literature as useful as it can be, laying the foundation for uniting humanity in a common intellectual conversation and quest for knowledge." The open access concept has parallels in the data world. In the atmospheric sciences community, an important consequence of the open sharing ideal has been the free flow of meteorological data across geographic boundaries, per World Meteorological Organization (WMO) Resolution 40, which commits the WMO to broadening and enhancing the free and unrestricted international exchange of meteorological and related data and products. The resulting sharing and free flow of data meteorological data has had a noticeable impact on education and research in atmospheric and related sciences, as illustrated in a companion article by Yoksas et al. (2005).

Another notable IT trend is the desire to integrate all information, including data and a variety of services behind a single entry point or a portal. Portals often include personalization features allowing users a tailored view into the information. The customization permits: (a) a single point of authentication to validate access permissions and enable links to available resources and (b) the ability to design a customized view of available information.

The open access, open source, and open standards are inter-related concepts that are gaining momentum and developers of data service are aggressively rethinking how they might both contribute to and benefit from these trends toward "openness." The benefits of open access, open source, and open standards are numerous and when they are combined the benefits can be even greater.

Open source software is software that includes source code and is usually available at no charge. The open source model for software has many benefits, as articulated in a collection of essays by Raymond (1999). For instance, it has the advantage of harnessing the collective wisdom, experiences, expertise, and requirements of large communities. Drawing upon the Linux development experience and based on his successful open source software project, *fetchmail*, Raymond makes a compelling argument for the proposition that, "Given enough eyeballs, all bugs are shallow" and for the importance of treating users as co-developers. Additional features and benefits include scalability, extensibility, and customizability. For example, people using a wide variety of hardware platforms, operating systems, and software environments can test, modify, and run software on their system to test for portability. Successful open source development efforts also do not start from scratch but rather try to adapt and build on top of existing code base, using the community process for refinement and reuse.

In the data services area, many excellent examples of open source software that are highly reliable and supported by a large community exist. They include Linux, Apache, MySQL, and similar projects. Network Common Data Form (netCDF), Open-source Project for a Network Data Access Protocol (OPeNDAP), Thematic Realtime Environmental Distributed Data Services (THREDDS, Domenico et al., 2002) are leading examples of open source software in the geosciences data infrastructure area. Because of its free and open source nature, netCDF software has been incorporated into over

50 other open source software packages and 15 commercial packages, resulting in its widespread use and status as a de facto standard for data format in atmospheric and related sciences. Likewise, the OPeNDAP software, which was originally called Distributed Ocean Data System, has found wide use outside the core oceanography data community where it originated. Open source software also increases opportunities for software reuse, adaptation to different hardware and software environments, and customization to user needs. The best example, perhaps, is the use of Linux in a wide range of electronic and computer systems, including videogame consoles, mobile phones, personal digital assistants, and personal, mainframe computers, and massively parallel high-performance computing systems. The large-scale availability of access to the Internet and Internet applications, coupled with widespread use of Linux in academia and the availability of inexpensive, commodity microprocessors, and storage devices, has had a democratizing effect on data provision to and access by the geosciences community. As an example, today over 160 colleges and universities worldwide are participating in the Unidata Internet data distribution system and they are receiving, sharing and distributing data, and integrating them into their education and research using inexpensive computers and freely-available, Linux-based open source applications.

The use of open standards models for middleware, a special kind of software between client and server processes to ensure consistency and interoperability, is particularly important for developing new data services. For example, an open standards-based middleware provides opportunities for the provision of a stable, consistent interface to a wide variety of applications, on a broad set of platforms and enable their inter-operability. In the process, it decouples data service providers from users, allowing end users with multiple clients to access the same services. This can accelerate the migration of data services to new and diverse platforms. Furthermore, it facilitates the "wrapping" of legacy systems in standard interfaces, giving them the ability to integrate with other distributed components and systems. Given the demand for standards-based, open systems that easily integrate, the open source development process provides a significant advantage over proprietary approaches to software development and use.

Interfaces based on open standards are by definition publicly documented and based on an explicit or de facto standard. There is evidence that well developed open standards for data formats are less likely to become quickly obsolete and are more reliable and stable than proprietary formats. Having access to the file format also allows users and developers to create data conversion utilities into other formats. File formats that use open standards can assist in long-term archiving because they allow for software and hardware independence. Open standards also allow for greater flexibility and easy migration to different systems and interoperability of diverse systems. Open access, open source software models, and open standards each offer a number of significant benefits in the provision of data services.

The XML, is a simple, highly flexible, text-based framework for defining mark up languages. This standard for classifying, structuring, and encoding data allows organizations and services to exchange information more easily and efficiently. Although originally developed to facilitate web-based publishing in a large scale, XML has since rapidly gained acceptance and usage in the exchange of a wide variety of data on the web. An important emerging standard for interoperability of data systems is in the metadata area, which can use XML to share descriptions of underlying datasets.

The ability of XML to organize data into a computer-interpretable format that is also easy to code and read by humans is quickly making XML the lingua franca for business services and electronic commerce and also rapidly becoming a widely used standard in the data services world. Because of its simplicity and elegance, XML has radically transformed the provision of data services in the scientific community. Some of its principal benefits include: (a) ability to delineate syntactic information from semantic information; (b) allows the creation of customizable markup languages for different use cases and application domains; (c) platform independence. For example, XML makes it possible for providers of data services to send information about data sets, metadata, in a form completely separate from the presentation of the underlying data. Furthermore, service providers can present the same information in multiple forms or views using XML style sheets, customized to the needs of particular users. For example, really simple syndication (RSS) is a lightweight XML format for sharing news and bulletins, and it has been used successfully by the US National Weather Service to disseminate weather information such as local forecasts, watches, and warnings to the Internet users. The same technology can also be used in the data services context to notify users when new data becomes available in a data system.

Web services, based on XML and HTTP, the two open standards that have become ubiquitous underpinnings of the web, are emerging as tools for creating next generation distributed systems. Besides recognizing the heterogeneity as a fundamental ingredient, web services, independent of platform, and development environment, can be bundled, published, shared, discovered, and invoked as needed to accomplish specific tasks. Because of their building-block nature, web services can be deployed to perform either simple, individual tasks or they can be chained to perform complicated business or scientific processes. As a result, web services, implemented in a service oriented architecture (SOA) or framework, are quickly becoming a technology of choice for deploying cyberinfrastructure for data services. By wrapping existing applications and their components as web services in a SOA, the traditional obstacles to interfacing legacy and packaged applications with data systems are being overcome through loosely coupled integration. Such an approach to lightweight integration affords an easier pathway to interoperability amongst disparate systems and distributed services. The new software architectures based largely on web services standards are enabling whole new service-oriented and event-driven architectures that are challenging traditional approaches to data services. In a series of articles, Channabasavaiah et al. (2003) present a persuasive case for developing and deploying a SOA, as the level of complexity of traditional architectures increases and approaches the limit of their ability. They also provide a realistic plan for migrating existing applications to a SOA, one that leverages existing assets and allows for incremental implementation and migration of those assets. Several efforts are underway within the geosciences community to apply web services and SOAs to both migrate existing, stove-piped data systems as well as in the development of common architectures for future data systems. For example, the strategic plan for the US Integrated Earth Observation System, the US contribution to Global Earth Observation System of Systems (GEOSS), calls for the implementation of GEOSS services within a web-enabled, component-based architecture in its overall data management strategy so that the value of Earth observations data and information resources is maximized (Hood, 2005; IWGEO, 2005). Likewise, the Integrated Ocean Observing System and the NOAA Group on Earth Observations Integrated Data Environment (GEOIDE) are both planning to use a SOA/web services approach for providing data services to their respective communities.

Another computing model that is beginning to transform how resources are applied to solve complex scientific problems is grid computing, a term that originated in the 1990s as a metaphor for making distributed computer power as easy to use as an electrical power grid. While many definitions of grid computing exist (Wikipedia: The Free Encyclopedia, 2006), the most definitive and widely used one is by Foster and Kesselman (1997). According to them, the grid refers to "an infrastructure that enables the integrated, collaborative use of high-end [and distributed] computers, networks, databases, and scientific instruments owned and managed by multiple organizations." Grid applications often involve large amounts of data and/or computing and often require secure resource sharing across organizational boundaries. Grid computing and the science enabled by it, eScience, are two major trends in distributed computing. A key advantage of grid computing over historical distributed computing systems is that the grid concept permits the virtualization of computing resources such that end-users have the illusion of using a single source of "computing power" without knowing the actual location where their computations are performed. The use of digital certificates to access systems on behalf of a user and third-party file transfer between grid nodes authenticated via certificates are specific examples of how grid technology enables resource allocation and virtualization. Grid Services, which implement web services in a grid architecture, are in still in their infancy, although several proof-of concept test beds have been deployed in a number of disciplines, including Earth and atmospheric sciences, high energy physics, and biomedical informatics. Although the distinction between traditional web services and grid services is subtle, grid services, ideally, should enable virtualization for building and running applications that span organizations and share resources and infrastructure in a seamless way. Gannon et al. (2005) and Foster (2002) provide the distinguishing characteristics of a grid service and specify what is needed for a web service to qualify as a grid service.

DATA SERVICE ATTRIBUTES

As articulated by Cornillon (2003), the ultimate objective of a data system or service is to provide requested data to the user or user's application (e.g., analysis or visualization tool) in a transparent, consistent, readily useable form. The users do not care as much about the technology behind those systems or services, but do about transparency and usability. The key to achieving Cornillon's two objectives is through interoperability of components, systems and services, via the use of standards.

In the opinion of this author, an ideal data service should have the following attributes:

- User-friendly interface
- Transparency (format, protocol, etc.)
- Customization and personalization of services
- Capability for server-side operations (e.g., subsetting, sub-sampling, etc)
- Aggregation of data and products
- Provision of rich metadata
- Integration across data types, formats, and protocols
- Intelligent client-server approaches to data access and analysis
- Interoperability across components and services
- Flexibility, extensibility, and scalability
- Ability to chain services via workflows
- Support an array of tools for access, processing, management, and visualization

As a result of the aforementioned trends, the last decade has seen an evolution of data systems like Earth Observing System Data and Information System (EOSDIS) towards a more layered and open architecture, while new data systems have been built and deployed using many open source and standards-based technologies (e.g., the NOAA National Operational Model Archive and Distribution System (NOMADS; Rutledge et al., 2002), Community Data Portal (Middleton, 2001) and Earth System Grid (Foster et al., 2002); and data system at the British Atmospheric Data Centre (Lawrence, 2003) However, the transition has not been without challenges for a number of reasons, including:

- Heterogeneity and complexity of distributed observing, modeling, data, and communication systems
- Nature of data coverage: diversity and multiple spatial and temporal scales
- Data systems using both legacy components alongside contemporary applications, creating integration challenges
- A lack of standards and interoperability
- Non-monolithic user community
- Political, technological, and cultural and regulatory barriers, especially in global sharing of and access to data

Given the very high data rates from current and future generation observing systems such as GOES-R and NPOESS satellites, the user community will need a hybrid solution that couples a satellite-based data reception system with a terrestrial, Internet-based data access system. Both local and remote data access mechanisms will be required to deal with the large volumes of data. Both push systems for distributing data (e.g., Unidata Local Data Manager) or just notifications (using RSS feeds) and pull systems for remote access (e.g., THREDDS and OPeNDAP) will be required.

Broad Data Categories

While far too many data categories exist to describe in detail, typical data systems in atmospheric sciences must provide a seamless, end-to-end services for accessing, utilizing, and integrating data across the following data types:

- Real-time data
- Archived data
- Field and demonstration project data
- Episodic or case study data
- Data from related disciplines (hydrology, oceanography, cryosphere, chemical and biosphere—soil, vegetation, canopy, evapotranspiration)
- GIS databases

The first four categories, and to a lesser extent the fifth one, include data from *in-situ* and remote sensing observations, and output from models. Even though each discipline within the geosciences is unique and has different data needs depending on the use or application, the geoscience disciplines do share a common interest in accessing data of the types listed above. For instance, the first four data types are important for many applications in atmospheric sciences, oceanography, hydrology, geologic subdisciplines of seismology, and volcanology. Another common attribute is the need for georeferencing and integration with information contained in GIS databases. This brings up an important area of ongoing research, namely, the development of a common data model for the geosciences, as they share a common representation of data in their spatial and temporal representations. A discussion of data models for geosciences is beyond the scope of this chapter, however.

Data Deluge, Data Mining, and Knowledge Discovery

Advances in computing, modeling, and observational systems have resulted in a veritable increase in the volume of data. These data volumes will continue to see exponential growth in the coming years. For example, data from current and future observing systems will result in a 100-fold increase in volume in the next decade. The GOES-R satellite, scheduled for launch in 2012, will have a hyperspectral sounder with approximately 1,600 channels. In contrast, the current generation GOES satellite sounders have 18 thermal infrared channels. Similarly, each NPOESS satellite when fully deployed will have raw data rates of nearly 1 Terabyte each day. Hey (2003) previews the imminent data deluge from the next generation of simulations, sensors, and modeling systems and experiments, and discusses the importance of metadata and the need to automate the process of converting raw data to useful information and knowledge and implications for grid middleware architecture.

The data deluge clearly requires extraction of higher level information useful to users. The process of extracting higher level information is referred to as data mining. Data mining is a key step toward data reduction and knowledge discovery. Graves (1996) and Ramachandran et al. (1999) offer a methodology to efficiently mine and extract content-based metadata from Earth Science datasets and describe the capabilities of the ADaM (A Data Miner) tool, which enables phenomena-oriented data mining by incorporating knowledge of phenomena and detection algorithms in the system. Their methodology provides a meaningful solution needed by users to convert data to knowledge and cope with the data deluge. An ideal data system or service should include algorithms and facilities for data mining that can be applied to data sets as needed by users. Future success will depend on how well users are served by such discovery and mining tools and services pertaining to data integration.

Geographic Information Systems

The GIS have become indispensable tools for geoscientific exploration, commerce, and for decision making in environmental and social sciences (Morss, 2002). In general, GIS involves a broad array of computer tools for mapping, and managing geographically referenced data, and for spatial analysis. Regardless of the application, all the geographically related data can be input and prepared in a GIS such that users can display the specific information of interest, or combine data contained within the system to produce additional value-added information that might help answer a specific problem. The extraordinary gains in computer performance over the past 2 decades have seen a parallel growth in GIS applications. These applications have not only been growing in number but also in their diversity. An important reason for the proliferation of GIS use is that it provides a convenient framework for multidisciplinary analysis and synthesis, which is becoming increasingly important as researchers explore the frontiers of science. As the demand for innovative GIS applications, services, and know-how grows, GIS is expected to play a pivotal role in shaping the cyberinfrastructure and data services in the geosciences.

The GIS's powerful capability is to integrate spatially referenced data from disparate sources into a single environment. An important application of GIS is linking remotely sensed data from a variety of instruments with various socioeconomic data and biophysical datasets in a common framework. For example, GIS can be used to integrate radar, lightning, and satellite data with land use and population data to study how deforestation and urban development affects the occurrence and frequency of forest fires. The ability to integrate observations and model data from several geoscience disciplines with socioeconomic and biophysical data in a common framework permits not only multidisciplinary analysis and synthesis, but also provides a pathway to approach geoscience problems and processes from an ESS perspective.

While the atmospheric science community has a rich tradition in developing specialized scientific analysis and visualization tools, which can be loosely characterized as SIS, to process, analyze, and display atmospheric data, the field has only recently begun to embrace the use of GIS in education and research. One reason for the slow adoption of GIS by the atmospheric science community is that current GIS frameworks, due to their limitations in data models and lack of conceptual and physical interoperability of proprietary GIS applications, are not suited to the management and analysis of dynamic, multidimensional atmospheric datasets. Research is needed to identify new frameworks and methodologies to integrate database constructs of GIS with SIS datasets. For instance, combining real-time and forecast weather information with GIS databases of population and infrastructure has significant potential for greatly improving weather related decision support systems. The utility and integration of "off-the-shelf" GIS tools and scientific applications in the context of climate change and disaster research, mitigation and response should also be of high priority in the development of future cyberinfrastructure in the atmospheric sciences.

The trend of GIS applications shifting from operational support tools to strategic decision support systems, as described above, is followed by the demand for the incorporation of more powerful analysis techniques. It is into this context that the need for basic research in the operation, development and use of spatial data analytical techniques and spatial modeling should be identified as an important focus in the future evolution of data services.

Another recent trend of GIS is to make it accessible via the Internet, allowing easier exchange of data and functionality. The GIS applications were historically built as stand-alone tools, often based on proprietary architectures. However, with the increasing availability of geospatial information from diverse sources and their disparate applications in different settings, there is a growing recognition that standards are the key to the interoperability and wider use of GIS tools and services. Organizations like the Open GIS Consortium (OGC) and International Standards Organization are aiming to addresses these interoperability and connectivity issues based on open standards, interfaces, and protocols. In fact, one of the stated goals of the OpenGIS specifications is to make it easy to integrate, superimpose, and render for display geospatial information from different sources and perform in spatial analyses even when those sources contain dissimilar types of data. For instance, the OGC "Geo-interface for Atmosphere, Land, Earth, and Ocean netCDF" (GALEON) Interoperability Experiment supports open access to atmospheric and oceanographic modeling and simulation outputs and it will implement a geo-interface to netCDF datasets via the OpenGIS Web Coverage Server protocol specification. The interface will provide interoperability among netCDF, OPeNDAP, ADDE, and THREDDS client/server and catalog protocols.

CONCLUSIONS

It is important to recognize that we are in the midst of a revolution in data services and the underlying information technologies. This revolution is far from complete. The data services we see today, though advanced, are still evolving and their evolution toward more complex and sophisticated systems is expected to continue for the foreseeable future.

This chapter has presented a brief overview of many issues reshaping geoscience education and research and it provided a survey of many of the technological trends that have contributed to new approaches to data provision and their integration in ESS education and research. The new approaches and services are transforming how students, faculty, and scientists use data services in their daily work The imminent data deluge from a new generation of remote sensing satellite instruments, next generation models, and experiments will have a profound impact on the scientific community and data infrastructure, and it calls for new ways to exploit these and other IT trends for the development of new approaches to scientific data services.

KEYWORDS

- **Cyberinfrastructure**
- **Earth system science**
- **Information technology trends**
- **Geographic information systems**

Chapter 2

Health Benefits of Geologic Materials and Geologic Processes

Robert B. Finkelman

INTRODUCTION

The reemerging field of medical geology is concerned with the impacts of geologic materials and geologic processes on animal and human health. Most medical geology research has been focused on health problems caused by excess or deficiency of trace elements, exposure to ambient dust, and on other geologically related health problems or health problems for which geoscience tools, techniques, or databases could be applied. Little, if any, attention has been focused on the beneficial health effects of rocks, minerals, and geologic processes. These beneficial effects may have been recognized as long as 2 million years ago and include emotional, mental, and physical health benefits. Some of the earliest known medicines were derived from rocks and minerals. For thousands of years various clays have been used as an antidote for poisons. "Terra sigillata," still in use today, may have been the first patented medicine. Many trace elements, rocks, and minerals are used today in a wide variety of pharmaceuticals and health care products. There is also a segment of society that believes in the curative and preventative properties of crystals (talismans and amulets). Metals and trace elements are being used in some of today's most sophisticated medical applications. Other recent examples of beneficial effects of geologic materials and processes include epidemiological studies in Japan that have identified a wide range of health problems (such as muscle and joint pain, hemorrhoids, burns, gout, etc.) that may be treated by one or more of nine chemically distinct types of hot springs, and a study in China indicating that residential coal combustion may be mobilizing sufficient iodine to prevent iodine deficiency disease.

Medical geology—the impacts of geologic materials and geologic processes on animal and human health—has been enjoying resurgence in recent years. Several books on this subject have been published within the past few years (Komatina, 2004; Selinus, in press; Skinner and Berger 2003), numerous articles on medical geology have appeared in various journals (Berger et al., 2001; Bowman et al., 2003; Bunnell, 2004; Ceruti et al., 2001; Dissanayake, 2005; Dissanayake and Chandrajith, 1999; Earthwise, 2001; Finkelman et al., 2001; Selinus et al., 2004) and a number of symposia (Natural Science and Public Health, 2003) have been devoted to this topic. Nearly all of these books, articles, and symposia have focused on the health problems caused by geologic materials such as exposure to elements such as arsenic, mercury, lead, fluorine, selenium, and uranium; minerals such as asbestos, quartz, pyrite; and geologic processes such as earthquakes and volcanic eruptions. The list of geologic materials and geologic processes that adversely impact human health seems endless. However, there is a kinder, gentler side of nature.

Although relatively little acknowledgment has been given to the health benefits provided by geologic materials and processes (Limpitlaw, 2004), there is a long, rich and varied record of these benefits. This is not new knowledge; rather it is knowledge that has, to some degree been lost. It has been pushed aside for a more "modern" way of healing. As tribes and languages die out, knowledge of the ways minerals, fossils, and rocks were used to cure vanishes (Min, 2004). Awareness of the beneficial health effects of rocks and minerals may have occurred more than 2 million years ago. Abrahams (2005) cites the discovery of powdered clays at a homo habilis site in Africa that is about 2 million years old. The most logical explanation is that these early humanoids used the powdered clays to aid in digestion or as an antidote for upset stomachs—the same uses that these clays are put to today! For eons primitive tribes throughout the world have used various types of clays for nutritional and therapeutic purposes (Price, 2000). As long as 3,000–5,000 years ago, ancient civilizations (Mesopotamia, China, India, Egypt, etc.) used minerals for their therapeutic value. "Terra sigillata," (Earth that has been stamped with a seal) was described in the first century AD (Abrahams, 2005) and may have been the first patented medicine (Figure 1).

Figure 1. Terra Sigillata; Adapted from Reinbacher (2003).

The health benefits of geologic materials and processes can be organized into several broad categories:

- Source of essential nutrients
- Pharmaceuticals and health care products
- Hot springs and geysers
- Talismans and amulets

This chapter will briefly discuss these health benefits providing a few examples from each category. It is not our intention to offer a comprehensive discussion of the health benefits. The purpose of this chapter is to provide the readers with an awareness of this oft-neglected aspect of medical geology.

SOURCE OF ESSENTIAL NUTRIENTS

All living organisms require certain naturally occurring elements (essential nutrients) for their metabolism to function efficiently. For humans there are 14 elements that are believed to be essential. These are: calcium, chromium, copper, fluorine, iodine, iron, magnesium, manganese, molybdenum, phosphorus, potassium, selenium, sodium, and zinc (Price, 2000). The ultimate source of these elements is the natural environment. From the dawn of time human beings have obtained the essential nutrients by eating plants that accumulated these elements from the soil; by eating meat from animals that accumulated the elements from plants or from animals that had obtained the elements from plant eating organisms lower in the food chain; by drinking water in which the nutrients are suspended or dissolved; or by directly ingesting soil (Geophagea). Today many people rely on vitamin pills to augment their daily requirements of the essential elements but the elements in vitamin pills are ultimately derived from rocks, minerals, or natural waters.

Geophagy or geophagea is a Greek word meaning "eating earth". It refers to the practice of people in almost every society to eat soil as part of their diet (Figure 2). Pica is a similar term that is commonly applied to the tendency of children to intentionally eat unnatural substances such as soil. The craving to eat soil, principally clay, may be physiological in nature or may have derived from observation of animals that eat soil, clay, pebbles, and so on. Soil may constitute 10% or more of what grazing animals ingest (Abrahams, 2005). Certain birds are known to ingest clays prior to eating toxic fruits. The clays apparently act to detoxify the fruits. In Africa, wild yams, an important source of nutrients during time of famine, can be eaten if detoxified through ingestion of clay (Abrahams, 2005). American Indians have also been known to mix soil with acorns, tubers, and berries to render them edible.

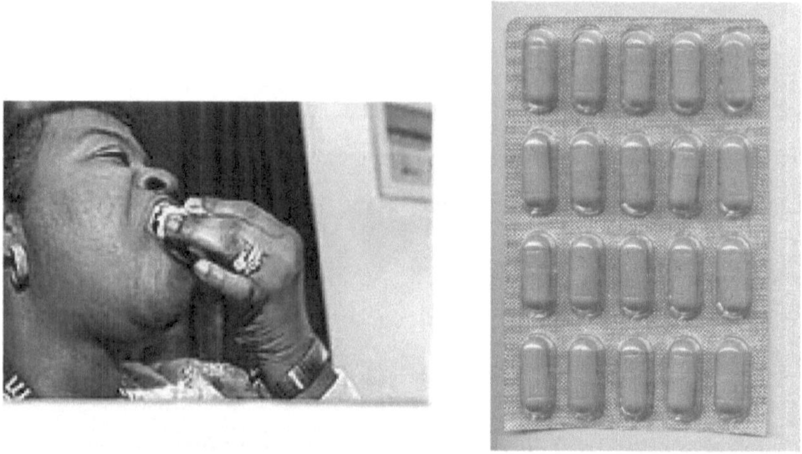

(a) (b)

Figure 2. (a) Eating clay; (b) Loesss (wind-blown sand) pills produced in Germany.

There is no question that people who eat unprocessed soil are at risk for several significant health problems (exposure to harmful levels of chemical elements, exposure to soil pathogens, intestinal blockages, etc.). There is, however, a lively debate as to whether there are other health benefits derived from geophagy. Reinbacher (2003) uses the term "geopharma"—eating Earth as a medicine for this facet of geophagy. Several hypotheses have been proposed for the health benefits of geophagy. The most commonly cited benefit is that the soil provides essential nutrients. In Africa and elsewhere, pregnant woman commonly eat red clays. The iron in these clays that may offset iron-deficiency associated with the pregnancy. There is a question as to whether the iron in these clays is bioavailable. The calcium in soils and minerals are also believed to aid in lactation of nursing women (Abrahams, 2005).

Reinbacher (2003) suggests that eating clay may augment an insufficient food supply, thus staving off the feeling of hunger (Figure 2). He cites workers in Germany who used "stone butter", talc, on their lunch sandwiches. There are also reports that ingesting clay may cure certain diseases. He cites the example of people in Cameroon who eat baked clay to cure hookworm.

There is general agreement that ingestion of clays can cure certain stomach disorders such as stomachaches, acid indigestion, nausea, and diarrhea. Kaolin and smectitic clays are the principle ingredients in some of the most commonly used products for treating stomach disorders. There are many books (Knishinsky, 1998, for example) that advocate eating clay as a safe and effective way to detoxify, aid digestion, relieve nausea, treat allergies, soothe ulcers, counteract poisoning, and so on.

An unusual source of an essential element was recently identified. Residential coal combustion that can cause severe health problems (Finkelman et al., 2002) may help to prevent iodine deficiency disorders (IDDs) in northwestern Guizhou Province, P.R. China (Wang et al., 2008). The coals from this region have anomalously high iodine contents (~8 ppm versus ~1 ppm for the world average). The IDD is rare in the northwest part of the province where coal is plentiful and people use it to cook, heat their homes, and dry crops hanging from the rafters. In the southeastern part of the province where wood is the primary fuel, IDD is endemic. Inhalation of iodine volatilized by the coal combustion and ingestion of foods enriched in iodine by drying over the high-iodine coal fires may be providing sufficient iodine to ward off the goiter, cretinism, and mental retardation caused by IDD.

PHARMACEUTICALS AND HEALTH CARE PRODUCTS

Most people would readily agree that "minerals" are essential for their health and wellbeing. They are, of course, referring to the essential elements (calcium, potassium, sodium, etc.) that are required for good health and not to minerals in the strict sense. Minerals are defined as a naturally occurring crystalline solid with a definite but not fixed chemical composition (Hurlbut, 1961). Minerals are the fundamental building blocks of rocks, are a major component of all soils, and are the ultimate source of the essential elements. The term mineral as used in nutrition is actually a misnomer resulting from the fact that the essential elements were derived from minerals; but minerals in their own right have many beneficial health properties.

They are used in a wide range of everyday healthcare products. Common products such as Kaopectate (kaolin), Tums (calcite), Milk of Magnesia (magnesite), talcum powder (talc), toothpastes (quartz, rutile), anti-perspirants (bauxite), calcium supplement (ground up coral) all rely on minerals for their beneficial health properties. In addition, various clays are used in antacids, gypsum in plaster of paris. Pumice is the abrasive in natural pumice stone and lava soap, extracts from coal are used in dandruff shampoos, fluorine in fluoridated water and toothpastes, and salt (halite) has been used as an antibacterial agent and as a food preservative for millennia. Epsom salt (epsomite) is found in most every household. Baking soda (trona) is a well-known antacid and is also used intravenously (Figure 3a–h).

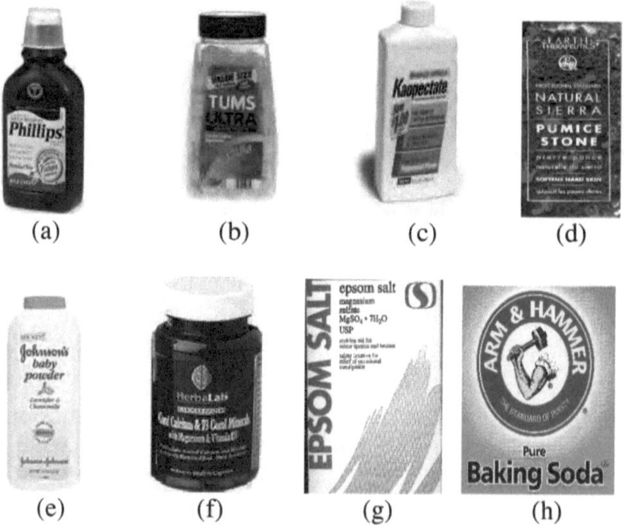

Figure 3. Minerals in everyday use. (a) milk of magnesia, (b) tums antacid, (c) kaopectate, (d) natural pumice stone, (e) baby powder, (f) calcium supplement, (g) epsom salt, and (h) baking soda. *******Note*: Illustration of these products does not constitute an endorsement.

Clays, the mineral group most widely used, and likely the longest used for heath care by both humans and animals, deserves special recognition (Reinbacher, 2003). One of the most important life saving uses of clay is its curing effect of buruli ulcers or flesh eating disease (De, 2002).

Zeolites have become a valuable medical resource. QuickClot (Figure 4), a product not available to the public at this point, has saved many lives with its blood coagulating properties. Emergency personnel and the military have it readily available. Heavy bleeding that occurs when femoral arteries are severed, can be stopped by ground up and sterilized zeolite (Alam et al., 2004). Furthermore, studies are under way to confirm zeolite's affinity for radionucleides. Zeolites are already being used in low level nuclear radiation clean ups and are now being tested on humans to counter the effects of nuclear radiation exposure. Clinical trials are being conducted to investigate zeolite's anticancer properties (Pavetic et al., 2000).

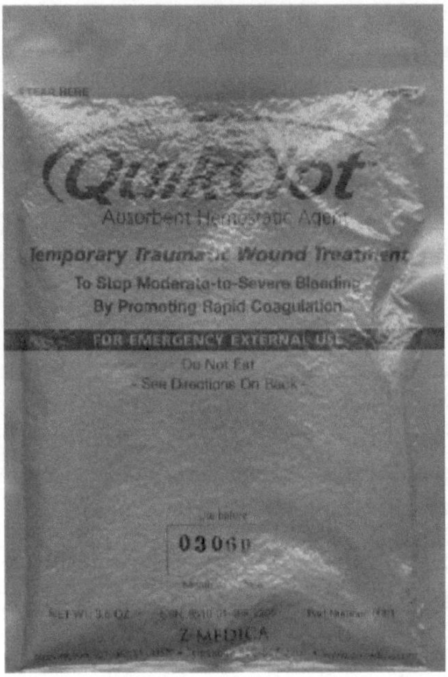

Figure 4. Ground up zeolite sold as QuickClot.

Elements such as arsenic (Trisenox), mercury (mercurochrome, dental amalgam), boron (boric acid), selenium (dandruff shampoos), sulfur (drugs), barium (enemas, X-radiography), bismuth (Pepto Bismol), zinc (skin ointments), are the active ingredients in various drugs and pharmaceuticals. Some of the more exotic elements such as cesium, lithium, indium, samarium, gold, platinum, titanium, and gallium, are used in an ever increasing array of sophisticated medical treatments and products (Figure 5).

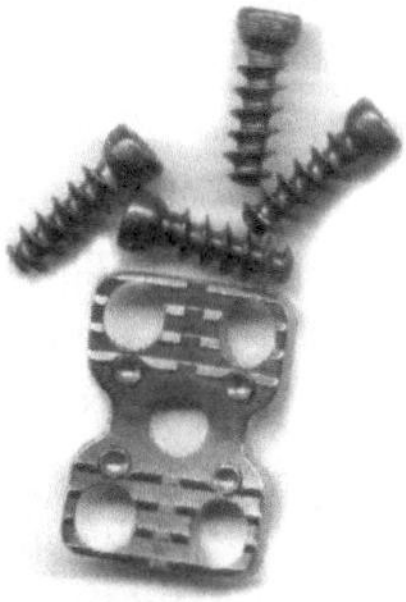

Figure 5. Surgical titanium plate and screws.

While modern day America uses a few common minerals for medical purposes, other countries manufacture tons of pharmaceuticals using a variety of minerals and fossils (Aschoff and Tashingang, 2001). More than 90 minerals, mineraloids, and rocks have been used for beneficial purposes with clay minerals leading the list, followed by quartz, amber, hematite, pearl, and malachite. Various illnesses and maladies have been treated with these Earth materials, such as malachite and clays for infections, clays and pearls for gastrointestinal problems, and amber for alcoholism and to strengthen the immune system.

In ancient cultures the only available healing materials came from plants, animals, or minerals. Often the color of a mineral was believed to be related to the condition of an illness such as blood disorders being treated with a red mineral like hematite or ruby. Reasons for use of these Earth materials range from actual effectiveness to the placebo effect.

Little research exists as to the actual curative and palliative qualities of minerals. However, the actual healing signature of one mineral that had been used for over 5,000 years (Nunn, 1996) was demonstrated in an *in vitro* experiment. The purported anti-microbial properties of malachite were verified in a preliminary study with *Staphylococcus aureus* and *Pseudomonas aeruginosa* (Limpitlaw, unpublished data). This study showed that knowledge of healing properties of certain minerals existed and was utilized for millennia.

HOT SPRINGS AND RELATED GEOLOGIC PHENOMENA

Visiting hot springs and thermal baths for prevention and cure of a variety of illnesses, once popular in the USA, has fallen out of favor. This practice, however, is still popular in many other cultures. For example, in some European countries, going to a hot springs "cure" is included in medical insurance.

The study of the therapeutic benefits of naturally occurring mineral waters is known as balneology. In the US, this science is not well known, and is seldom practiced. However, throughout Europe and Japan, balneology and hot springs therapy is very much a part of routine medical care. Medical prescriptions are given by licensed doctors for the treatment of a wide range of conditions, and utilizing mineral waters as a part of preventative medicine is widely recognized and encouraged.

Advocates claim a large number of health problems can be cured by hot spring therapies. In Japan nine different chemical classes of hot springs have been recognized (simple, chloride, hydrogen carbonate, sulfate, carbon dioxide, iron, sulfur, acid, and radioactive).

Each class of hot spring is reputed to be effective in treating a wide range of health problems such as burns, hypertension, diabetes, gout, muscle aches, hemorrhoids, and so on. (Mio Takeuchi, 2002, personal communication).

There appears to be little in the way of epidemiological support for such claims in the US medical literature. A brief review of the National Library of Medicine abstracts of articles dealing with the therapeutic effects of hot springs revealed that most authors are from Japan. Other authors were from Germany, Turkey, China, Portugal, Greece, and Israel. The one article by a US author reported on the historic use of hot springs to effectively treat venereal disease.

In addition to hot springs, geologic materials such as mud and sand have been used for therapeutic purposes. The therapeutic powers of Dead Sea black mud are mentioned in the Bible (Abdel-Fattah, 2004) and the mud is currently marketed worldwide. Beach sand baths in Porto Santo Island, Portugal are popular therapeutic practice. Gomes and Silva (personal communication, 2004) believe that essential elements in the carbonate and clay rich sands are incorporated into the bodies of the people covered by the sand by percutaneous mechanisms.

TALISMANS AND AMULETS

Talismans and amulets, commonly naturally occurring crystals, are believed by some to have positive health effects but, unlike the rock-derived pharmaceuticals, do not have scientifically proven health benefits. Talismans have been used for thousands of years and were believed to ward off evil and heal certain medical conditions. Geinger (2003) identifies the unique spiritual, emotional, mental, and physical healing powers of nearly 100 rocks and minerals.

Since 1996 a scientific group in Germany has been doing empirical research on the health effects of crystals. Seventy different study units are investigating four crystals each year. As of July 2004, 40 different minerals had been tested. Their findings are published in a magazine called *Opalitho*. All reports include physical, spiritual, and emotional effects (Opalitho et al., 2005).

A recent survey on a college campus found that the student populations coming from tribal backgrounds such as Native Americans, Hawaiian, and Polynesian all have standing traditions of using talismans and amulets (Limpitlaw, 2005). These traditions are being continued to this day.

Finally, it is an undisputed fact that a certain group of minerals exert wonderfully positive effects when presented to another person—these minerals are commonly called gems.

CONCLUSIONS

The health benefits derived from geologic materials and geologic processes are commonly ignored in formal treatises, seminars, and research agendas. However, the list of benefits is extensive, varied, and often significant. Moreover, there still exists an exciting potential for finding novel health benefits of geologic materials and processes.

KEYWORDS

- **Geologic materials**
- **Geologic processes**
- **Iodine deficiency disorders**
- **Medical geology**
- **Pharmaceuticals**

Chapter 3

Cloud Statistics from Spaceborne Lidar Systems

S. Berthier, P. Chazette, J. Pelon, and B. Baum

INTRODUCTION

The distribution of clouds in a vertical column is assessed on the global scale through analysis of lidar measurements obtained from three spaceborne lidar systems: LITE (Lidar In-space Technology Experiment, NASA), GLAS (Geoscience Laser Altimeter System, NASA), and CALIOP (Cloud-Aerosol LIdar with Orthogonal Polarization). Cloud top height (CTH) is obtained from the LITE profiles based on a simple algorithm that accounts for multilayer cloud structures. The resulting CTH results are compared to those obtained by the operational algorithms of the GLAS and CALIOP instruments. Based on our method, spaceborne lidar data are analyzed to establish statistics on the CTH. The resulting columnar results are used to investigate the interannual variability in the lidar CTHs. Statistical analyses are performed for a range of CTH (high, middle, low) and latitudes (polar, middle latitude, and tropical). The probability density function (PDF) of CTH are developed. Comparisons of CTH developed from LITE, for 2 weeks of data in 1994, with International Satellite Cloud Climatology Project (ISCCP) cloud products show that the cloud fraction observed from spaceborne lidar is much higher than that from ISCCP. Another key result is that ISCCP products tend to underestimate the CTH of optically thin cirrus clouds. Significant differences are observed between LITE-derived cirrus CTH and both GLAS and CALIOP-derived cirrus CTH. Such a difference is due primarily to the lidar signal-to-noise ratio that is approximately a factor of three larger for the LITE system than for the other lidars. A statistical analysis for a full year of data highlights the influence of both the Inter-Tropical Convergence Zone and polar stratospheric clouds (PSCs).

One of the most challenging objectives of current climate research programs is in understanding the impact of clouds on the global energy budget and hydrological balance. Indeed, clouds have a significant influence on the Earth's radiative balance and induce various climatic feedbacks that are not well known (e.g., Forster et al., 2007; Stephens, 2005). One important issue is the cloud spatial and vertical distribution (e.g., Rossow and Schiffer, 1991). The vertical distribution of the cloud layers in an atmospheric column can lead to very different assumptions of cloud overlap in numerical models. Clouds influence the heating rates and the radiative energy budget. Feedbacks due to cirrus clouds are an important issue in climate modeling as they have a significant radiative impact which largely depends on their characteristics. Such feedbacks become more complex if lower-level clouds are present. To properly model overlapping cloud layers, it is necessary to know the thermodynamic phase of each cloud layer in addition to the other properties such as height, optical thickness, and effective particle size. Multilayered, overlapping clouds are presently poorly modeled because their life cycle implies dynamical processes at scales much smaller than those used in general circulation model (GCM) calculations, and also because of their complex microphysics (Flatau et al., 1989). A better knowledge of the horizontal and vertical distribution of all cloud layers is required to improve cloud parameterization in existing climatic models and better assess their feedbacks. Numerous previous studies have been performed using spaceborne passive instruments to infer the vertical distribution of clouds, for example combination of ISCCP and SSM/I (Yeh and Liou, 1983), NIMBUS 7 (Stowe, 1984), TOVS (Susskind et al., 1987), 3DNEPH and RDNEPH (Hugues and Henderson-Sellers, 1985), ISCCP (Rossow et al., 1985), combination of AVHRR and HIRS/2 (Baum et al., 1995), AVHRR (Heidinger and Pavolonis, 2005; Pavolonis and Heidinger, 2004), MODIS published by Copernicus Publications on behalf of the European Geosciences Union.

Berthier et al. made comparisons of cloud statistic from spaceborne lidar system (Baum et al., 2003; Chang and Li, 2005; Nasiri and Baum, 2004), SAGE-II (Kent et al., 1993), and HIRS/2 (Jin and Rossow, 1997; Wylie et al., 2005). Even with techniques now available that show some facility in detecting the occurrence of multiple, but overlapping, cloud layers in passive radiometric data, it is still problematic to infer the properties of each cloud layer. Lidar offers the opportunity to better determine the presence of optically thin ice clouds and to detect lower-level stratiform systems (e.g., Winker et al., 1998). Active measurements may also be used to mitigate biases in CTHs that arise in complex situations, such as in the polar regions. New spaceborne backscatter lidar missions (ESA: Ingnann et al., 2008 and http://www.esa.int/export/esaLP/index.html, NASA: http://www.veg3dbiomass.org/VolzVeg3Dworkshop.pdf) are currently underway or in preparation to give further insight on the spatial and vertical distribution of both clouds and aerosols in the troposphere, on a continuous observational basis as required by the models. However, compared to ground-based systems, spaceborne lidar systems provide an atmospheric backscattered signal with a relatively weak signal to noise ratio (SNR), thus requiring significant signal processing (e.g., Chazette et al., 2001).

In this study, we develop and apply a methodology to derive the PDF of cloud layer structures from lidar profiles obtained during the LITE, (Winker, 1996) mission in September, 1994. This pioneering mission provides an opportunity to estimate the cloud spatial distribution with a high spatial resolution under a given satellite footprint. The PDF retrieved from LITE data are compared with those calculated from the new spaceborne lidar missions, such as GLAS, (Palm et al., 2002) and CALIOP (Winker et al., 2002). The GLAS and CALIOP data are processed two ways: (1) with the methodology developed for the LITE profiles and (2) the methodology used for the operational products. For the LITE time period in 1994, we perform comparisons to the ISCCP cloud products (Rossow and Schiffer, 1991). A full year of products generated from CALIOP is used to analyze the impact of the ITCZ latitudinal position and the occurrence of PSCs on the intra-and inter-annual lidar signal variability.

METHOD AND SPACEBORNE OBSERVATIONS

Different approaches have been developed to retrieve the CTH from spaceborne lidar systems. Such approaches have been used for the operational algorithms for GLAS (Palm and Spinhirne, 1998) and CALIPSO (Cloud-Aerosol Lidar Pathfinder Satellite Observation), (Vaughan et al., 2004) missions. Based on spaceborne lidar modeling performed by Chazette et al. (2001), we suggest an alternative methodology to retrieve the CTH for both semitransparent and dense clouds and apply this methodology to the LITE data.

Algorithm to Retrieve the Cloud Top Heights

We adapt the method developed by Chazette et al. (2001) to infer the CTHs of scattering layers in the atmosphere from simulated spaceborne lidar signals with low SNRs (~3) to actual spaceborne lidar measurements. The method will be called the "Local Method" hereafter.

To determine the existence of a peak (i.e., a cloud) in a lidar calibrated and attenuated backscatter signal *S* at any altitude level *i* requires an ability to discriminate between an actual signal and signal noise. Here, the discrimination is performed by determining a threshold value *F*. The value of *F* is proportional on the signal noise, which is used to define the variance *Var* as follows:

$$Var(k)=\frac{1}{2n+1}\sum_{i=k-n}^{k+n}\left[\frac{S[i]-\overline{S}}{\sigma_N}\right]^2>F \qquad (1)$$

where (2n+1) is the number of points of the filtering window. A constant size is assumed for the windows, with n equal to 3, corresponding to a window size equivalent to 7 pixels. The $\overline{S}$ and σ_N are respectively the mean value of the detected signal and the noise standard deviation in an altitude range where only noise is expected to be present (i.e., between 19 and 20 km height).

Figure 1 provides an example of the determination of *F* for GLAS. This figure gives the mean cloud depth as a function of *F*. The method has been optimized based on the depth of the scattering layers so that the only cloud structures considered have a geometrical depth larger than 100 m. This approach attempts to minimize the number of false alarms.

For a value of *F* in the interval of 1–10, most of the values of *Var* are greater than *F*. This means that for an individual altitude level, the noise dominates the measurement and it is not possible to discern the presence of a cloud. If this situation occurs at every altitude, the lidar shot is not used in our analysis. As values of *F* increase between 10 and 1,000, fewer lidar signals are misclassified as cloudy structures. The misclassifications between noise and an actual cloud structure are further reduced in the case of the LITE data by using a median filter. For values of $F > 1{,}000$, the lidar signal from a cloudy structure is more certain to not be caused by signal noise.

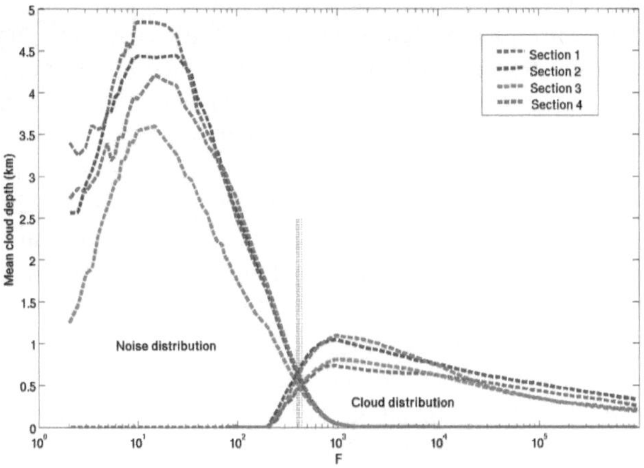

Figure 1. Distributions of both the mean cloud geometric thickness and the signal noise against the value of the threshold F for GLAS measurements. The Glas/ICESat data used for this study are from October 13, 2003. Data are collected between 16:39 and 16:59 (GMT) for Section 1, 16:59 and 17:19 (GMT) for Section 2, 17:19 and 17:39 (GMT) for Section 3, and 18:20 and 18:40 (GMT) for Section 4. Here, F has been assessed to be equal to 395, 441, 413, and 411, for respectively Sections 1, 2, 3, and 4.

Two distributions are thus retrieved; the first is associated with the noise and the second is associated with the scattering layers themselves. A value for F is inferred from the intersection of these two distributions to minimize the error probability that is a function of the no detection and false alarm probabilities (Chazette et al., 2001).

From all the structures identified after the first step of the algorithm, we further discriminate between clouds and aerosols. However, this operation can be quite difficult for lidar data, primarily for the case of dust aerosol that is denoted by a low Angstrom exponent (Grant et al, 1997). Different classification approaches have been suggested for GLAS (Palm et al., 2002) and CALIOP (Liu et al., 2004; Vaughan et al., 2004) lidar profiles. We use the GLAS prototype algorithms to separate clouds and aerosols in the LITE data processing between the ground level and the altitude of 8 km. The lidar signal is explained in term of the attenuated volume backscatter coefficient $\beta(r)$ (Platt et al., 1998), that is, to the calibrated, range-corrected lidar signals within each layer. The discriminator used here is based on the threshold relation given by $P = \beta'_{max} |\Delta\beta' / \Delta z|_{max} > X$. β_{max} is the maximum attenuated backscatter of the layer and $|\Delta\beta' / \Delta z|_{max}$ is the maximum vertical gradient magnitude within the layer. X defines the thresholds previously defined. Layers with values of P larger than X are interpreted as cloud whereas the others are classified as aerosol. After a statistical study, we determined the value of X to be 3.10^{10} $m^{-3}.sr^{-2}$.

The altitude range of the resulting CTHs is classified following the ISCCP approach: low (L), middle (M), and high (H) clouds corresponding to pressure levels of 1,000–680 hPa, 680–440 hPa, and 440–50 hPa, respectively (Rossow and Schiffer, 1991).

Spaceborne Datasets

The LITE data were recorded on board the Space Shuttle Discovery during the NASA space shuttle mission STS-64 in September 1994. Over 11-day period of the mission, the LITE instrument accumulated 53 hr of data (i.e., 70 available orbits) with a pulse repetition rate of 10 Hz, at ~240 km height, within a few degrees of nadir at three wavelengths: 355, 532, and 1,064 nm. The vertical resolution is 15 m and the horizontal sampling is 700 m along the footprint. Only the data at 532 nm are used in this study because of its better SNR, which is close to nine in the planetary boundary layer (PBL). The LITE database contains only Level 1 (calibrated and geolocated lidar backscatter profiles) data that can be accessed at http://www-lite.larc.nasa.gov.

The GLAS instrument was on the satellite platform called ICESat (Ice, Cloud, and land Elevation Satellite, http://icesat.gsfc.nasa.gov/). The ICESat was launched on January 13, 2003, had an inclined orbit of about 94°, and an altitude of about 590 km at the equator. The vertical and the horizontal resolution of the collected data are respectively 76.8 m and 175 m. Despite a pre-launch goal for the lidar of 3 years continuous operation, the GLAS Operation Center needed to reduce the energy and the time period of the lidar activities because of a technical malfunction of the lidar (Abshire, 2005; Thome et al., 2004). In this study, we use data from laser 2A that was recorded before a temperature anomaly occurred, and for the same season as when LITE was in operation, that is between the September 25, 2003 and October 3, 2003. Raw data at 532 nm (named Level 1) and the Level 2 official GLAS cloud product are used in this study to assess the accuracy of our algorithm. The GLAS data are characterized by a lower SNR than that of LITE, close to 1.5 in the PBL, due to the higher altitude of the satellite and the lower energy emitted by the instrument. The GLAS database includes Level 1 and Level 2 (derived products such as CTHs, optical depth, …) data (available at http://nsidc.org/data/icesat.data.html).

The CALIOP instrument is on the CALIPSO satellite platform. The CALIPSO satellite was inserted in the A-Train constellation (http://www-calipso.larc.nasa.gov/about/atrain.php) behind Aqua on April 28, 2006. This satellite, which resulted from a collaboration between NASA and CNES, began to collect data in June 2006. This database is the first to provide more than 1 year of spaceborne lidar ob-seasonal variations may be studied with a high vertical resolution. The mean altitude of the satellite is 705 km, resulting in vertical and horizontal resolutions of 30 and 330 m, respectively. The inclination of the satellite is about 98.2°, and thus covers the polar regions. The technology of the GLAS and the CALIOP instrument are quasi-similar, but operational modes are different. The SNR values are similar to that of the GLAS instrument (~2.1). The CALIOP Level 1 and Level 2 data can be accessed at http://eosweb.larc.nasa.gov/PRODOCS/calipso/tablecalipso.html or at http://www.icare.univ-lille1.fr.

The ISCCP database (Schiffer and Rossow, 1983), collects analyzed infrared (11 μm) and visible (0.6 μm) radiances measured by the operational geostationary and polar-orbiting weather satellites. The ISCCP products provide a detailed description of the horizontal variations of cloud top pressure and optical thickness. The horizontal resolution of the ISCCP-DX data, used in this work, is 30 × 30 km^2 (http://climserv.lmd.polytechnique.fr). The ISCCP dataset is built from Meteosat (http://www.eumetsat.

int), GOES (Geostationary Operational Environmental Satellites, http://www. goes. noaa.gov/), and TOVS (TIROS-N Operational Vertical Sounder, http://www.class. noaa.gov/) measurements.

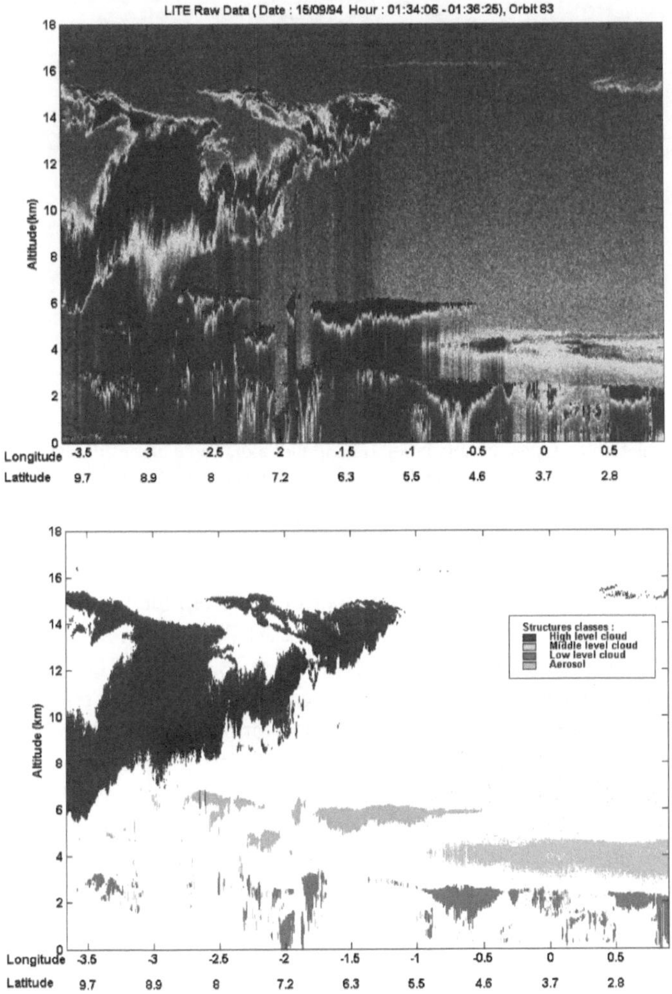

Figure 2. Raw LITE data for September 15, 1994 (orbit 83) are given in the upper figure. The result of the classification of clouds and aerosol layers is given in the lower figure.

CLASSIFICATION OF CLOUDS AND AEROSOLS

A case study illuminates the classification of clouds and aerosols in the lidar data. Figure 2 gives an example of the classification results for the LITE orbit 83 in September 15, 1994. The layers of clouds and aerosols are quite distinct. In particular, the layers of low, middle, and high clouds are quite separate.

In the case of LITE data, there is no existing Level 2 cloud product so we will use the results derived from the local algorithm.

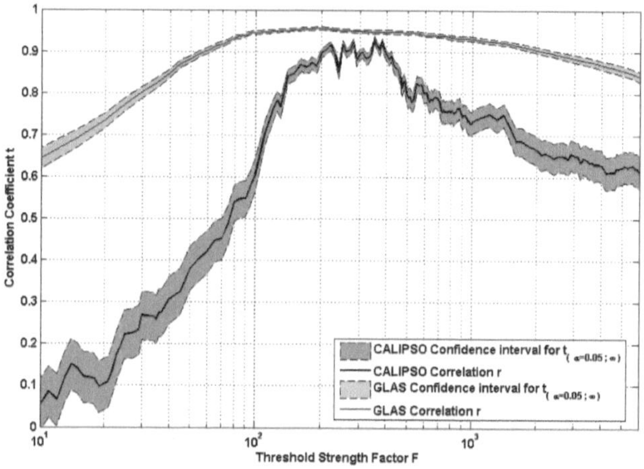

Figure 3. Correlation between the local method and the operational algorithm of GLAS and CALIOP lidar against the value of *F*. Considering the GLAS data, the segment that are been considered are the same that ones used to construct the Figure 1. For the CALIPSO data, the used data come from the day of the June 26, 2006, at 20:52 GMT.

To compare the cloud classifications performed from the previous algorithm and the operational algorithms of the GLAS and CALIPSO missions, it is necessary to evaluate the coherence between their respective results. For the CTH of only the uppermost structure on each lidar shot, we calculated the coefficients of correlation between each operational algorithm and the local method for cloudy scenes observed by GLAS and CALIOP. Figure 3 shows the results for GLAS and CALIOP according to the value of *F*. The maximum level of correlation is reached for a value of *F* close to 400 for both GLAS and CALIOP. These values correspond to the thresholds that we will use in subsequent analyses. The correlation is high with values of 0.95 and 0.93 for GLAS and CALIPSO, respectively. The variations observed are related primarily to the non-detection of cloudy structures associated with low optical thickness (lower than 0.1 at the wavelength of 532 nm) by both GLAS and CALIOP. Indeed, the detection of the semitransparent scattering structures is less sensitive with GLAS and CALIOP because the SNR is weaker than that of LITE. The limits of cloud detection that result from different SNR values have implications for the statistics of global cloud cover presented later in this chapter.

LIDAR-DERIVED CLOUD TOP HEIGHT

Now that some understanding has been gained of the coherence between the various algorithms for the identification of cloudy structures, the associated statistical distributions of the CTHs can be compared.

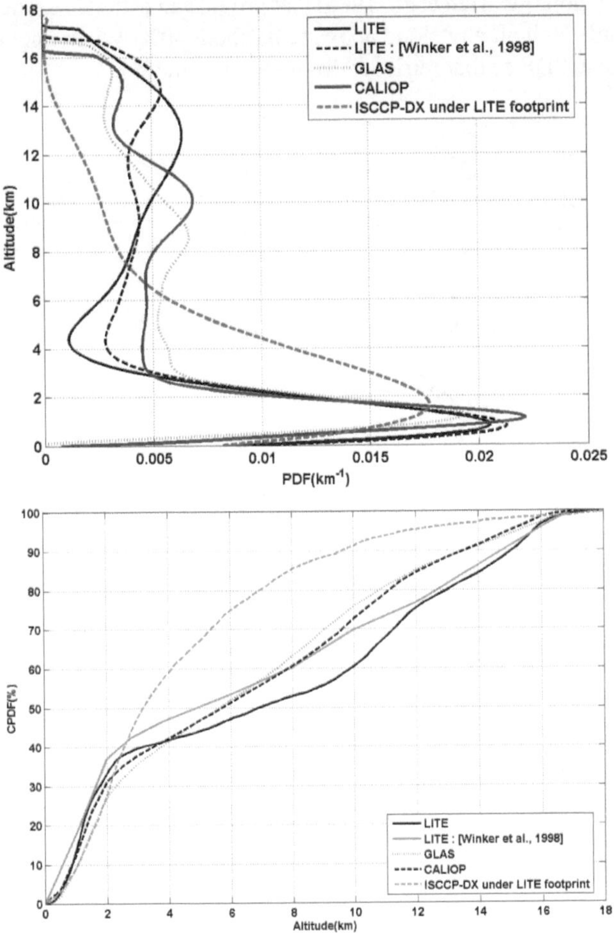

Figure 4. Cloud top height probability density functions (Figure 4a) and cumulative PDF (CPDF) (Figure 4b) for the highest cloud structure established from: LITE (local method applied on September, 1994), GLAS (operational algorithm, applied on last week of September and first 2 week of October, 2003), and CALIOP (operational algorithm, applied on September, 2006. The CPDF previously retrieved by Winker et al. (1998) with LITE data and the CDPF for ISCCP on the footprint of the LITE orbits on September, 1994 are also given.

Cloud Top CPDF

Figure 4a and 4b shows respectively the top of cloud PDF, and cumulative PDF (CPDF) as obtained from the datasets of LITE (local method), GLAS (operational algorithm) and CALIOP (operational algorithm). These CPDF indicate good agreement for low-level clouds up to an altitude of approximately 3 km. There are very few differences between GLAS and CALIOP for all CTH, whereas the LITE CPDF shows a stronger sensitivity to high clouds. Since LITE has a higher value of the SNR, it is better able to detect high altitude semitransparent cloud structures. Note that for the low level clouds (CTH <~3 km), the results in Figure 4b are in good agreement with surface observations (Warren et al., 1985) that provide an estimate of the percentage of low level clouds without any cloud above to be about 40%. Winker et al. (1998) show about the same CPDF as that retrieved from the local method.

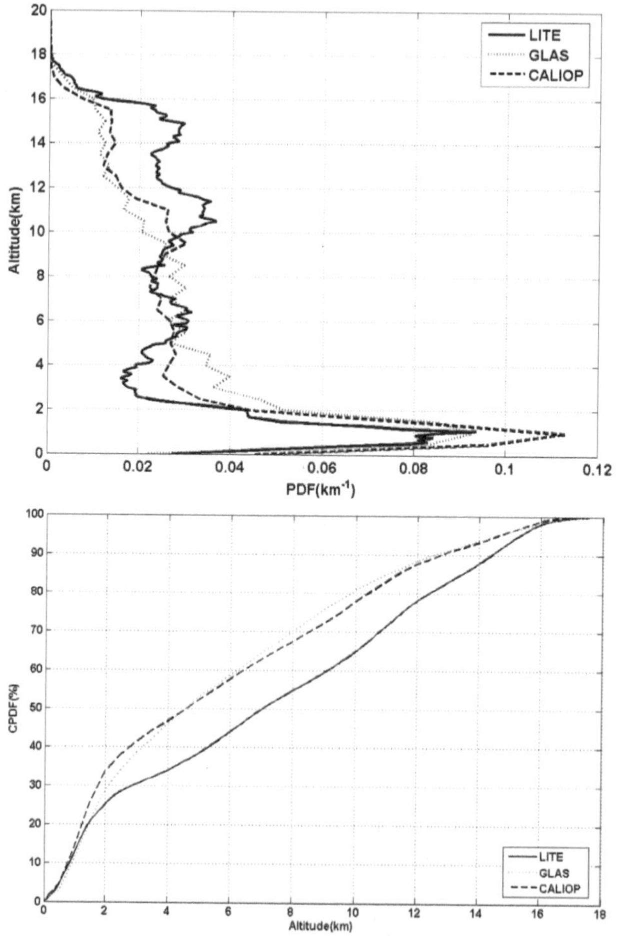

Figure 5. Cloud top height probability density functions (Figure 5a) and cumulative PDF (CPDF) (Figure 5b) for all the cloud structure established from LITE, and CALIOP at the same date as for Figure 4a and b.

When the upper cloud structure is semitransparent, the lidars offer the possibility of identifying lower-level scattering layers. The efficiency of lower scattering layer detection is directly linked to the SNR. Figure 5a and b gives the PDF (CPDF) for all the cloud structures for each spaceborne lidar. They are similar for GLAS and CALIPSO with a slight increase of middle-altitude clouds. The difference is more important for LITE where more low level cloud structures are detected. In this last case, the CPDF reach about the same value (~45%) at an altitude of approximately 6 km.

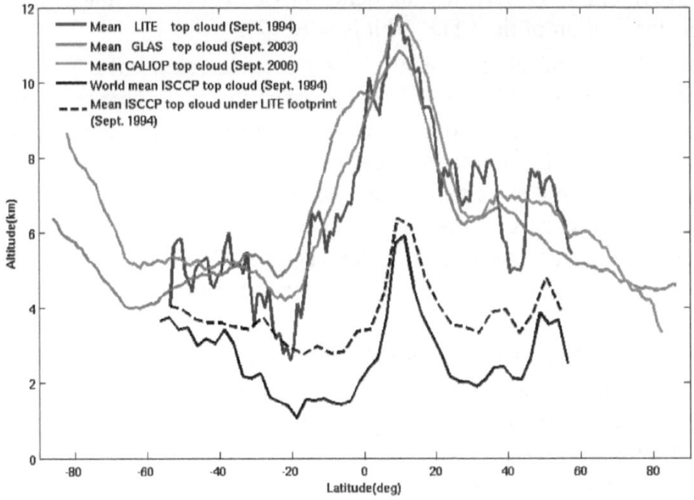

Figure 6. Mean cloud top height against the altitude of the highest structure for LITE (local method applied on September, 1994), GLAS (operational algorithm applied on the last week of September and the first 2 weeks of October, 2003), and CALIOP (operational algorithm applied on September, 2006). The mean cloud top height given for both the global coverage of ISCCP data and the ISCCP data under the LITE footprint on September, 1994 is also given.

Latitudinal Cloud Top Height Distributions

Mean Cloud Top Height

Figure 6 gives the mean CTH for the highest-level structure as a function of the latitude for LITE, GLAS, and CALIOP. The acquisition periods are different. Ten days are considered for the LITE mission on September, 1994. A full month (September, 2006) of CALIOP is analyzed, and for GLAS the time period encompasses the last week of September and the first 2 weeks of October, 2003. The reduced number of days for LITE mission may contribute to the greater variability of the data. No significant difference in the shape was observed on GLAS data when splitting the data into weekly sequences.

Between 60°S and 60°N, there is good agreement between the observations of the various missions, which is fortunate considering difference in the acquisition periods and the potential for interannual variability to influence the results. The northern maximum near 10°N corresponds to the position of the ITCZ (Inter Tropical Convergence Zone) in September (Waliser et al., 1993). The ITCZ moves seasonally, following the solar heating and the warmest surface temperatures: it moves toward the Southern Hemisphere from September through February and reverses from March through August.

Mean CTHs are calculated for clouds ranging between 6 and 11.5 km in the tropical latitudes. Minima are found near 20°S and near 30°N, corresponding to the descending circulation of the Hadley cell. In the latitude range [−60°; −20°] and [30°; 60°] corresponding to the mid-latitudes, the mean CTH varies between 4.5 and 6.5 km.

Only the GLAS and CALIOP measurements provide data at latitudes higher than 60° because the inclination of the LITE orbit is only 57°.

Between 60° and 82°N, the mean CTH tends to decrease linearly from 6 to 4 km (5–4.5 km) height from CALIPSO (GLAS) measurements. There is no significant difference between the two spaceborne lidar observations. On the contrary, between 60° and 82°S, the mean CTH tends to increase from 4 to 6.3 km height for GLAS measurements, and from 5.1 to 8.6 km height for CALIPSO measurements. Such a difference may be due to the presence of PSC. It is interesting to note that Hopfner et al. (2007) showed a stronger occurrence of PSC in 2006 compared to 2003 using MIPAS/Envisat (Michelson Interferometer for Passive Atmospheric Sounding) instrument (Burkert et al., 1983).

Winker et al. (2007) demonstrated the capabilities of CALIOP measurements to provide high resolution vertical profiles to a latitude of 82°S. Winker et al. (2007) showed that clouds observed over the East Antarctic plateau in the middle of the Antarctic Winter were relatively tenuous PSC extending up to ~25 km. Pitts et al. (2007) analyzed such PSC structures from the CALIPSO database and document the occurrence of extensive PSC over large regions of Antarctica throughout the 2006 Austral winter. They show that the 2006 season is very similar to the cold 1987 season, with a higher probability of occurrence under 16 km in September. Furthermore, in a personal communication, C. David, PI of the LOANA (Lidar Ozone and Aerosol for NDSC in Antarctica) ozone lidar, based at the French NDSC Station of Dumont d'Urville, confirms that there was a high occurrence of PSCs in 2006. The influence of PSC on the CTH will be discussed again in a later.

Distributions as Functions of Latitude and CTH

Figure 7 shows the two-dimensional histograms as a function of latitude and altitude of the CTH. The occurrences are given for the LITE, GLAS, and CALIPSO missions, respectively. They show strong similarities but with a noisier pattern for the LITE-derived CTH. However, this variability may be caused by a sampling issue, with LITE providing a lower number of observations compared to GLAS and CALIOP measurements.

In the following discussion, low-level clouds are defined as having cloud-top pressures ranging from 1,000 to 680 hPa; middle-level and high-level clouds ranging respectively from 440 to 680 hPa and 440 to 50 hPa.

For the three cases, a low cloud pattern is observed between 0 and 2 km with a maximum in the frequency of occurrence at a height close to 1.5 km. A minimum in the frequency of low cloud occurrences is observed near the location of the ITCZ due to the presence of high optically dense clouds that mask the potential presence of low-and mid-level cloud structures. For all cases a strong occurrence of high clouds is clearly highlighted between 13 and 18 km height in the ITCZ region, that is between [10°S; 20°N], corresponding to regions of deep convection, and the occurrence of cumulonimbus and cirrus clouds. Generally, one finds higher occurrences of cloud near the tropopause, in particular for the northern middle latitudes and the polar latitudes (>60°). The altitude of the maximum cloud occurrence decreases with increasing latitude.

The higher frequency of occurrence of high clouds is mainly due to the presence of optically thin cirrus clouds.

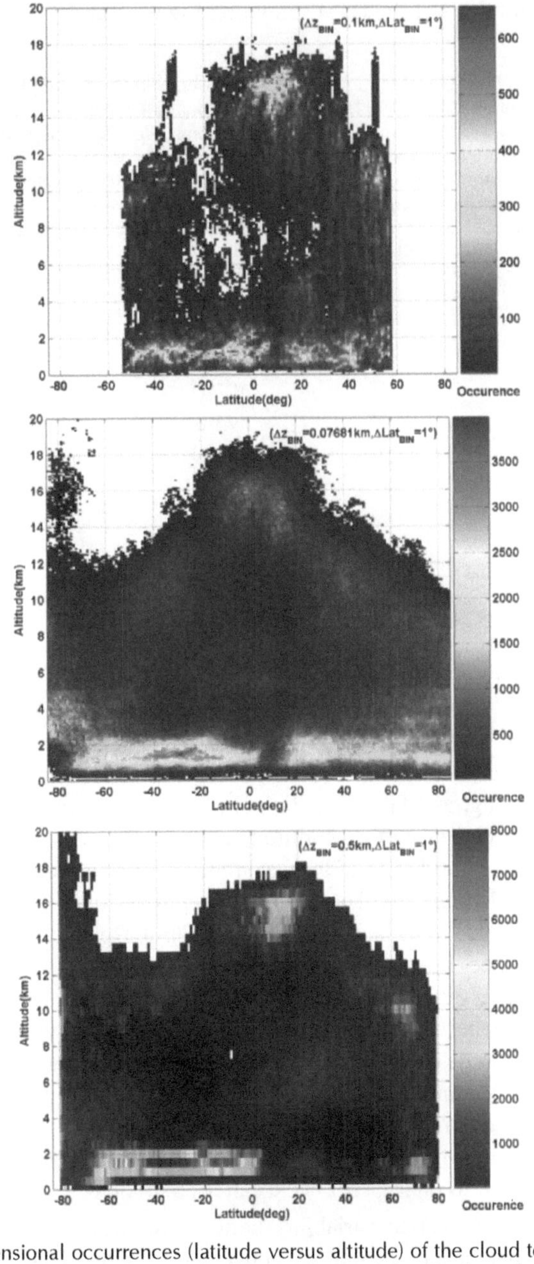

Figure 7. Two-dimensional occurrences (latitude versus altitude) of the cloud top height from LITE (September 10–19, 1994), GLAS (last week of September and the first 2 weeks of October, 2003), and CALIOP (September, 2006) measurements.

To summarize the previous results, lidar-derived cloud statistics based on only the uppermost cloud layer on each lidar shot from LITE, GLAS, and CALIOP are given in Table 1. Cloud statistics are calculated for the latitude intervals defined by [60°S; 20°S], [20°S; 30°N], and [30°N; 60°N]. These intervals were chosen from the two-dimensional distributions of the CTH shown on Figure 6 and 7.

Table 1. Separation of lidar-based cloud rerievals into low, middle, and high cloud classes for LITE (local method, September 10–19, 1994), GLAS (operational algorithm, last week of September and first 2 weeks of October, 2003), and CALIOP (operational algorithm, September–October, 2006) measurements as determined from the ISCCP classification against low, middle, and high clouds. The bottom group gives the cloud cover ratio. The first column gives the global cloud statistics, and the other columns the statistics on the three latitude intervals [–60°; –20°], [–20°; 30°], and [30°; 60°].

		All latitude	[-60°; -20°]	[-20°; +30°]	[+30°; +60°]
	LITE	52.4%	37.6%	57.4%	56.1%
High	GLAS	46.5%	47.5%	48.3%	42.0%
	CALIOP	45.0%	33.9%	53.3%	48.6%
Middle	LITE	9.0%	10.2%	6.6%	12.4%
	GLAS	18.8%	19.3%	18.3%	19.2%
	CALIOP	13.9%	14.9%	10.9%	17.7%
Low	LITE	38.6%	52.2%	36.0%	3 1.5%
	GLAS	34.7%	33.2%	33.4%	38.8%
	CALIOP	41.1%	51.2%	35.8%	33.7%
Cover Ratio	LITE	69.8%	72.8%	72.5%	64.0%
	GLAS	69.2%	76.5%	62.4%	70.5%
	CALIOP	70.5%	78.6%	67.5%	64.52%

The global cloud fractions are similar for the three datasets with values of 69.8%, 69.2%, and 70.5% for LITE, GLAS, and CALIOP measurements, respectively. This is somewhat surprising, as LITE should be able to detect more clouds due to its better SNR. A strong occurrence of high altitude clouds is highlighted (52%, 46.5%, and 45% for LITE, GLAS, and CALIOP measurements, respectively) representing half of the detected cloud structures. The LITE detects more high clouds than GLAS and CALIPSO, probably due to its better sensitivity (SNR). Detected low cloud fraction is comparable in the three datasets. However, middle-altitude clouds are more frequently detected by GLAS and CALIPSO than by LITE (9.8 and 4.9%, respectively). Besides a difference in atmospheric cloud structure due to inter-annual variability, this may be also be attributed to some extent to the larger multiple scattering (MS) impact in LITE measurements. The MS increases signal at altitudes below an upper cloud layer, and may prevent the detection of a second lower cloud layer with our algorithm. Lower level clouds tend to be composed of liquid water and are less perturbed by MS from upper cloud layers due to their larger backscatter coefficient.

There is less cloud fraction variability between the tropical and middle latitudes. However, we find that the latitude interval between [60°S; 20°S] has a higher proportion of low clouds that may exceed 50% although it is lower in the case of GLAS (~33%).

The relatively high frequency of occurrence of high cloud structures at latitudes greater than 60° for the GLAS and CALIOP instrument may be related to the tightening of the orbit footprints for these latitudes. Some caution must be exercised with the interpretation of high clouds in polar latitudes. There may be over-representation of the high cloud structures of greater horizontal expansion.

CROSS-COMPARISON BETWEEN LIDAR AND PASSIVE SPACEBORNE INSTRUMENTS

In operational cloud retrievals from passive radiometric measurements, the assumption is made that only one cloud layer is present in a satellite pixel (i.e., Platnick et al., 2003; Stubenrauch et al., 2006b). In the case of multilayered cloud structures, this assumption will introduce biases in the determination of CTH, especially for the case of cirrus overlying a lower-level water cloud. Fortunately, spaceborne lidar can provide insight to the vertical distribution of clouds. Moreover, the determination of the CTH is a direct measurement, and the multilayer cloud distribution can be assessed as long as the lidar signal does not attenuate in the uppermost cloud structure.

We now compare the CTH distributions given by passive and active instruments. The coherence between the approaches to retrieve the CTH from spaceborne lidar measurements has been demonstrated. With the increased vertical resolution of the LITE lidar and its better SNR, these data have been retained for the comparisons with passive instruments.

From use of the surface echo for LITE measurements, we have separated the contribution of high semitransparent clouds from that of optically thick clouds. The result is presented in Table 2 for the latitude intervals that have been discussed earlier. We note that the occurrence of semitransparent clouds approaches 70% for all latitude bands.

Table 2. Proportion of optically thin clouds compared to the total cover of high clouds for LITE, MODIS (Chang and Li, 2005), and TOVS PATH-B (Stubenrauch et al., 2006b).

Instrument	All latitudes	[-60 -20]	[-20 +30]	[+30 +60]
LITE	70.6%	71.3%	72.4%	69.4%
ISCCP	87.2%	84.62%	87.6%	86.0%
MODIS	73.8%	72.2%	64%	80.5%
VS PATH-B	91.9%	90%	94.7%	89. 1%

Comparison to ISCCP Database

A first comparison between the CPDF of the CTH is given in Figure 4b. Only the ISCCP measurements within the footprint of the LITE orbit have been considered. As specified before, only the highest cloud structure has been considered. For clouds at altitudes of less than 3 km (i.e., low level clouds), the ISCCP and the LITE CPDF are very similar. The ISCCP CPDF shows a cumulative probability of about 47% of cloud detected at up to 3 km height, which is in agreement with of the surface observation analysis by Warren et al. (1985). For clouds over 3 km height, significant differences occur in the number of detected CTH. The LITE CTH tends to place clouds at higher altitudes than ISCCP. The CPDF retrieved from LITE profiles reaches 95% at an altitude close to 16 km whereas that from ISCCP reaches the same value at an altitude close to 11 km. It seems that detection of optically thin high cloud layers is problematic for ISCCP. This leads to notable differences when comparing mean CTH retrieved from ISCCP and LITE datasets. The mean altitude of the cloud layers is much lower for ISCCP, more than 30% on average, as shown on Figure 6 and 7. The difference is less marked in the southern hemisphere but is about 2 km under 20°S and increases toward the northern latitudes.

A different way of comparing LITE and ISCCP CTH is given in Figure 8, which divides the CTH into the cloud classes defined by ISCCP: high (H), middle (M), and low (L) clouds. For the LITE CTH, each class is sub-divided considering the potential overlap from the other cloud layers. Our goal is to use data from active instruments to highlight the occurrence of cloud overlap.

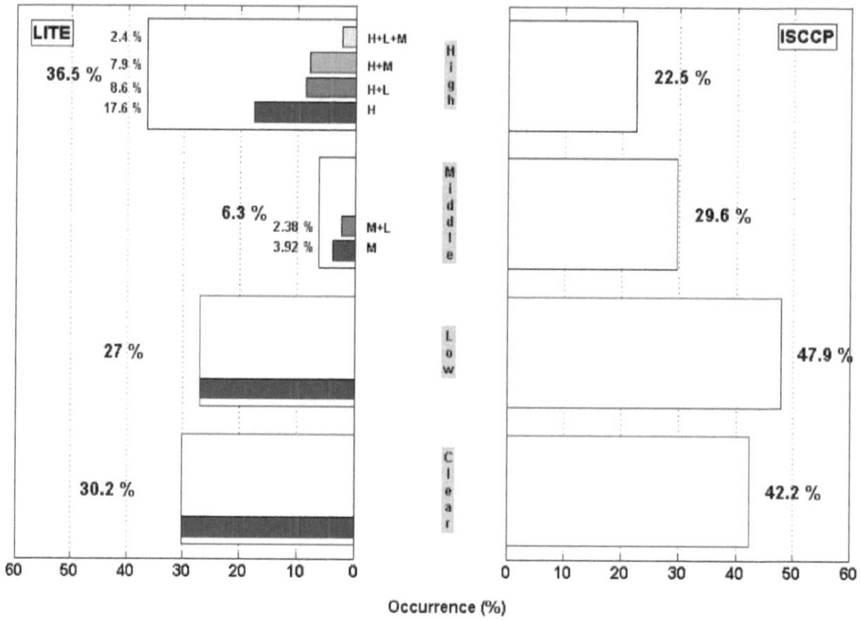

Figure 8. Distribution of each cloud class determined from LITE measurements following the classification of ISCCP in term of high (H), middle (M), and low (L) clouds (left figure). For each class the percentage corresponding to the multilayered clouds by the other cloud classes is given. The bottom percentage gives the occurrence of the clear sky. The ISCCP-DX classification is given on the right figure.

The largest difference between LITE and ISCCP statistics (14%) is observed in the occurrence of the high clouds. Such a difference may result from two causes. First, ISCCP may not detect all the high semitransparent clouds. In particular, situations with optically thin cirrus clouds may be classified by ISCCP as being clear sky. This could explain in part the difference in the clear sky occurrence between the two types of measurements (12%). Also, note that ISCCP assumes that if a cloud is present, there is a single cloud structure in the atmospheric column. The inference of CTH for optically thin cirrus is problematic and such a cloud may be assessed as being a lower-level cloud. The assumption of a single cloud structure in a satellite imager pixel can lead to biases in CTH when multilayered cloud structures exist. It is interesting to note that the multilayered cloud classes (H+M), (H+L+M), and (H+L) retrieved from LITE profiles are redistributed in the middle and low cloud classes of the ISCCP climatology. This could explain the great percentage of middle and low cloud structures in the ISCCP products. Moreover, Evan et al. (2007) demonstrated that the long term global trends in the cloudiness from the ISCCP record are influenced by artifacts associated with satellite viewing geometry. This study underlines a non-physical decrease of the total cloud amount as given by ISCCP of about 6% between 1987 and 2000. This can also explain partially the differences observed in the proportion of clear sky observations.

The lack of optically thin cirrus clouds in the ISCCP products is also described in Jin et al. (1996). They assessed the latitudinal spread of the thin cirrus cloud fraction, in a comparison of the ISCCP and high resolution infrared radiation sounder (HIRS) high cloud products. In all the cases and considering all the latitude range, ISCCP seems to underestimate the fraction of cirrus cloud type. In particular, we note a larger difference in the ITCZ region.

Comparison to Cloud Statistic from MODIS Measurements

Cloud products are available from the MODerate-resolution Imaging Spectrometer (MODIS; Platnick et al., 2003) and provide global CTH. The approach assumes only a single-layer cloud in the entire atmospheric column, and effort is underway to include multispectral approaches for the detection of multilayered clouds, specifically for the case of cirrus overlying a lower-level water cloud.

High-level and mid-level clouds are analyzed with the CO_2-slicing method developed by Menzel et al. (2002); this approach infers cloud top pressure and effective cloud amount (emittance multiplied by cloud fraction) for clouds at pressures lower than approximately 700 hPa. The MODIS results are similar to those presented in this study for the spaceborne lidar systems. In particular, the ratio of semitransparent clouds to the total cover of high clouds shown in Table 2 is closer to the results deduced from LITE dataset.

Comparison to Clouds Statistics from TOVS and Others

Other cloud products derived from passive sensors, such as TOVS Path-B (Stubenrauch et al., 2006a, b), or from atmospheric infrared sounder (AIRS), identify more cirrus clouds than ISCCP. From analysis of TOVS Path-B products from 1987 to 1995, the percentage was ~4% in the middle latitudes (30°–60°N and 30°–60°S), and up to 20% in the Tropics (15°N–15°S). For the AIRS L2 data analyses, the ITCZ high cloud amount is retrieved as being 10% larger than that from ISCCP (Stubenrauch et al., 2006a). As for MODIS, the ratio of semitransparent clouds to the total cover of high clouds shown in Table 2 indicates that a higher proportion of optically thick clouds than that derived from LITE measurements for all the latitudes.

Impact of the Seasonal Variation on the CALIOP Derived Parameters

We find a very strong difference in the mean CTH as a function of latitude between the passive and active measurements, particularly for the tropical areas (see Figure 6). This region plays an important role in the hydrological and radiation energy balance (Forster et al., 2007) and the differences observed can influence our understanding of potential feedback mechanisms. Earlier, we associated the maximum of the CTH with the position of the ITCZ for September. From the annual observations of CALIOP, we are able to further study this aspect and to analyze the interannual variability of the CTH. Figure 9 shows the temporal evolution of the CTH distribution from June, 2006 to May, 2007.

The mean values of CTH are comparable for both the middle and the northern polar latitudes. Large differences appear for both the tropical and the southern polar latitudes. We discuss hereafter the causes of these differences that can have a considerable impact on the cloud climatology.

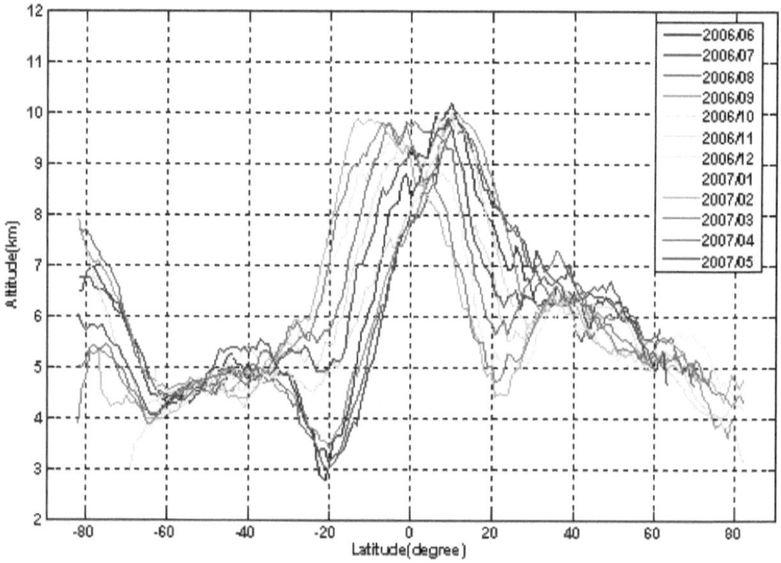

Figure 9. Monthly average of the mean cloud top height against the latitude. Only the highest cloud structure retrieved from the operational algorithm applied on CALIOP measurements has been used between June, 2006 and May, 2007.

Tropical Variability of the Mean CTH

As shown in Figure 9, the maximum of the CTH distribution moves from north to south of the equator with the ITCZ between February and August. Figure 10 shows the location of the ITCZ retrieved by Waliser et al. (1993) with the CTH distribution. We note that the correspondence between the ITCZ location and the mean CTH retrieved from CALIOP is very similar. This confirms that the maximum of the median value of the CTH can be associated with the mean position of the ITCZ. The extreme positions of ITCZ are 12°N during boreal summer (August, 2006) and 7°S during boreal winter (February, 2007).

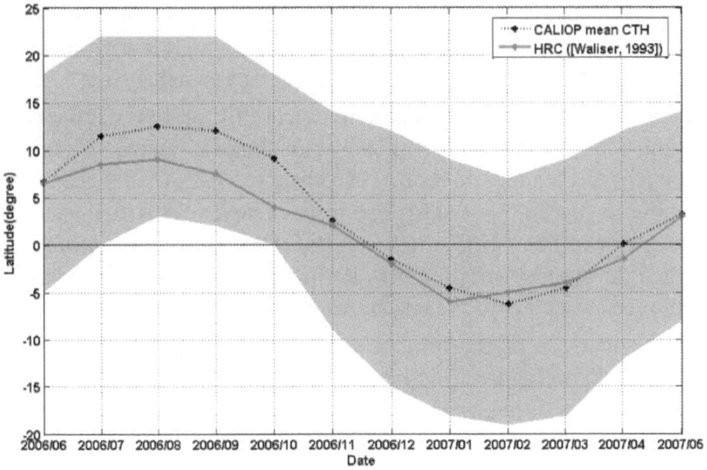

Figure 10. Seasonal variation of the mean CTH retrieved from CALIOP measurements performed between June, 2006 and May, 2007 against the latitude (dotted line). The gray area gives the peak width of tropical CTH distribution (Figure 9) taken arbitrary at 8 km height. The black full line gives the annual cycle of the ITCZ computed from 17 years of HRC (Hightly Reflective Cloud) data (January, 1971 to December, 1987) by Waliser et al. (1993).

The ITCZ tends to reside longer in the Northern hemisphere than in the Southern hemisphere (7 months and 5 months, respectively). The width of the tropical CTH distribution at an arbitrary height of 8 km is shown as the gray area in the Figure 10. This belt seem to be largest between November and March (close to 7°S), and minimal between July and October (close to 12°N).

This asymmetry, which could seem in contradiction with the symmetry of the solar radiation to the equator, has been well documented, and modeled by numerous authors such as Philander et al. (1996), Li (1997), and Hu et al. (2007). The fundamental cause of the asymmetry in the eastern Pacific is the tilt of the western coast of the Americas, which perturbs the sea surface temperature in the vicinity of the coastal region through a coastal wind-upwelling mechanism. The asymmetry in the Atlantic results from the land—ocean thermal contrast between the bulge of northwestern Africa and the ocean to the south. The ocean—atmosphere interactions act as an amplifier to enhance the asymmetry that is set up by the continental or coastal morphology (Li, 1997).

Influence of Polar Stratospheric Clouds

We highlighted the potential impact of the PSC on the CTH statistics for the southern polar latitudes. PSC are both high altitude and optically thin clouds but they can be detected from lidar measurements (see Figure 7 under 60°S). They play a major role in stratospheric chemistry and in particular on ozone depletion (i.e., Solomon, 1999).

Figure 9 shows also the mean CTH of all the clouds retrieved for southern polar latitudes (between 70°S and 90°S). The mean CTH significantly evolves during the year with the higher values between June and October (~7 km at 80°S) and smaller values between February and April (~5 km) at 80°S. July is an intermediate situation with a mean CTH close to 6 km. In December and January the detection of cloud structures is not efficient due to the duration of the day. The SNR in lidar measurements for optically thin clouds is lower due to the influence of sky radiance at 532 nm.

There are very few instruments able to detect PSC and very few existing studies about the climatology of this type of optically thin cloud. In a previous study, David et al. (1998) showed a percentage of about 70% of detected PSCs between 11 and 26 km height. Hopfner et al. (2007) used MIPAS instrument analyses to investigate the inter-annual occurrence of PSC. They show that PSC mainly occur from May to the end of October when the stratospheric temperature is very cold, with a maximal occurrence retrieved between mid-July to the end of August.

Hence, the increase of the mean CTH observed between GLAS and CALIOP measurement periods (Figure 6) is likely due to a more important presence of PSC during year 2006 than during year 2003.

CONCLUSIONS

To investigate the statistical distribution of the CTH, we use the observations from three spaceborne lidar missions: LITE, GLAS, and CALIPSO. We developed a methodology to infer the CTH from the lidar measurements. This methodology was compared with the operational algorithms of GLAS and CALIOP missions and proved to be quite powerful. One way of comparing the CTH from these lidars is through the CPDF. Optically thin cirrus clouds are better identified from LITE profiles because these measurements have a higher SNR than with the other lidars. The better SNR is mainly due to the altitude of the shuttle (~240 km) compared to the GLAS and CALIOP satellites (~590 and ~705 km, respectively).

Important variations are noted from a comparison of the CTH statistics from LITE and those from ISCCP, although these comparisons cover a very short time period. These differences are noted especially for the high clouds but also on the mean CTH. Low clouds are well identified by the two types of instrument (i.e., active and passive remote sensing sensors). We note that a similar comparison using MODIS cloud products based on the CO_2 slicing method led to results more similar to those deduced from the lidar measurements.

Natural causes of variability can be related in the tropical areas to the position of the ITCZ and in the southern polar regions with the monthly and inter-annual cycles of the PSC. In particular, important differences can be recorded from 1 year to another as between September, 2003 and September, 2006 when the mean CTH increases from 6 to 9 km, respectively. The first year of measurements was obtained from GLAS and the second by CALIOP. The observed differences are not thought to be caused by differences in the instruments but from changes in the atmosphere itself.

Such a study has highlighted that the use of spaceborne lidar observations performed at the global scale is a potentially powerful way of assessing critical parameters of the cloud distribution that may significantly change under the influence of human activities.

KEYWORDS

- **Cloud probability density function**
- **Cloud top height**
- **Geoscience laser altimeter system**
- **Inter tropical convergence zone**
- **Lidar in-space technology experiment**
- **Signal to noise ratio**

ACKNOWLEDGMENTS

The authors would like to thank Alcatel-Space, now Thales-Alenia Space and CNES, for the granting of one of them (S. Berthier). This study was supported by CEA and CNRS. The authors note their deep respect for their esteemed colleague Pierre Couvert, now deceased, who took part in this work.

Chapter 4

Pre-earthquake Signals

F. T. Freund

INTRODUCTION

Earthquakes are feared because they often strike so suddenly. Yet, there are innumerable reports of pre-earthquake signals. Widespread disagreement exists in the geoscience community how these signals can be generated in the Earth's crust and whether they are early warning signs, related to the build-up of tectonic stresses before major seismic events. Progress in understanding and eventually using these signals has been slow because the underlying physical process or processes are basically not understood. This has changed with the discovery that, when igneous or high-grade metamorphic rocks are subjected to deviatoric stress, dormant electronic charge carriers are activated: electrons and defect electrons. The activation increases the number density of mobile charge carriers in the rocks and, hence, their electric conductivity. The defect electrons are associated with the oxygen anion sublattice and are known as positive holes or pholes for short. The boundary between stressed and unstressed rock acts a potential barrier that lets pholes pass but blocks electrons. Therefore, like electrons and ions in an electrochemical battery, the stress-activated electrons and pholes in the "rock battery" have to flow out in different directions. When the circuit is closed, the battery currents can flow. The discovery of such stress-activated currents in crustal rocks has far-reaching implications for understanding pre-earthquake signals.

Seismologists use earthquakes as "flash lights" to illuminate the interior of the Earth. Information extracted from the propagation of seismic waves has produced great insights into the hidden structure of our dynamic planet. Unfortunately, earthquakes are erratic "flash lights" that seem to go off at unpredicted times and at unpredicted places. Since they often lead to destruction and death, it is understandable that seismologists have endeavored to find ways to predict—within limits as narrow as possible—time, place, and magnitude of major seismic events (Gokhberg et al., 1995; Lomnitz, 1994; Milne, 1899, Rikitake, 1976 #452; Sykes et al., 1999; Turcotte, 1991; Wyss and Dmowska, 1997). However, using the "flash lights" is different from understanding when and where a "flash" might go off. The two require different skills and different tools.

Though the seismological models have become ever more sophisticated and tend to take into account ancillary information (Holliday et al., 2005; Keilis-Borok, 2002; Rundle et al., 2003), the tools of seismology are blunt when it comes to recognizing the build-up of stress before the rupture.

At the same time it has been known for a long time that the Earth sends out a bewildering array of non-seismic signals before major events (Tributsch, 1984). Understanding how these signals are generated and what information they may provide has remained an elusive goal (Bernard et al., 1997; Hough, 2002; Kagan, 1997; Kanamori, 1996; Knopoff, 1996; Park, 1997; Uyeda, 1998).

Nature of Pre-earthquake Signals

The dilatancy theory (Brace et al., 1966) fits a simple mechanical concept. It is based on the observation that, when rocks are stressed, they expand normal to the stress vector and change their pore volume. This can account for bulging of Earth's surface and for changes in the resistivity due to pore water (Brace, 1975).

Earthquake Lights

Earthquake lights have been observed since ancient times (Derr, 1973; Galli, 1910; Mack, 1912; Tributsch, 1984). Based on over 1,500 reports, Musya (1931) stated: "The observations were so abundant and so carefully made that we can no longer feel much doubt as to the reality of the phenomena" (Terada, 1931). Nonetheless, doubts persisted in the scientific community even beyond 1960s when EQLs were photographed during an earthquake swarm at Matsushiro, Japan (Yasui, 1973). Hedervari and Nosczticzius (1985) covered many observations in Europe. St-Laurent (2000) evaluated reports of luminous phenomena sighted at the time of the Saguenay earthquake in Canada. Tsukuda (1997) reported luminous phenomena associated with the January 16, 1995 Kobe earthquake. Similar observations were made in Mexico (Araiza-Quijano and Hernandez-del-Valle, 1996) and other seismically active regions (King, 1983; Lomnitz, 1994).

Low Frequency Electromagnetic Emissions

Low frequency electromagnetic (EM) emissions possibly related to pre-earthquake activity have attracted attention over the past 10–20 years (Fujinawa and Takahashi, 1990; Gershenzon and Bambakidis, 2001; Gokhberg et al., 1982; Molchanov and Hayakawa, 1998; Nitsan, 1977; Vershinin et al., 1999; Yoshida et al., 1994; Yoshino and Tomizawa, 1989). Other authors report local magnetic field anomalies (Fujinawa and Takahashi, 1990; Gershenzon and Bambakidis, 2001; Kopytenko et al., 1993; Ma et al., 2003; Yen et al., 2004; Zlotnicki and Cornet, 1986) or increases in radio-frequency noise (Bianchi and al., 1984; Hayakawa, 1989; Martelli and Smith, 1989). Mercer and Klemperer (1997) modeled the EM emissions prior to the 1989 $M = 7.1$ Loma Prieta earthquake (Fraser-Smith et al., 1990) assuming streaming potentials caused by the movement of water along the fault plane. Sometimes no EM emissions are recorded, which has caused considerable consternation in the science community (Karakelian et al., 2002).

Ionospheric Perturbation

The ionosphere marks the transition from the Earth's atmosphere to the vacuum of space. It is a highly dynamic region where the solar radiation creates a plasma of ions and free electrons, which partly decays during the night. Prolonged ionospheric perturbations were observed before the great 1960 Chilean earthquake (Warwick et al., 1982) and the 1964 "Good Friday" earthquake in Alaska (Davis and Baker, 1965). Changes of the Total Electron Content (TEC), observed days before major events suggest the presence of a positive charge on the Earth's surface to which the ionosphere responds (Liperovsky et al., 2000; Liu et al., 2000, 2001; Naaman et al., 2001). A discussion of the pre-earthquake effects is found in Pulinets and Boyarchuk (2004) who favor release of radon at the Earth's surface as the cause of the reported ionospheric perturbations.

"Thermal Anomalies"

Non-stationary areas of enhanced infrared (IR) emission, linked to impending (Gornyi et al., 1988; Qiang et al., 1990, 1991; Srivastav et al., 1997) with apparent land surface temperature variations on the order of 2–4°C. The effect has become known as "thermal anomalies". The cause has remained enigmatic (Ouzounov et al., 2007; Srivastav et al., 1997; Tramutoli et al., 2005; Tronin, 2002, 2004). The rapidity of the temperature variations rules out a flow of Joule heat from deep below. Several alternative processes have been invoked: Rising well water levels and changing moisture contents in the soil; near-ground air ionization due to radon emission leading to the condensation of water vapor and the release of latent heat; emanation of warm gases (Gornyi et al., 1988), in particular of CO_2 (Quing et al., 1991; Tronin, 1999, 2002).

Other Pre-earthquake Signals

There are claims of other pre-earthquake phenomena such as differences in ground potentials (Varotsos, 2005; Varotsos et al., 1993, 1986), low-lying fog and unusual clouds (Tsukuda, 1997), and of course the rich folklore of abnormal animal behavior (Tributsch, 1984).

Common Traits Among Non-seismic Pre-earthquake Signals

Many pre-earthquake signals require transient electric currents in the Earth's crust. Electric currents arising from streaming potentials are well known (Bernabe, 1998; Jouniaux et al., 2000; Merzer and Klemperer, 1997; Morrison et al., 1989). Currents due to piezoelectric voltages generated in quartz-bearing rocks have been invoked to explain pre-earthquake low-frequency EM emissions (Gershenzon and Bambakidis, 2001; Ogawa and Utada, 2000; Sasai, 1991). However, no consensus of opinion has emerged.

EXPERIMENTAL

Important properties of rocks have been profoundly misunderstood or misinterpreted in the past, specifically the electrical properties of igneous and high-grade metamorphic rocks, which make up the bulk of Earth's crust in the depth range where most earthquakes occur, about 7–35 km.

At the root lies the fact that, in the geosciences, electrical conductivity of rocks is typically, often exclusively discussed in terms of ionic conductivity (high temperatures, partial melts) or electrolytical conductivity (low temperatures, fluids). However, from a solid state physics viewpoint there may be other mechanisms that can contribute significantly. One of these mechanisms arises from the fact that not all oxygen anions exist in their common 2-valence state but in the 1-valence state.

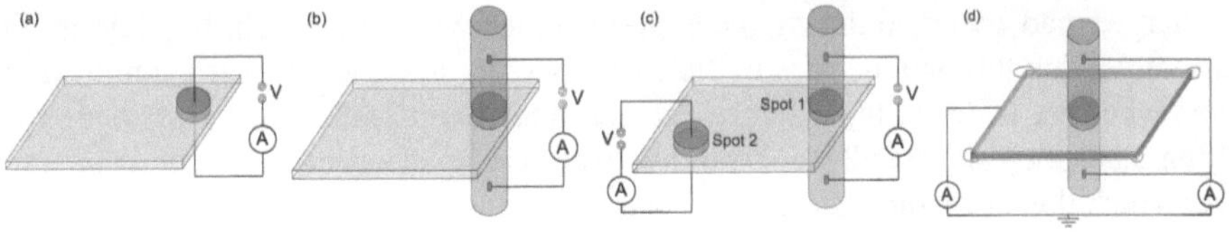

Figure 1. Different ways to measure electrical conductivity. (a) standard procedure, no load; (b) under load; (c) measuring at two places, one under load; (d) measuring a battery current.

Though valence fluctuations on the oxygen anion sublattice may appear to be of interest only to the narrowest of the specialists, they can have far-reaching consequences the capability of rocks to generate electric currents and, hence, to generate electric or EM signals.

To provide some background I start with a tutorial about electrical conductivity measurements. The electrical conductivity is typically measured with a set-up such as depicted in Figure 1a: a voltage is applied between a pair of electrodes on opposite surfaces of a flat sample and the current is recorded with an ammeter. If the effect of stress is to be studied, two pistons may be used to apply a force as shown in Figure 1b. Figure 1c depicts a set-up, where the conductivity is measured at two places, spot 1 between the pistons and spot 2 where no stress is applied. Conventional wisdom suggests such an experiment will lead to nothing new: the conductivity across spot 2 should not change when stress is applied to spot 1. Figure 1d shows yet another set-up: a circuit without a voltage source but a pair of steel pistons at the center, in electrical contact with the rock and connected to ground through one ammeter plus a Cu stripe as a contact along the periphery, connected to ground through a second ammeter. This circuit does not apply a voltage across the rock sample and, hence, will not measure electrical conductivity. Instead it will measure currents that flow out of the stressed rock under a self-generated voltage differential between the pistons and the periphery of the rock.

Figure 2a shows a typical current-voltage (I-V) plot recorded with a set-up as in Figure 1a. The sample is a dry gabbro. Except for the low voltage region, the I-V plot is linear as it should be for an Ohmic response. Figure 2b gives an example for the conductivity of dry granite measured with a set-up as shown in Figure 1b. During the first 7 min, the current is recorded without applying stress to the rock. The conductivity before loading is 0.7×10^{-6} S/m. Upon loading, the conductivity rises sharply, increasing by a factor of 3–4 to about 2.5×10^{-6} S/m and then continues to increase slowly. Stress-induced increase in the conductivity is commonly explained by better point-to-point contacts between grains under load (Glover and Vine, 1992). However, the next experiment will show that we are dealing with a more complex and much more interesting phenomenon.

The data shown in Figure 2c were obtained with a set-up as depicted in Figure 1c. We measured simultaneously the currents at spot 1 and at spot 2, that is between a pair of electrodes away from the stressed rock volume. The bold line shows the current at spot 1, the thin line the current at spot 2. After about 6 min, when we began applying the load, the current flowing through the stressed rock increased very rapidly, similar to what we had observed during the experiment shown in Figure 2b. In addition, the current flowing through the rock in the unstressed portion also increased significantly. Even when we widened the distance between spot 1 and 2, the effect remained, indicating that spot 1 and spot 2 were "talking to each other electrically" beyond the range over which the mechanical stress was transmitted.

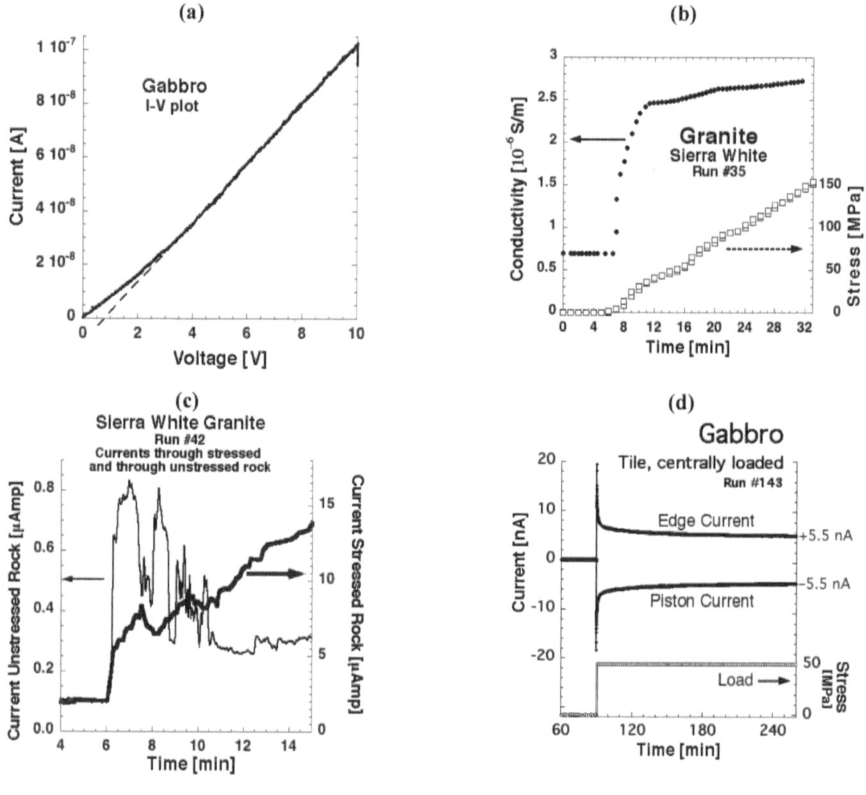

Figure 2. Exemplary results obtained with dry rocks using the four different circuits depicted in Figures 1a–d. (a) current-voltage plot indicating essentially a classical ohmic behavior; (b) significant increase in the electrical conductivity when stress is applied and strongly non-linear current response; (c) bold line: conductivity changes during application of stress through the stressed rock volume; thin line: instant change in the electrical conductivity of the unstressed rock; (d) demonstration of the stressed rock volume turning into a battery (see text).

To find out more about this "cross talk" we used the circuit in Figure 1d where we simply run a wire from the pistons at the center to the Cu contact along the edges. Initially no current flows. When a load is applied to the center, both ammeters in the circuit record a current as shown in Figure 2(d). The first ammeter records electrons flowing from the pistons to ground. The second ammeter records electrons flowing from ground into the rock. The two currents are of the same magnitude, suggesting that they are in fact the same current.

What we have here is, in fact, a battery with electrons flowing out from the stressed rock volume through the pistons into the external circuit and reentering the unstressed rock along the edges. A current through the external circuit implies a current of the same magnitude through the rock.

Dormant Electronic Charge Carriers

What Figure 2d demonstrates is a mechanism, previously unknown, to generate electrical currents in a rock subjected to deviatoric stress (Freund, 2002, 2003). The mechanism is fundamentally different from piezoelectricity. It is based on the fact that a small but non-zero number of the oxygen anions in the minerals that make up these rocks are not in their usual 2-valence state (O^{2-}) but have converted to the 1-state (O^{-}).

From a semiconductor perspective an O^{-} in a matrix of O^{2-} represents a defect electron or hole, also known as positive hole or phole for short. The O^{-} normally form positive hole pairs, PHP, chemically equivalent to peroxy links, O_3Si-OO-SiO_3. In the form of PHPs the O^{-} are electrically inactive or "dormant". During deformation, dislocations are generated in large numbers. When they intersect a PHP, the peroxy bond breaks, "waking up" a phole. The phole then becomes a highly mobile electronic charge carrier.

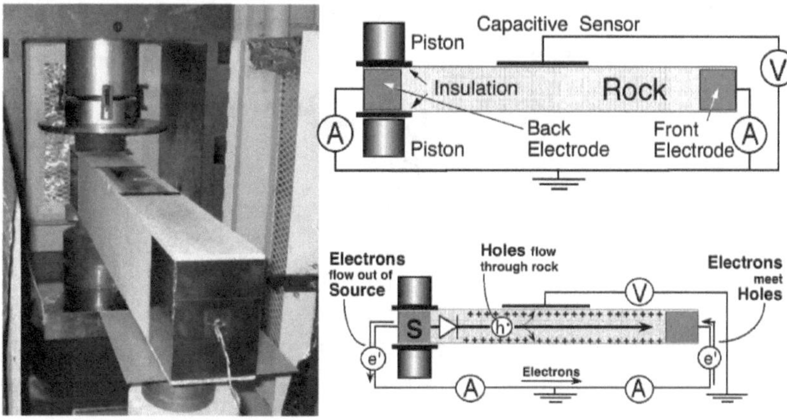

Figure 3. Experimental set-up with a slab of granite, one end in a press, Cu contacts attached to front and the back (left); electric circuit with ammeters to measure currents, one between each Cu contact and ground, capacitive sensor to measure the surface charge (upper right); circuit showing the flow of electrons through the outer circuit and the flow of holes through the rock (lower right).

When we stress a portion of a block of igneous rock, a grayish-white Sierra Nevada granite, 10 × 15 × 120 cm^3, insulate it from the press and ground, as shown in Figure 3 (left), the stressed portion turns into a battery (Freund et al., 2006). Figure 3 (right, top) shows the electric circuit similar to the one in Figure 1d. Figure 3 (right, bottom) shows the current flow: electrons, e', leave the stressed rock, the "source" S, and re-enter the rock at the unstressed end. To close the circuit a current of equal magnitude must flow through the rock. This current is carried by pholes. Some are trapped at the rock surface generating a positive surface charge relative to ground. This surface charge is recorded by means of a capacitive sensor visible in the photograph in Figure 3 (left) as a flat metal plate 0.8 mm above the rock surface.

The stressed rock volume becomes the "source" in which electrons and pholes are co-activated (Freund et al., 2006). There are two points to be stressed: (i) inside the stressed rock volume the number of charge carriers that is available to transport electric current has gone up, allowing the rock to transport more current than before, in the unstressed state; (ii) There are two kinds of charge carriers, both electronic but one kind (the electrons) representing a negative charge and the other kind (the pholes) representing a positive charge. For both the number density inside the stressed rock volume is higher than outside the stress filed. Both would "like" to spread out of the stressed rock volume, if they can.

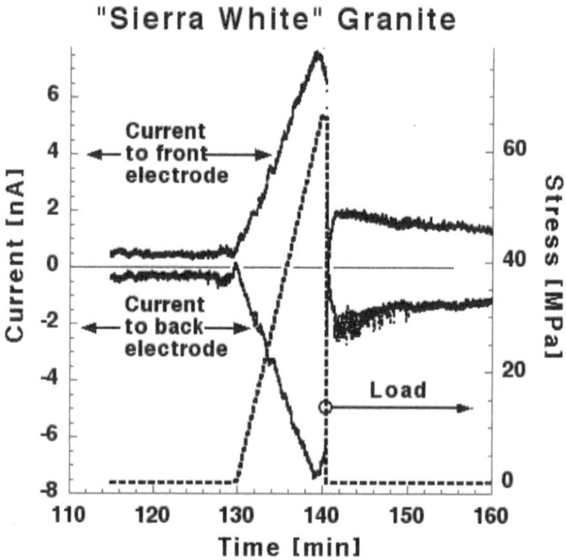

Figure 4. As one end of the granite slab is loaded (dotted line) electrons flow out of the stressed rock volume to ground (lower curve) and from ground into the unstressed rock (upper curve). The two currents are legs of one and the same circuit.

We conjecture that the boundary between stressed and unstressed rock acts like a Schottky barrier: it lets holes pass but rejects electrons. This barrier function is illustrated in Figure 3 (lower right) by the symbol of a diode. Indeed, as shown in Figure 4, a current carried by positive charge begins to flow as soon as a load is applied. A positive charge layer builds up on the rock surface (not shown), due to trapping of pholes.

In the case of the granite slab, a linear system, the currents increased approximately linearly with applied stress. In Figure 5, for a planar geometry, the currents increase in a non-linear fashion, very steep at first with strong fluctuations, and then saturate before the modest stress level of 48 MPa is reached. The cause of these fluctuations is unknown. They may be due to coupling, mediated by an internal electric field, between the electron and phole currents flowing out of the stressed rock volume in opposite directions.

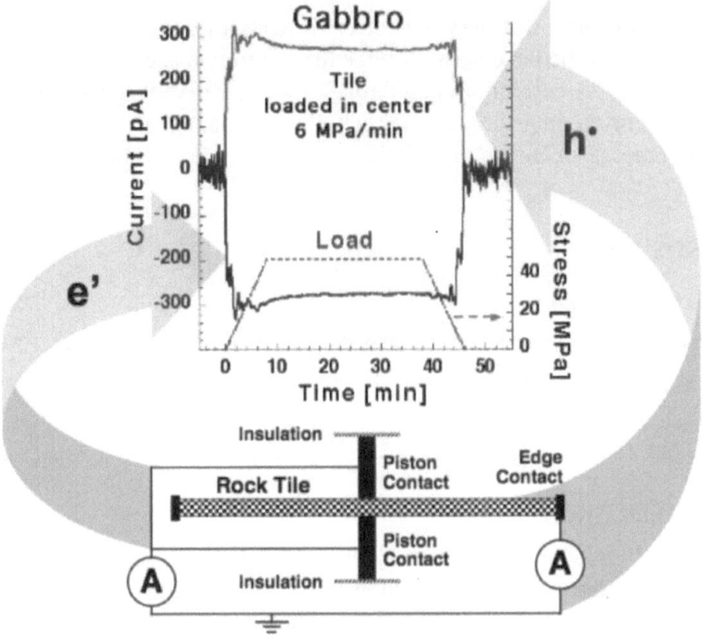

Figure 5. Outflow currents from a gabbro tile, 30 x 30 x 1cm³, loaded at the center (~10 cm³) from 0 to 48 MPa and kept at constant load for 30 min. Currents increase rapidly at beginning, reach a steady state value and decay slowly. Upon unloading currents return to zero.

Another interesting feature in Figure 5 is that, when stress is removed slowly, the currents return to zero. The process is repeatable. We have applied the same load-hold-unload-hold program 22 times over a period of 12 hr and observed that the currents always return to the same level.

Rocks Turn into Batteries

The most important conclusion to be draw from this series of experiments is that, when rocks such as granite or gabbro are subjected to deviatoric stress, they behave as if they were batteries. When stresses are applied, electronic charge carriers are activated, both electrons and pholes. In other words, the application of stress charges the battery.

Removing the stress causes the battery to return to an inactive state, ready to be charged again. So far we have not reached a limit of how often we can repeat the process. So long as the stress is well below the threshold of macroscopic damage, in our case about 1/4th the failure strength of the unconstrained rock, the process is repeatable many times.

How much current can be delivered per unit rock volume? In the case of gabbro the rock volume between the pistons in about 10 cm^3. We measure typically 300 pA. If the rock volume were 1 km^3, the outflow current would be on the order of 30,000 A. By stressing the rock tiles faster, the steady-state outflow currents were found to increase to 50,000 A, plus an initial spike that can rise to 100,000 A/km^3.

In an electrochemical battery the current flowing through the outer circuit is carried by electrons, while the current through the electrolyte is carried by cations. In the case of the rocks, the current flowing through the outer circuit is carried by electrons, while the current running through the rock, the internal current, is carried by defect electrons or holes, also known as "positive holes" or pholes for short at briefly outlined above.

KEYWORDS

- **Earthquake lights**
- **Low frequency electromagnetic**
- **Pre-earthquake phenomena**
- **Seismological models**

ACKNOWLEDGMENTS

This work has been supported in part by the NASA Ames Research Center Director's Discretionary Fund and by a GEST Fellowship from the NASA Goddard Space Flight Center, Planetary Geodynamics Laboratory. I thank my students and my coworkers A. Takeuchi (supported by a JSPS fellowship) and Dr. Bobby WS Lau (supported through a grant from NGA).

Chapter 5

Using Earthquakes to Uncover the Earth's Inner Secrets

C. Nostro, G. Cultrera, P. Burrato, A. Tertulliani, P. MacrM, A. Winkler, C. Castellano, P. Casale, F. Di Felice, F. Doumaz, A. Piscini, P. Scarlato, M. Vallocchia, A. Marsili, L. Badiali, A. Bono, S. Stramondo, L. Alfonsi, E. Baroux, M. G. Ciaccio, and A. Frepoli

INTRODUCTION

The Educational & Outreach Group (E&O Group) of the National Institute of Geophysics and Volcanology (INGV) designed a portable museum to bring on the road educational activities focused on seismology, seismic hazard, and Earth science. This project was developed for the first edition of the Science Festival organized in Genoa, Italy, in 2003.

The museum has been mainly focused to school students of all ages and explains the main topics of geophysics through posters, movie, and slide presentations, and exciting interactive experiments. This new INGV museum has been remarkably successful, being visited by more than 8,000 children and adults during the 10 days of the Science Festival. It is now installed at the INGV headquarters in Rome and represents the main attraction during the visits of the schools all year round.

Italy is a land prone to high seismic and volcanic hazards. About 60% of the country is classified as seismic on current hazard maps and has suffered more than 200 damaging earthquakes in the last century. Large cities such as Naples and Catania are located near the two major active volcanoes of Europe, Mt. Vesuvius and Mt. Etna, respectively. Italy is also the land where, historically, the first scientific observations and studies on seismic events and volcanic eruptions were made (e.g., De Dolomieu, 1785; Mallet, 1862; Plinius Caecilius Secundus, 1968 first century AD), and where many words now used in the scientific language were invented. Nevertheless, in spite of this long-lived cultural heritage, subjects such as seismology and volcanology are neglected in today schoolbooks and school programs. For this reason many schoolteachers feel the need to request visits to academic and scientific institutions with their class, for geophysical talks and exhibitions. They also need to gather up-dated educational material to upgrade school programs.

The INGV is currently the largest European scientific institution dealing with research and real-time surveillance, early warning and forecast activities in geophysics and volcanology. Five years ago, the E&O Group was created in the INGV headquarters of Rome to promote, develop, and disseminate Earth science programs and geophysical knowledge. Every year more than 4,000 students from primary, secondary, and high schools visit the INGV center in Rome during more than 65 open days, and demand increases each year. This group also participates in exhibitions and outreach projects organized by several public and private institutions.

In the last few years, the attention of scientific word to outreach in geophysics has grown, as demonstrated by many international projects and other activities performed by research institutes (Burrato et al., 2003; Hamburger et al., 2001; Johnson, 1999; Tertulliani and Donati, 2000; Virieux, 2000) and by the special sessions held at the international conferences as AGU meetings, EGU General Assembly, and ESC General Assembly (DiDa Working Group, 2002; Nostro et al., 2004; Tertulliani et al., 2004). Moreover, in 2003 Genoa has been the capital of the scientific outreach with the first edition of Science Festival ("Festival della Scienza") in Italy. The Festival lasted for 10 days and included more than 180 events, such as scientific and technological exhibits, educational activities, lectures, and videos. The large success of the initiative includes the Festival in the European Scientific Week, a European Union network of the VI Framework Program (http://www.festival.infm.it/it/home.php).

In this chapter, we describe the educational project developed by the E&O group for the 2003 Genoa Science Festival: a portable museum designed to bring on the road educational activities focused on seismology, seismic hazard, and Earth Sciences. The exhibits include activities that allow visitors to play back famous historical earthquakes, understand where and why earthquakes happen, discover the relationships between earthquake locations and plate boundaries, produce their own earthquakes, and track recent earthquake activity.

Figure 1. (a) the 3-D magnetic Plate Tectonic Puzzle; (b) children reconstructing the magnetic puzzle to learn the subdivision in plates of the Earth's lithosphere.

THE EXHIBIT ELEMENTS

The visitors follow a guided tour that starts with a short movie showing the location of the largest Italian earthquakes during the last 1,000 years, conceived to illustrate the most seismic prone areas in Italy. In the first section, PC programs and posters introduce the visitors to the Plate Tectonics theory. Here an interactive program working on three desktop PCs illustrates where and when earthquakes and volcanic eruptions occurred from 1960 to the present. In the next exhibit the visitors can put down and reconstruct a 3-D magnetic plate tectonic puzzle. This exercise helps to better understand the Earth's surface configuration and constrains the concept of the Earth's lithosphere subdivision in plates. Another 3-D Earth model shows the inner structure of our planet and the secret engine driving the plates.

After having been introduced to the plate tectonics' world, the visitors become familiar with earthquake geology and seismology. They play with a 3-D strike-slip fault model that can produce earthquakes of different sizes and, finally, produce their own earthquakes jumping close to an S-13 seismometer connected to a drum recorder. The tour ends with an exciting visit to the reproduction of the INGV National Seismic Network Center, watching seismic data in real-time and how seismologists locate the earthquakes.

Let's explore together the interactive museum.

When and Where did Earthquakes Happen in Italy?

The entrance of the museum is a small movie-theatre where a 3-min movie shows the location of the largest Italian earthquakes during the last 1,000 years, conceived to illustrate the most seismic areas in Italy. In the last 10 centuries Italy was struck by more than 200 earthquakes with magnitude greater than 5.5 (CPTI) (Catalogo Parametrico dei Terremoti Italiani, 1999) and many cities were completely destroyed, sometimes abandoned, and re-built in other places. In Italy even moderate size earthquakes can be damaging, and so it is important to study this level of seismicity.

Figure 2. A 3-D Earth model shows what drives the plates and the inner structure of our planet.

Looking at the magnitude and location of historical events and considering the effects of major earthquakes in different parts of our country, the visitors can learn that earthquakes and damage do not occur by chance, and Earth often shakes where it has already shaken in the past.

Where do Earthquakes Happen in the World?

After the film the visitors move on to three computer stations connected to a database of earthquakes and volcanic eruptions, covering a time span from 1960 to present. The educational computer program was developed by Alan Jones (see also Jones et al., 2003) and was translated into Italian and modified to add the largest Italian earthquakes. When the program starts, the user is presented with a map of the world where he can choose the area and the number of earthquakes and volcanic eruptions which are to be displayed each second. Plates boundaries can be shown to point out the relationships between earthquake locations and the plate boundaries. The visitors are generally impressed by the huge amount of earthquakes occurring all over the Earth and discover that the earthquakes and eruptions location are not random.

What About Earth's Lithosphere?

The third step is a 3-D magnetic Plate Tectonic Puzzle (Figure 1). Visitors can position several magnetic plates on a sphere of 1-meter in diameter, representing the Earth globe. The magnetic pieces illustrate the eight largest tectonic plates with their boundaries and can remain on the globe only if positioned in the right location. Thanks to this exhibit visitors can touch by hands and experience that the Earth's lithosphere is presently subdivided into single plates whose boundaries and positions are clearly identified. This was one of the favorite "games" of the younger children, who spent lots of time trying to set up the puzzle.

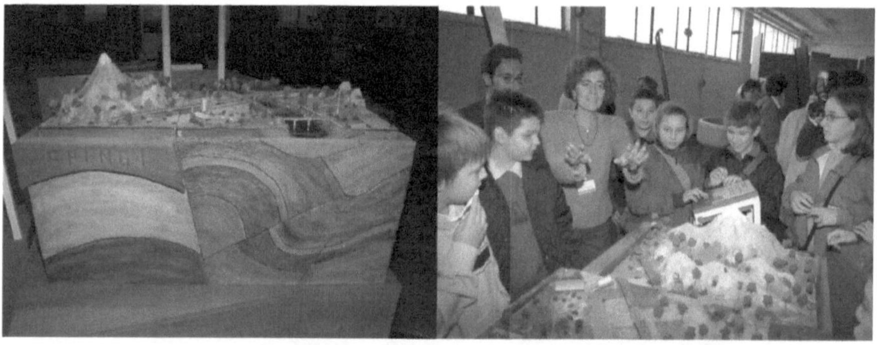

Figure 3. An interactive 3-D electro-mechanical model of a 1 meter-long strike-slip fault.

The Earth's Interior and the Plates Engine

After learning the subdivision of the lithosphere in plates, the visitors go to a 3-D Earth model where the inner structure of our planet is shown (Figure 2). The model shows also the mantle convection that moves the plates. This exhibit explains that the Earth is made up of many layers characterized by different temperatures, densities, and viscosities, and that the convection currents are the driving force behind the movement of the plates.

How do Earthquakes Occur?

The previous exhibits have explained where and why seismicity occurs. The visitors now enter the section where it is shown how earthquakes happen and what kind of damage they can produce. A 3-D reconstruction of a fault zone reproduces a rupture that causes a damaging earthquake (Figure 3). This exhibit is a 3-D electro-mechanical model of a 1 meter-long strike-slip fault made by two blocks representing the hanging-wall and foot-wall of the fault. The surface is realistically rendered with buildings, factories, bridges, and trees. When one of the two blocks is pushed, it slips, causing the rupture along a vertical plane. The stronger the push, the larger the damage at the surface: trees and constructions fall down and can be recovered by pressing a button.

With this experience it is possible to understand how different forces produce different-magnitude earthquakes and that the level of damage is also related to the energy of the earthquake.

JumpQuake

Another way to experience the energy released by an earthquake is to produce a ground motion and look at it through seismometer records. By jumping on the ground in front of an S-13 seismometer, people create their own earthquakes that are recorded mechanically on a large rotating drum (Figure 4). This simple interactive exhibit introduces the concept of seismic waves, their amplitudes, and duration, displaying the trace of ground motion directly on paper. Moreover, it shows the instrumentation used by the National Seismic Network and how it works. This exhibit was particularly appreciated by children, who loved jumping higher and higher to induce the greatest earthquake of all. The 3-D model of the fault and the seismometer is made simple to make visitors understand the relationships between the force applied (stress) and the effect (earthquakes and permanent deformation of the surface).

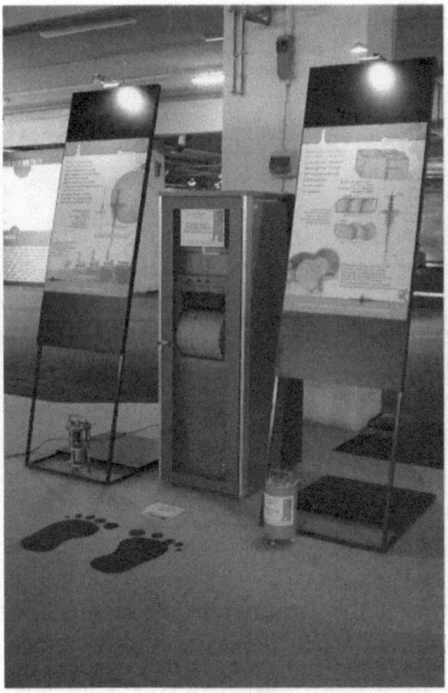

Figure 4. By jumping in front of a S-13 seismometer connected to a drum recorder, people induce their own earthquakes.

Figure 5. The reproduction of the INGV National Seismic Network Center shows the seismic records in real-time and how seismologists locate the earthquakes.

The INGV National Seismic Network Center

The tour ends with an exciting visit to the reproduction of the INGV National Seismic Network Center, where it is possible to watch the seismic records in real-time and how seismologists locate the earthquakes (Figure 5). Several computers are linked with the acquisition system of the National Seismic Network run in different headquarters of the INGV.

The time series of the ground velocity recorded in real-time by the Italian Seismic Network stations are shown on six large monitors. Visitors can also see the earthquake waveforms as they occur and how the earthquake location is computed using a software developed at INGV (Bono and Badiali, 2005).

People discover the large number of earthquakes recorded everyday and they are amazed because earthquakes are usually thought as not frequent but always damaging. They learn that earthquakes are common phenomena on the Earth and they are the signals of the geological forces of our planet.

CONCLUSIONS

One of the main goals of our exhibition was to familiarize people with earthquakes, following a journey from their origin to the monitoring and study of those phenomena. This is important for natural disaster preparedness in a country with a high exposure, like Italy. The public reactions were in general of wonder to know how many earthquakes hit Italy in historical time (they usually just remembered the last two or three earthquakes occurring in the past 20 years), and of curiosity to see how the seismic monitoring is performed.

We choose to use very few computer animations in designing our tour, preferring hands-on exhibits to provide the opportunity to directly interact with the phenomena. We think that in this way the curiosity is much more stimulated, and the number of visitors and their feedback seems to demonstrate the success of this experience. The 3-D mechanical displays and the real-time earthquake information help to understand what links the processes ruling the dynamic Earth and seismicity and volcanic activity.

To give an idea of the expenses sustained for the creation of this traveling mini-museum and of the effort for the participation to an event such as the 2003 Genoa

Science Festival, let's try to give some figures. The cost of the design and construction completely financed by INGV, was about 150,000 EUR. The ongoing costs for transportation and maintenance were around 4,000 EUR to move it from Rome to Genoa. The number of people needed to staff the exhibits is variable according to the attendance, but on average with a visit every 15 min eight people were necessary to guide and illustrate the exhibits. Each visit lasted at least 1 hr. We organized the staff in several watches of 4–5 days, and the total number of researchers and technicians who dedicated their time was 30 people.

Our future plans are to follow this experience and design new exhibits dealing with other fields of geophysics (such as Geomagnetism and Volcanology) to provide an all-around comprehension of the planet Earth. Up to now, we designed for the 2004 Genoa Science Festival a new exhibit tour dealing with the Earth's magnetic field (Winkler et al., 2004).

KEYWORDS

- **3-D Earth model**
- **Earthquakes**
- **Educational & Outreach Group**
- **Seismometer**

Chapter 6

Ground Water Desalination

William M. Alley

INTRODUCTION

Desalination is a water-treatment process that removes salts from water. Concerns about the sustainability of freshwater supplies, as well as rapid advances in membrane and other water-treatment technologies, are fostering a renewed interest in desalination as a partial solution to increased water demand. It is projected that more than 70 billion dollars will be spent worldwide over the next 20 years to design and build new desalination plants and facilities (Sandia National Laboratories, 2002).

Desalination presents the possibility of providing freshwater not only from the ocean, but also from saline ground water. The potential importance of desalination is exemplified, for example, by New Mexico where approximately 75% of ground water is too saline for most uses without treatment (Reynolds, 1962, p. 91). Earth-science issues related to desalination of ground water include the distribution of saline ground water resources, the chemical characteristics and suitability of ground water for desalination, effects of extraction of saline ground water on connected freshwater resources, and disposal of residual products. These issues are discussed by posing a series of basic questions.

WHAT IS SALINE GROUND WATER?

Salinity is a term used to describe the amount of salt in a given water sample. It usually is referred to in terms of total dissolved solids (TDSs) and is measured in milligrams of solids per liter (mg/l). Water with a TDS concentration greater than 1,000 mg/l commonly is considered saline. This somewhat arbitrary upper limit of freshwater is based on the suitability of water for human consumption. Although water with TDS greater than 1,000 mg/l is used for domestic supply in areas where water of lower TDS content is not available, water containing more than 3,000 mg/l is generally too salty to drink. The US Environmental Protection Agency has established a guideline (secondary maximum contaminant level) of 500 mg/l for dissolved solids. Ground water with salinity greater than seawater (about 35,000 mg/l) is referred to as brine.

WHERE IS SALINE GROUND WATER LOCATED?

Little is known about the hydro-geology of the parts of most aquifers that contain saline water compared to the parts that contain freshwater, because the need to utilize saline ground water has been limited. Most ground water resource evaluations have been devoted to establishing the extent and properties of freshwater aquifers, whereas evaluations of saline water-bearing units have been mostly devoted to determining the effects on freshwater movement.

Much of the work to characterize saline ground water resources in the US was done in the 1950s, 1960s, and 1970s. Surveys of saline water resources of several states (e.g., Winslow and Kister, 1956), and of selected areas within states (e.g., Hood, 1963), were published in the late 1950s and early 1960s. Krieger et al. (1957) undertook a preliminary survey of the saline water resources of the US during this period. Later, Feth at el. (1965) prepared a generalized map of the depth to saline ground water for the conterminous US (Figure 1). This map provides a preliminary perspective on the location of saline ground water resources, but provides limited information about critical factors required to understand the development potential of the resources such as aquifer hydraulic conductivity and well yields.

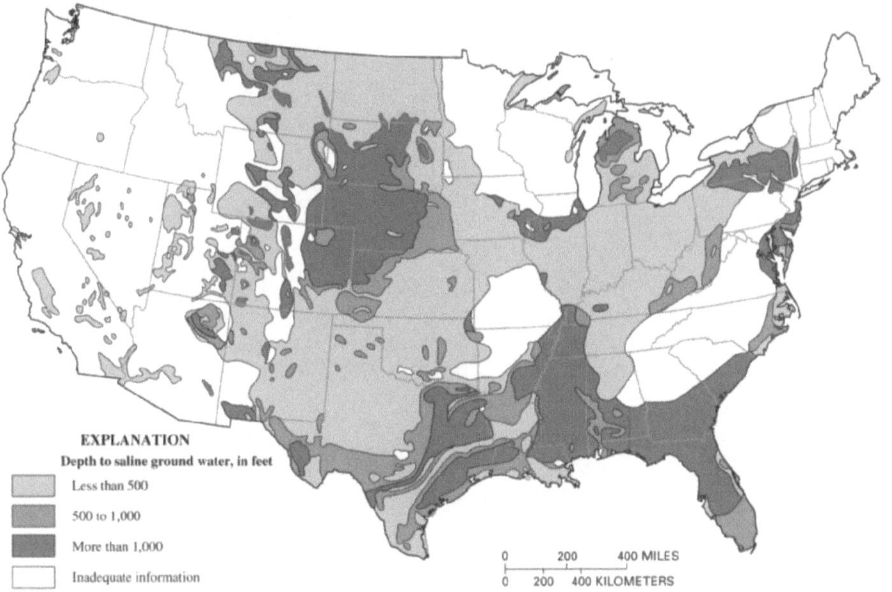

Figure 1. Depth to saline ground water in the US (generalized from Feth et al., 1965).

Reports published in US Geological Survey (USGS) Professional Paper Series 813 in the 1970s present very generalized analyses of ground water resources within each of the 21 Water Resources Regions of the US. Many of these reports contain a short summary statement about the saline water resources of a region, including use of saline aquifers as a source of treatable water, place to store freshwater, or place to dispose of liquid waste. Subsequently, the USGS developed quantitative assessments of 25 of the Nation's most important regional aquifer systems, as part of the Regional Aquifer-System Analysis (RASA) Program (USGS Professional Paper Series 1400–1425). Reports on the geochemistry of some of these aquifer systems and the Ground Water Atlas of the US (Miller, 2000) produced as a part of the RASA Program contain maps and data about dissolved-solids and specific ion concentrations for selected areas (Figure 2).

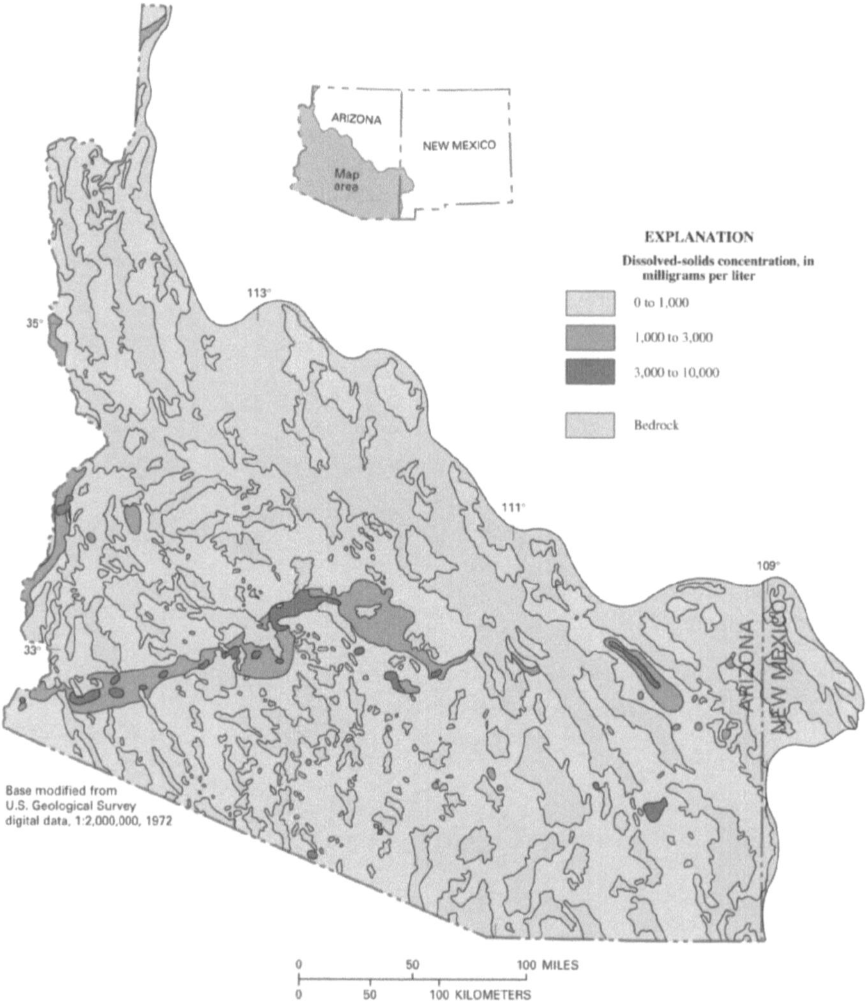

Figure 2. Dissolved-solids concentrations in basin-fill aquifers in Arizona and New Mexico are typically less than 1,000 mg/l, but can be much higher. Areas of higher concentrations include low parts of some basins where dissolved solids are concentrated by evapotranspiration from shallow ground water, basins with smaller fluxes of water naturally circulating through the ground water system, near deposits of saline minerals, and in some areas of salt buildup from intensive irrigation. (modified from Kister, 1973; Robson and Banta, 1995).

Subsurface data on water chemistry and hydrology typically are restricted to near-surface, freshwater aquifers and to sediments containing deeper, commonly highly saline, oil-field water. Few wells are located in intermediate areas that might be of interest for saline-water development. Thus, surface and subsurface geophysical techniques to supplement information from wells may prove useful tools for saline water-resource evaluation.

HOW DO WE CHARACTERIZE THE CHEMISTRY OF SALINE GROUND WATER WITH RESPECT TO SUITABILITY FOR DESALINATION?

The dissolved-solids concentration of saline ground waters typically is less than that of seawater. Saline ground water, however, has a greater tendency than seawater to precipitate sulfate (e.g., gypsum), carbonate (e.g., calcite), and silicate scales (Figure 3). The tendency of saline ground water to form scale is important to its suitability for use in desalination. Saline ground water also may contain some constituents, such as arsenic, elevated radioactivity, and dissolved organic material, at greater concentrations than in seawater. Although a relatively large body of information exists on the chemical behavior of solutes within desalination systems, relatively little research has been undertaken on the chemical suitability of water from various aquifers as sources for desalination.

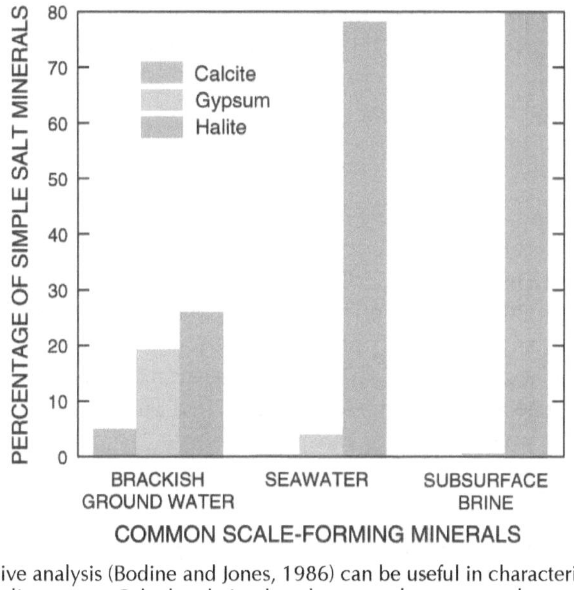

Figure 3. Normative analysis (Bodine and Jones, 1986) can be useful in characterizing the chemical composition of saline water. Calculated simple salt norms demonstrate the preferred association of cations and anions based on relative solubility of evaporite minerals. These are derived from the equilibrium salt assemblages obtained on complete evaporation of a water sample under ambient conditions. The resulting cation-anion associations (simple salts) readily show the relative abundance of principal scale-forming mineral phases. This is illustrated graphically by comparison of simple salt mineral percentages calculated from analyses of brackish ground water, seawater, and subsurface brine (data from Wood and Jones, 1990). In this way, a quick evaluation can be made of the magnitude and type of potential scaling problems for a given saline water resource.

WHAT ARE THE EFFECTS OF USING SALINE GROUND WATERS ON FRESHWATER RESOURCES AND THE ENVIRONMENT?

Although one might at first consider saline ground water as a "new" source of water, the parts of aquifers containing saline water commonly are connected hydraulically to parts of the same aquifer or aquifer system that contain freshwater. Thus, development of one resource affects the other, as well as potentially affecting the flow and quality of surface-water bodies connected to the ground water system. These effects may extend over many years or decades. Techniques are needed to predict the effects of saline ground water development in three dimensions and with time. Variable density numerical models have been used successfully to simulate the subsurface occurrence and movement of saltwater in coastal environments in many locations. The use of these models to quantify the movement of inland saline water resources would aid in understanding the impacts of utilizing saline water resources for desalination.

Disposal of the volumes of brine produced as a residual product of desalination is a primary environmental issue associated with desalination. Residual products from desalination of seawater are pumped to the sea. The expense of geologic disposal of residual products from inland saline ground water is a major factor that affects the cost-effectiveness of ground water desalination. Disposal of residual products of desalination requires careful consideration of their chemical composition. For example, disposal in deeper (higher temperature) saline aquifers may lead to precipitation of minerals, including calcite, and gypsum that have lower solubility at higher temperatures (Kharaka at al., 1997). Underground injection of saline water currently is done in the US to dispose of oil-field brines.

ARE THERE OTHER SOURCES OF TREATABLE GROUND WATER?

Other sources of ground water may be amenable to desalination. These possibilities, for which research continues, include treating water co-produced with oil and conventional natural gas (Kharaka et al., 1999) and water co-produced with coal-bed methane (Rice and Nuccio, 2000). Many oil- and gas-producing formations have associated water that is only slightly saline (less than 10,000 mg/l), notably in the intermountain basins of the western US. Coal-bed methane water in the Powder River Basin of Wyoming is generally less than 3,000 mg/l TDS. The likely presence of organic compounds in this water complicates treatment.

(Left photograph) A desalination plant for the City of Cape May, New Jersey, was built inside the brick building of the former Cape May Water Works. (Right photograph) The automated desalination plant can produce from 750,000 to 2 million gallons of water per day. (Photos: Jennifer Kopp, courtesy of CapeMay.com) Saltwater contamination has forced the closure of water-supply wells for the City of Cape May, New Jersey, and caused concerns about the future sustainability of ground water. Long-term monitoring of the resource and a numerical ground water model of the aquifer system of the Cape May Peninsula were key elements to help engineers design an approach to combine desalination of brackish ground water at some wells and reduce pumping at others to stabilize the saltwater front (Galloway et al., 2003; Lacombe and Carleton, 2002).

CONCLUSIONS

Desalinated ground water potentially represents an increasing component of the Nation's water supply; however, relatively little is known about saline ground water resources. An improved knowledge base is needed to better define their distribution and physical and chemical characteristics; to predict the effects of saline-water extraction on the environment; and to support proper disposal of waste products. This information is needed to support economic development of the resources, and to provide a scientific basis for regulatory and policy issues.

KEYWORDS

- **Desalination**
- **Environmental Protection Agency**
- **Total dissolved solids**
- **US Geological Survey**

Chapter 7

Deforestation Prediction for Different Carbon-prices

Georg E. Kindermann, Michael Obersteiner, Ewald Rametsteiner, and Ian McCallum

INTRODUCTION

Global carbon stocks in forest biomass are decreasing by 1.1 gigatonnes of carbon (GtC) annually, owing to continued deforestation and forest degradation. Deforestation emissions are partly offset by forest expansion and increases in growing stock primarily in the extra-tropical north. Innovative financial mechanisms would be required to help reducing deforestation. Using a spatially explicit integrated biophysical and socio-economic land use model we estimated the impact of carbon price incentive schemes and payment modalities on deforestation. One payment modality is adding costs for carbon emission, the other is to pay incentives for keeping the forest carbon stock intact.

Baseline scenario calculations show that close to 200 mil ha or around 5% of today's forest area will be lost between 2006 and 2025, resulting in a release of additional 17.5 GtC. Today's forest cover will shrink by around 500 million hectares, which is 1/8 of the current forest cover, within the next 100 years. The accumulated carbon release during the next 100 years amounts to 45 GtC, which is 15% of the total carbon stored in forests today. Incentives of 6 US$/tC for vulnerable standing biomass payed every 5 years will bring deforestation down by 50%. This will cause costs of 34 billion US$/year. On the other hand a carbon tax of 12 $/tC harvested forest biomass will also cut deforestation by half. The tax income will, if enforced, decrease from 6 billion US$ in 2005 to 4.3 billion US$ in 2025 and 0.7 billion US$ in 2100 due to decreasing deforestation speed.

Avoiding deforestation requires financial mechanisms that make retention of forests economically competitive with the currently often preferred option to seek profits from other land uses. Incentive payments need to be at a very high level to be effective against deforestation. Taxes on the other hand will extract budgetary revenues from the regions which are already poor. A combination of incentives and taxes could turn out to be a viable solution for this problem. Increasing the value of forest land and thereby make it less easily prone to deforestation would act as a strong incentive to increase productivity of agricultural and fuelwood production, which could be supported by revenues generated by the deforestation tax.

Deforestation is considered the second largest source of greenhouse gas (GHG) emissions amounting to an estimated 2 GtC per annum over the last decade (IPCC, 2001). It is a persistent problem. The UN Food and Agriculture Organization, in its recently released most comprehensive assessment of forests ever, puts deforestation at about 12.9 mil. ha per year (FAO, 2005). At the same time, forest planting, landscape restoration, and natural expansion of forests reduce the net loss of forest area. Net change in forest area in the period 2000–2005 is estimated at 7.3 million hectares per year (FAO, 2005). This reduces the annual GHG emissions to an estimated 1.1 GtC. In comparison, 7.3 GtC were emitted in 2003 by using fossil energy sources (Marland et al., 2006).

Deforestation has been difficult to tackle by governments, as its drivers are complex and many land uses yield higher revenues than those from forested land. Some see climate policy as a new opportunity to effectively reduce a major source of GHGs and biodiversity loss as well as to increase incomes of many people in rural areas whose livelihood depends on forests. The implementation of measures avoiding deforestation would require innovative financial mechanisms in the context of global climate policies. In this chapter we study the potential magnitude of effects of different financial mechanisms to help reduce deforestation, using a modeling approach.

To estimate the impact of financial incentives, to reduce deforestation and assuming profit maximizing behavior, we calculate differences in net present value of different land uses using a spatially explicit integrated biophysical and socio-economic land use model. Key model parameters, such as agricultural land use and production, population growth, deforestation, and forest product consumption rates were calibrated against historical rates. Land use changes are simulated in the model as a decision based on a difference between net present value of income from production on agricultural land versus net present value of income from forest products. Assuming fixed technology, the model calculates for each 0.5° grid cell the net present value difference between agricultural and forest land-uses in 1-year time steps. When carbon market prices, transferred through a financial mechanism, balance out differences between the net present value of agricultural land and forest-related income, it is assumed, consistent with profit maximizing behavior, that deforestation is avoided.

The net present value difference of forest versus other land uses can be balanced out through two mechanisms. One is to reduce the difference by adding costs to conversion through taxing emissions from deforestation, for example, through a land clearance tax and wood sales taxes. The other is to enhance the value of the existing forest by financial support when keeping the forest carbon stock, to be paid in certain time intervals. In both cases the value of forest carbon stock would be pegged to carbon market prices. The modeling results for different hypothetical tax or subsidy levels show the potential magnitude of avoided deforestation through financial incentive or disincentive mechanisms. The model results are annual, spatially explicit estimates of the forest area, and biomass development from 2000 to 2100, with particular focus on the period 2006–2025.

MATERIALS AND METHODS

The model is based mainly on the global afforestation model of (Benítez and Obersteiner, 2006) and calculates the net present value of forestry with equation (1–16) and the net present value of agriculture with equation (17–20). Main drivers for the net present value of forestry are income from carbon sequestration, wood increment, rotation period length, discount rates, planting costs, and wood prices. Main drivers for the net present value of agriculture on current forest land are population density, agricultural suitability, and risk adjusted discount rates.

These two values are compared against each other and deforestation is subsequently predicted to occur when the agricultural value exceeds the forest value by a certain margin. When the model comes to the result, that deforestation occurs, the speed of deforestation was constraint by estimates given by equation (24). The speed of deforestation is a function of sub-grid forest share, agricultural suitability, population density and economic wealth of the country.

All symbols and abbreviations in the following equations are explained in the "List of Symbols".

Net Present Value of Forestry

The net present value of forestry is determined by the planting costs, the harvestable wood volume, the wood-price, and benefits from carbon sequestration.

For existing forests which are assumed to be under active management the net present value of forestry given multiple rotations (F_i) over the simulation horizon is calculated from the net present value for one rotation (f_i) (equation 1). This is calculated by taking into account the planting costs (cp_i) at the begin of the rotation period and the income from selling the harvested wood ($pw_i \cdot V_i$) at the end of the rotation period. Also the benefits from carbon sequestration are included denoted as (B_i).

The planting costs (equation 3) are calculated by multiplying the planting costs of the reference country (cp_{ref}) with a price index (px_i) and a factor which describes the share of natural regeneration (pr_i). The ratio of plantation to natural regeneration is assumed to increase with increasing yield for the respective forests (equation 4). The price index (equation 5) is calculated using the purchasing power parity (PPP) of the respective countries. The stumpage wood price (equation 6) is calculated from the harvest cost free income range of wood in the reference country. This price is at the lower bound when the population density is low and the forest share is high and at the higher bound when the population density is high and the forest share is low. The price is also multiplied with a price index converting the price range from the reference country to the examined country. The population-density and forest-share was standardized between 1 and 10 by using equation (7) and equation (8) respectively.

The harvested volume (V_i) is calculated by multiplying the mean annual increment (MAI_i) with the rotation period length (R_i) accounting for harvesting losses (equation 9).

The rotation period length (equation 10) depends on the yield. Fast growing stands have a short and slow growing sites a long rotation length. In this study the rotation length is in the range between 5 and 140 years.

The mean annual increment (equation 11) is calculated by multiplying the estimated carbon uptake (ω_i) and a transformation factor which brings the carbon weight to a wood volume ($C2W_i$). The carbon uptake (ω_i) is calculated by multiplying the net primary production (NPP_i) with a factor describing the share of carbon uptake from the NPP (equation 12).

The benefits of carbon sequestration (equation 13) are calculated by discounting the annual income from additional carbon sequestration and subtracting the expenses incurred from harvesting operations and silvicultural production. At the end of a rotation period the harvested carbon is still stored in harvested wood products and will come back to atmosphere with a delay. This is considered in the factor (θ_i) which shares the harvested wood volume to short and long living products (equation 14).

The effective carbon price represents the benefit which will directly go to the forest owner. In equation (16) a factor describing the percentage of the transaction cost free carbon price is used. A factor $leak_i$ is calculated as the average of the percentile rank from "political stability," "government effectiveness," and "control of corruption" (Kaufmann et al., 2005).

$$F_i = f_i \cdot \left[1-(1+r)^{-R_i}\right]^{-1} \tag{1}$$

$$f_i = -cp_i + pw_i \cdot V_i + B_i \tag{2}$$

$$cp_i = cp_{ref} \cdot pr_i \cdot px_i \tag{3}$$

$$pr_i = \begin{cases} 0 & \text{MAI}_i < 3 \\ (MAI_i - 3)/6 & 3 \le MAI_i \le 9 \\ 1 & MAI_i > 9 \end{cases} \tag{4}$$

$$px_i = \frac{PPP_i}{PPP_{ref}} \tag{5}$$

$$pw_i = pw_{\min} - \frac{pw_{\max} - pw_{\min}}{99} + \frac{pw_{\max} - pw_{\min}}{99} \cdot SPd \cdot SNFs \cdot px_i \tag{6}$$

$$SPd = \begin{cases} 1 + \frac{Pd \cdot 9}{100} & Pd \le 100 \\ 10 & Pd > 100 \end{cases} \tag{7}$$

$$SNFs = 1 + (1 - Fs) * 9 \tag{8}$$

$$V_i + MAI_i \cdot R_i \cdot (1 - HL_i) \tag{9}$$

$$R_i = \begin{cases} 5 & MAI_i > 180/10 \\ \frac{600 - |MAI_i - 6| \cdot 50}{MAI_i} & \frac{10}{3} \le MAI_i \le \frac{180}{10} \\ 140 & MAI_i < 10/3 \end{cases} \tag{10}$$

$$MAI_i = \omega_i \cdot C2W \tag{11}$$

$$\omega_i = NPP_i \cdot CU \tag{12}$$

$$B_i = epc_i \cdot w_i \cdot (1 - b_i) \cdot \left\{r^{-1} \cdot \left[1-(1+r)^{-R_i}\right] - R_i \cdot (1-\theta_i) \cdot (1+r)^{-R_i}\right\} \tag{13}$$

$$\theta_i = \left(1 - \frac{dec_{llp} \cdot frac_{llp}}{dec_{llp} + r} - \frac{dec_{slp} \cdot frac_{slp}}{dec_{slp} + r}\right) \cdot (1 - frac_{sb}) + (1 - frac_{sb}) * frac_{sb} \tag{14}$$

$$frac_{slp} = 1 - frac_{llp} \tag{15}$$

$$epc_i = pc_i \cdot leak_i \tag{16}$$

Net Present Value of Agriculture

The net present value of agriculture (A_i) is calculated with a two-factor Cobb–Douglas production function (equation 17). It depends on the agriculture suitability and the population density. A high agriculture suitability and a high population density causes high agricultural values. The value ranges between a given minimum and a maximum land price. The parameters α_i and γ_i determine the relative importance of the agriculture suitability and the population density and v_i determines the price level for land. The agriculture suitability and the population density are normalized between 1 and 10.

$$A_i = v_i \cdot SAgS_i^{\alpha_i} \cdot SPd_i^{\gamma_i} \tag{17}$$

$$SAgS_i = \begin{cases} 10 & AgS_i \geq 0.5 \\ 1 + 9 \cdot AgS / 0.5 & AgS_i < 0.5 \end{cases} \tag{18}$$

$$\alpha_i = \frac{\ln(PL_{\max}) - \ln(PL_{\min})}{2 \cdot \ln(10)} \tag{19}$$

$$\gamma_i = \alpha_i \tag{20}$$

Decision of Deforestation

The deforestation decision is expressed by equation (21). It compares the agricultural and forestry net present values corrected by values for deforestation and carbon sequestration. For the deforestation decision the amount of removed biomass from the forest is an important variable. The agricultural value needed for deforestation increases with the amount of timber sales and its concomitant flow to the wood products pool (HWP) pool. On the other hand the agriculture value will be decreased by the amount of released carbon to the atmosphere. This mechanism is expressed by a deforestation value (DV_i, equation 22). The model also allows for compensation of ancillary benefits from forests. This additional income is modeled either as a periodical income or a one time payment and will increase the forestry value by (IP_i). If it is a periodic payment it has to be discounted, which has been done in equation (23).

$$Defor = \begin{cases} Yes & A_i + DV_i > F_i \cdot H_i + IP_i \\ & \wedge \, not \, Protected \\ No & A_i + DV_i \leq F_i \cdot H_i + IP_i \\ & \vee \, Protected \end{cases} \tag{21}$$

$$DV_i = BM_i \cdot \left\{ pw_i \cdot C2W \cdot (1 - HL_i) - epc_i \cdot \left[(1+r) \cdot \left(\frac{frac_{llp} \cdot dec_{llp}}{dec_{llp} + r} + \frac{frac_{slp} \cdot dec_{slp}}{dec_{slp} + r} \right) \cdot (1 - frac_{sb}) + frac_{sb} \right] \right\} \quad (22)$$

$$IP_i = (BM_i + BMP_i) \cdot pca_i \cdot \frac{(r+1)^{fr_i}}{(r+1)^{fr_i} - 1} \quad (23)$$

There exist several ways of how financial transfers can be handled. Two mechanisms are realized in equation (21). One is to pay the forest owner to avert from the deforestation, the other is to introduce a carbon price that the forest owner gets money by storing carbon and paying for releasing it. The introduction of a carbon price focuses the money transfer to the regions where a change in biomass takes place. Payments to avoid emissions from deforestation can be transferred to cover all of the globe's forests, target to large "deforestation regions" or individual grids.

Deforestation Rate

Once the principle deforestation decision has been made for a particular grid cell (i.e., the indicator variable $Defor_i = 1$) the actual area to be deforested within the respective grid is to be determined. This is done by the auxiliary equation (24–25) computing the decrease in forest share. We model the deforestation rate within a particular grid as a function of its share of forest cover, agricultural suitability, population density and gross domestic product. The coefficients c_1–c_6 were estimated with a generalized linear model of the quasibinomial family with a logit link. Values significant at a level of 5% were taken and are shown in Table 1. The parameters of the regression model were estimated using R (R Development Core Team, 2005). The value of c_0 was determined upon conjecture and directly influences the maximum possible deforestation rate. For our scenarios the maximum possible deforestation is set to 5% of the total land area per year. That means, a 0.5° × 0.5° grid covered totally with forests cannot be deforested in a shorter time period than 20 years.

$$Fdec_i = \begin{cases} 0 & Defor = No \\ FS_i & Ftdec_i > FS_i \wedge Defor = Yes \\ Ftdec_i & Ftdec_i \le FS_i \wedge Defor = Yes \end{cases} \quad (24)$$

Table 1. Coefficients for equation (25)—Deforestation speed.

Coef	Estimate	Std. Error	Pr(> ltl)
C_0	0.05	–	–
C_L	-1 .799e+OO	4.874e-O1	0.000310
C_2	-2.200e-O1	9.346e-02	0.019865
C_3	-1 .663e-O 1	5.154e-02	0.001529
C_4	4.029e-02	1.712e-02	0.019852
C_5	-5.305e-04	1.669e-04	0.001789
C_6	-1.282e-04	3.372e-O5	0.000206

$$Ftdec_i = \begin{cases} 0 & Fs_i = 0 \vee AgS_i = 0 \\ x_i & Fs_i > 0 \wedge AgS_i > 0 \end{cases} \tag{25}$$

$$x_i = \frac{c_o}{1 + e^{-\left(c_1 + \frac{c_2}{Fs_i} + \frac{c_e}{AgS_i} + c_4 \cdot Pd_i + c_5 \cdot Pd_i^2 + c_6 \cdot GDP_i\right)}} \tag{26}$$

The deforestation rates (Ft_{dec}) were taken from (FAO, 2005), where the forest area from 1990, 2000, and 2005 for each country was given. For the estimation of the model parameters the area difference between 1990 and 2005 was used to infer the deforestation rate. All values which showed an increase of the forest area have been set to 0, because the model should only predict the deforestation. Countries with an increasing forest area have a deforestation rate of 0. It should be mentioned that the change rate is based on the total land area in the grid i and not on the current forest area.

By using c_2/F_s the model can only be used on grid's where there is some share of forest. This makes sense, because on places where there is no forest, no deforestation can appear. The model will only be usable on grids where forests occur. Therefore, for parameterization, the average agricultural suitability and the population density of a country are also only taken from grids which indicate forest cover.

Development of Forest Share

After calculating the deforestation rate, the forest share has to be updated each year with equation (27) assuring that the forest share stays within the permissible range of 0–1.

$$Fs_{i,year} = \begin{cases} fsx_{i,year} & fsx_{i,year} \le 1 - (Bul_i + Crl_i) \\ 1 - (Bul_i + Crl_i) & fsx_{i,year} > 1 - (Bul_i + Crl_i) \end{cases} \tag{27}$$

$$fsx_{i,year} = Fs_{i,year-1} - F_{i,dec} \tag{28}$$

Aboveground Carbon in Forest Biomass

The model describes the area covered by forests on a certain grid. It can also describe the forest biomass if the average biomass on a grid is known and the assumption was made, that the biomass in forests on the grid is proportional to the forest area.

For this reason a global carbon map of aboveground carbon in forest biomass, was created, based on country values from (FAO, 2005). By dividing the given total carbon, for each country, with the forest area of the country, the average biomass per hectare can be calculated. Now the assumption was made, that the stocking biomass per hectare on sites with a higher productivity is higher than on sites with a low productivity. Not for every country with forests (FAO, 2005) gives values of the stocking biomass. So a regression, describing the relation between tC/ha and NPP, was calculated and the biomass of grids of missing countries have been estimated to obtain a complete global forest biomass map.

Simulations

In the simulations the effect of different carbon-prices and/or incentives, for keeping forest, have been tested. The simulation period started in the year 2000 and ends in 2100. The decision, whether deforestation takes place or not and how fast it goes on, was done in 1-year time steps. Scenario drivers, available on coarser time resolution (e.g., population density), have been interpolated linearly between the given years.

Outputs of the simulations are trajectoria of forest cover, changes in carbon stocks of forests, and financial resources required to cut emissions from deforestation under varying scenario assumptions.

Data

The model uses several sources of input data some available for each grid, some by country aggregates and others are global. The data supporting the values in Table 2 are known for each grid. Some of the values are also available for time series.

Table 2. Spatial dataset available on a 0.5° x 0.5° grid.

Value	Year	Source
Land area	2000	[11]
Country	2000	[12]
NPP	-	[10]
Population density	1990- 2015	[13]
Population density	1990-2100	[14]
GDP	1990-2100	[14]
Buildup	2010-2080	[15]
Crop	2010-2080	[15]
Protected	2004	[16]
Agriculture suitability	2002	[17]
Biomass	2005	Self
Forest area	2000	[11]

Beside the datasets, available at grid level, the PPP (World Bank, 2005) from 1975–2003, the discount rates (Benítez et al., 2004) for 2004, the corruption in 2005 (Kaufmann et al., 2005) and the fraction of long living products for the time span 2000–2005 (FAO, 2005) are available for each country (Table 3).

Table 3. Country level values.

	Source
Discount rate	[8]
Fraction of long living products	[2]
Corruption	[5]
ppp	[7]

The values of Table 4 are used globally. Monetary values are transformed for each country with their price index. Brazil was taken as the price-reference country as described in (Benítez et al., 2004) and (Obersteiner et al., 2006).

Table 4. Global values.

Baseline	0.1
Decay rate long	ln(2)/20
Decay rate short	0.5
Factor carbon uptake	0.5
Frequency of incentives payment	5 years
tC to m3	4
Harvest losses	0.3
Hurdle	1.5
Maximum rotation interval	140 years
Minimum rotation interval	5 years
Planting costs	800 $/ha
Carbon price	0-50 $/tC
Carbon price incentives	0-50 $ftC
Minimum Land price	200 $/ha
Maximum Land price	900 $/ha
Minimum wood price	5$/ha
Maximum wood price	35$/ha

In Figure 1 the net primary productivity taken from (Alexandrov et al., 1999) is shown. The values range up to 0.75 $gC/m^2/year$. The highest productivity is near the equator.

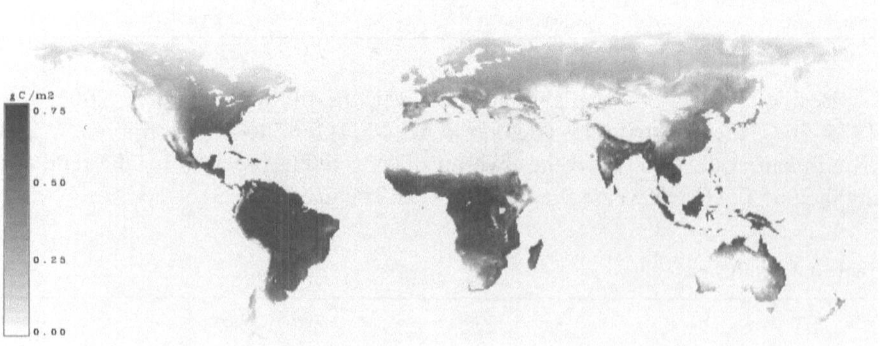

Figure 1. Deforested Area until 2100. Deforested Area under alternative assumptions. Incentives... periodic payments for standing biomass, Tax... payments for harvesting wood, Burn... felled wood is burned immediately, Sell... harvested wood is sold, Burn/Sell... share of the wood will be burned the other part sold.

In Figure 2 the population density in 2000 and in Figure 3 in the year 2100 is shown. It can be seen, that the highest population densities are reached in India and in south-east Asia. The densities are also quite high in Europe and Little Asia, Central Africa, and the coasts of America. The map of 2100 shows an increase in India and in south-east Asia.

Figure 2. Released Carbon from Deforestation until 2100. Released carbon from deforestation under alternative assumptions. Incentives... periodic payments for standing biomass, Tax... payments for harvesting wood, Burn... felled wood is burned immediately, Sell... harvested wood is sold, Burn/Sell... share of the wood will be burned the other part sold.

Figure 3. Avoided Carbon releases under different carbon prices during the next 100 years. Incentives... periodic payments for standing biomass, Tax... payments for harvesting wood, Burn... felled wood is burned immediately, Sell... harvested wood is sold, Burn/Sell... share of the wood will be burned the other part sold.

Figure 4 shows a map of the current forest, crop and buildup land cover. Large regions are covered by forests. Adjacent to the forests, large areas, used for crop production, can be seen.

Figure 4. Saved forest area under different carbon prices during the next 100 years. Incentives... periodic payments for standing biomass, Tax... payments for harvesting wood, Burn... felled wood is burned immediately, Sell... harvested wood is sold, Burn/Sell... share of the wood will be burned the other part sold.

In Figure 5 the suitability for agriculture is shown. Most of the high suitable land is used today for crop production (see Figure 4).

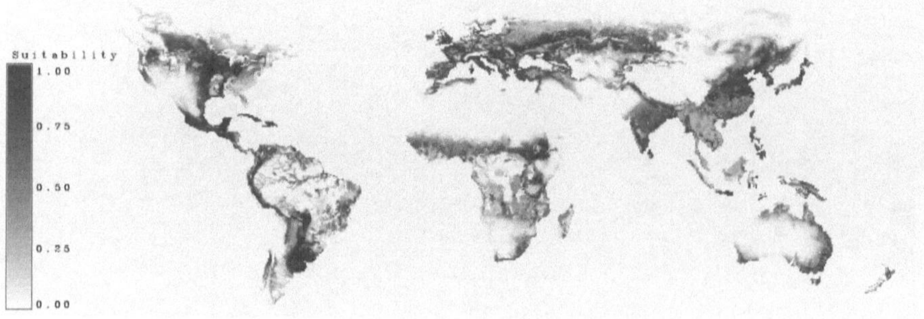

Figure 5. Income under different carbon prices. Tax... payments for harvesting wood, Burn... felled wood is burned immediately, Sell... harvested wood is sold, Burn/Sell... share of the wood will be burned the other part sold.

Figure 6 shows the carbon in forests. It can be seen, that the highest densities are located near the tropical belt. One reason for this is that, the biomass in tropical forests is high. Note that this picture shows the tons of carbon per grid and the grid size is 0.5° × 0.5° so the grid has its largest size near the equator.

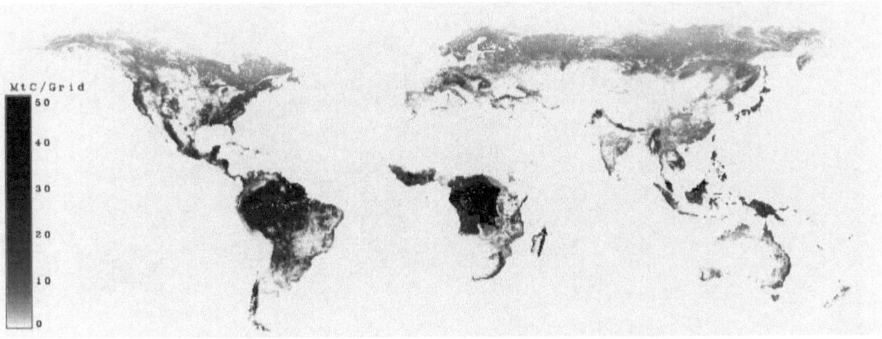

Figure 6. Expenditure under different carbon prices. Incentives... periodic payments for standing biomass, All... payments are done, without considering the effectiveness of the payment, in all regions, Region... payments are done in regions where the payments protect forest against deforestation, Affected... payments are done for forests where the payments protect them against deforestation.

Figure 7 shows the PPP which was used to calculate a price-index. It can be seen that the poorest countries are in Africa and the richest in North America, Europe, Australia, and Japan.

Figure 7. Cash flow until 2100 for different carbon prices. Incentives... periodic payments for standing biomass, Tax... payments for harvesting wood, Affected... payments are done for forests where the payments protect them against deforestation, Burn... felled wood is burned immediately, Sell... harvested wood is sold, Burn/Sell... share of the wood will be burned the other part sold.

Figure 8 shows the discount-rates given in (Benítez et al., 2004). Here also the richest countries have the lowest discount rates.

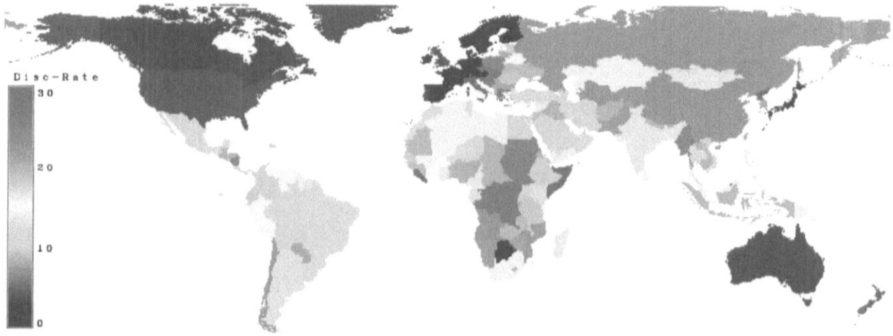

Figure 8. Expenditure until 2100 for different incentive payment strategies. Incentives... periodic payments for standing biomass, All... payments are done, without considering the effectiveness of the payment, in all regions, Region... payments are done in regions where the payments protect forest against deforestation, Affected... payments are done for forests where the payments protect them against deforestation.

Figure 9 shows the effectiveness of the carbon incentives. In low risk countries nearly all of the spent money will be used for maintaining forest sinks in risky countries not all of the money will come to the desired sink.

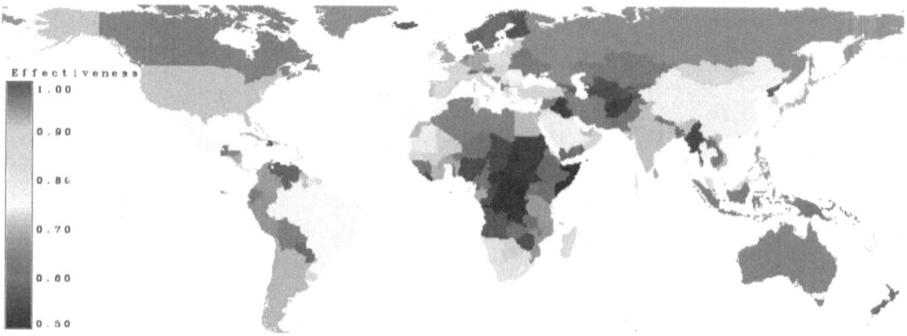

Figure 9. Removed biomass without a carbon price. Green areas show grids where nowadays forests can be found. Red areas indicate grids where deforestation will occur in a scenario without carbon prices.

Figure 10 shows the proportion of harvested wood entering the long living products pool (FAO, 2005).

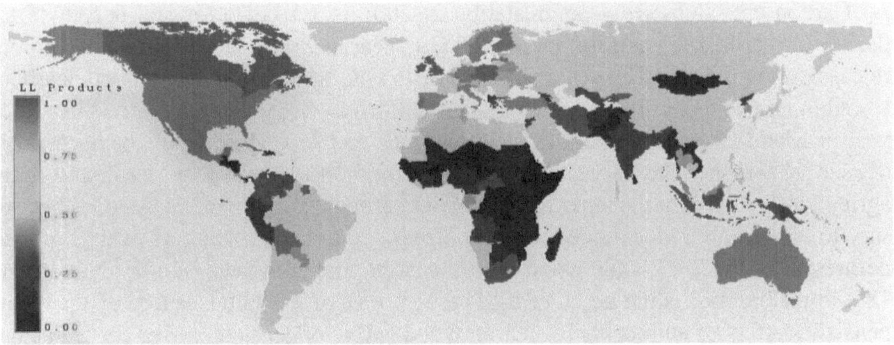

Figure 10. Saved biomass by 12$/tC (burn sell). Light green show grids where nowadays forests can be found. Dark green areas indicate grids where forest biomass can be saved by introducing a carbon price of 12$/tC compared to the baseline scenario. Red areas indicate grids where there will still be deforestation.

RESULTS AND DISCUSSION

Baseline Deforestation 2000–2100 and Effects of Financial Mechanisms Aiming at Cutting Emissions in Half

Baseline scenario calculations (i.e., a carbon price of 0 US$/tC is assumed) show that close to 200 mil ha or around 5% of today's forest area will be lost between 2006 and 2025, resulting in a release of additional 17.5 GtC to the atmospheric carbon pool. The baseline deforestation speed is decreasing over time, which is caused by a decreasing forest area in regions with high deforestation pressure. In the year 2025 the annual deforested area decreases to 8.2 million hectares, compared to 12.9 million hectares in 2005. By the year 2100 deforestation rates decline to some 1.1 million hectares. According to the base line scenario, today's forest cover will shrink by around 500 million hectares or by more than 1/8 within the next 100 years (Figure 11).

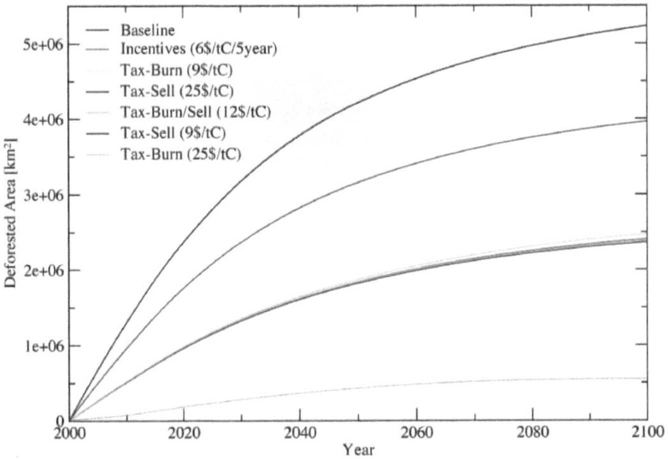

Figure 11. Net primary production (NPP). Areas with a high increment have a high net primary productivity and are indicated by dark gray. Sites with low productivity are indicated by light gray.

Carbon emissions from deforestation in 2005 is 1.1 GtC/year and decreases to 0.68 GtC/year in 2025 and further to 0.09 GtC/year in 2100. The accumulated carbon release during the next 100 years amounts to 45 GtC which is 15% of the total carbon stored in forests today. To bring deforestation down by 50%, incentives of 6 US$/tC/5 year or a land clearance tax of between 9 US$/tC and 25 US$/tC would be necessary, depending whether the harvested wood is burned on the spot (e.g., slash-and-burn agriculture) or sold. In the latter case, a higher carbon tax of up to 25 US$/tC is necessary to effectively reduce incentives to deforest, to a degree that cuts overall global deforestation by 50%. If the wood is further used and converted into products, only 18% of the biomass could be saved by a carbon price of 9 US$/tC, caused by the compensating effect of an income by selling wood and a longer time-period for releasing carbon. On the other hand, if the carbon price is 25$/tC and the wood is assumed to be slash burned, the reduction of deforestation calculated to be 91% (Figure 11 and 12). On a first sight it seems, that incentive payments might be more effective, than taxation. However, incentives payment contracts have to be renewed every 5 year for the actual standing biomass and the change of biomass has to be known to detect a breach of the contract, while a deforestation tax will be payed once for the harvested biomass once detected by targeted earth observation systems (see Figures 13 and 14). In the latter, transactions costs for implementing avoided deforestation are small.

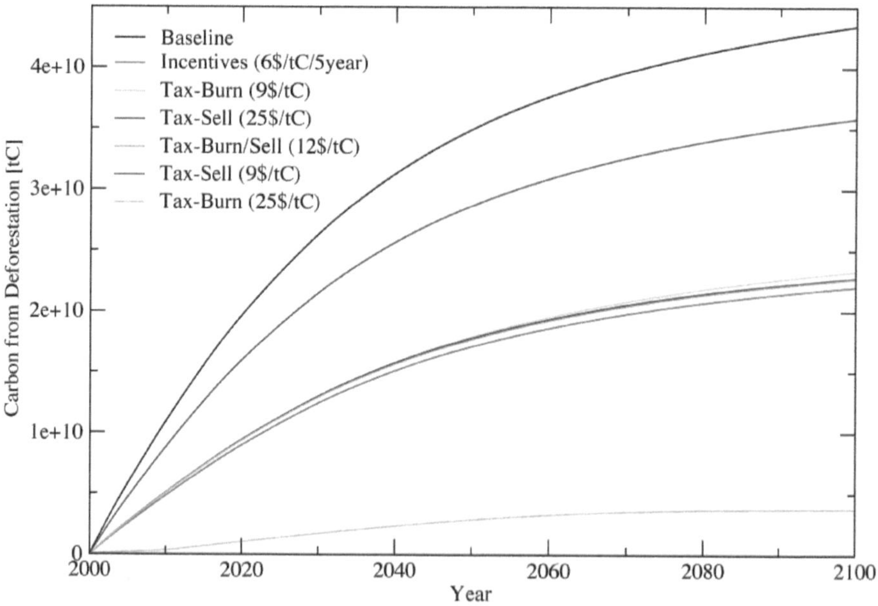

Figure 12. Population density in year 2000. Grids with few people are given in white. A rising population density is marked by gray up to high population densities (≥1,000 people/km²) which are indicated by black.

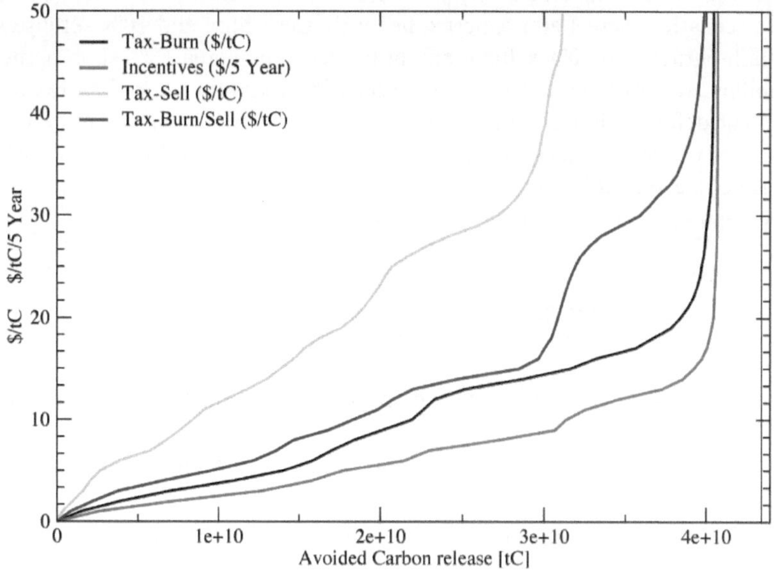

Figure 13. Population density in year 2100. Grids with few people are given in white. A rising population density is marked by gray up to high population densities (≥1,000 people/km^2) which are indicated by black.

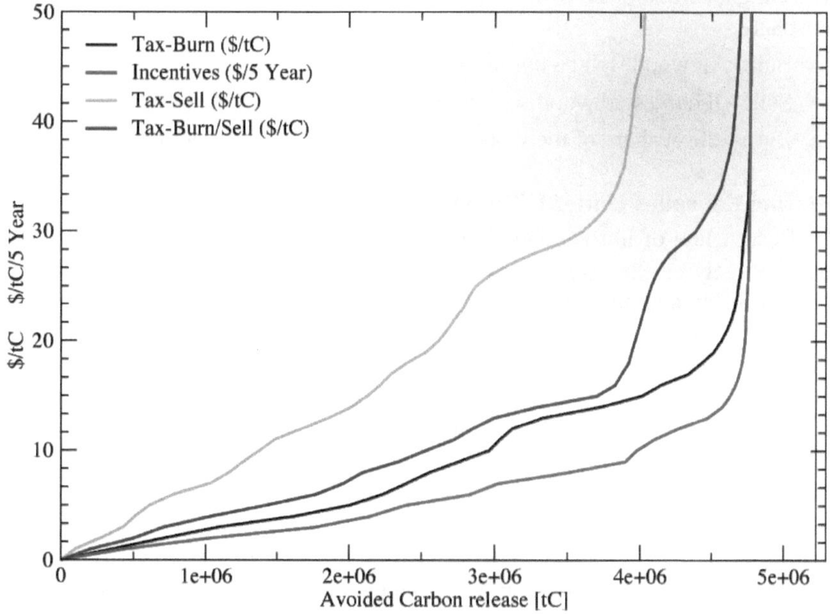

Figure 14. Forest, crop and buildup land cover.

The assumption, that either only slash burn or all wood will be sold is unrealistic. Thus, a scenario where Latin America has 90% slash burn and 10% selling, Africa 50% slash burned and 50% selling, and in the remaining area 10% slash burned and 90% selling, was examined. Under such scenario assumptions a carbon tax of 12 $/tC will cut deforestation in half. Also the assumption, that a carbon price will stay constant over time may not be close to reality but it can be used to see the long-term influence of a given carbon price.

We differentiate between the following cases:

Baseline

Introducing no carbon price.

Incentives

Introducing a carbon price which will be payed periodic for the carbon stored in the standing forest biomass.

- All: Payments are done, without considering the effectiveness of the payment, in all regions.
- Region: Payments are done in regions where the payments protect forest against deforestation.
- Affected: Payments are done for forests where the payments protect them against deforestation.

Tax

Introducing a carbon price which has to be paid for releasing the stored carbon to the atmosphere.

- Burn: All wood will be burned immediately.
- Sell: All harvested wood will be sold.
- Burn/Sell: A share of the wood will be burned and the other part sold.

Costs and Revenues Under Different Carbon Prices

The effectiveness of introducing a carbon price to influence deforestation decisions depends largely on the levels set for carbon prices, apart from considerations of political feasibility and implementability. Low prices have little impact on deforestation rates. During the 21st century carbon tax schemes of 9 US$/tC for slash burn and 25 US$/tC for situations when removed wood enters a harvested HWP would generate some 2–5.7 billion US$/year respectively when emissions from deforestation are to be cut in half. For the variant of 12 US$/tC, with regionally differentiated slash burn and HWP assumptions, the average annual income for the next 100 years are calculated to be around 2.7 billion US$. These tax revenues decrease dramatically over time mainly due to the declining baseline deforestation rate. Tax revenues are computed to be 6 billion US$ in 2005, 4.3 billion US$ in 2025 and 0.7 billion US$ in 2100. This indicates the magnitudes and their temporal change of funds generated from a deforestation tax scheme aiming at a 50% emission reduction (Figures 15 and 17).

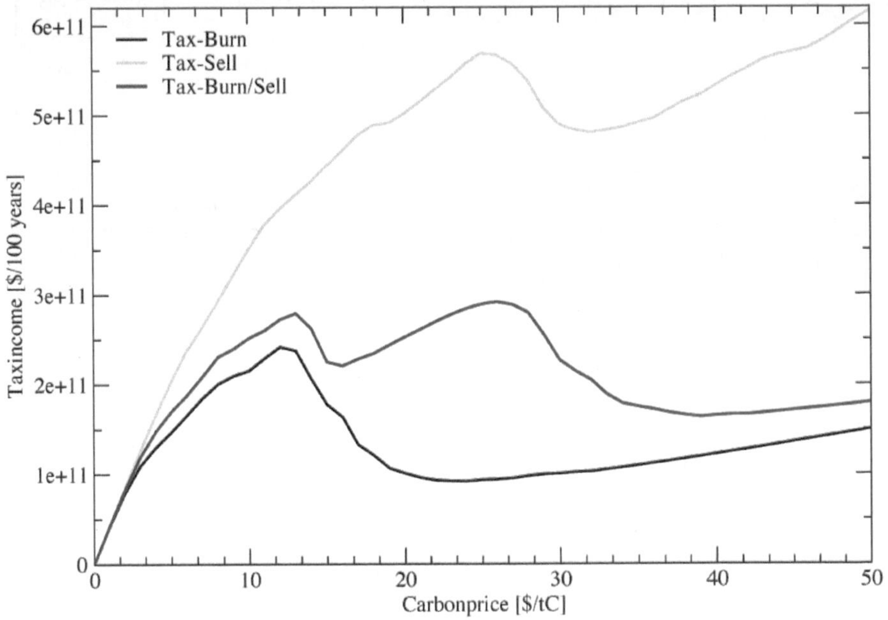

Figure 15. Agriculture suitability.

In the alternative incentive scheme, the amount of funds necessary, is depending on the strategy of payments, either increasing, staying constant, or decreasing over time. If incentives are paid only for those forest areas that are about to be deforested, and with a global target of cutting deforestation by 50%, a minimum payment of 6 US$/tC/5 year or 0.24 billion US$ in 2006 would be required. This amount rises to some 1.2 billion US$ in 2010, 4.1 billion US$ in 2025, and 10 billion US$ in 2100 caused by the increasing area of saved forest area. As precise information of forests about to be deforested is absent, incentive payment schemes would have to focus on regions under deforestation pressure. Given that incentives are only spent on regions of 0.5° × 0.5° where they can effectively reduce deforestation in an amount that they will balance out the income difference between forests and alternative land use up the 6 US$/tC/5 year, this would come at a cost of 34 billion US$/year (Figures 16 and 18). It should be noted that the tax applies only on places currently deforested while the subsidy applies to larger areas depending on how far it is in practice possible to restrict the subsidy to vulnerable areas. All figures above are intentionally free of transaction costs. Transaction costs would *inter alia* include expenditure for protecting the forests against illegal logging by force and expenditures monitoring small scale forest degregation. Governance issues such as corruption and risk adjustment, depending on the country are, however, considered in the analysis to the extent possible.

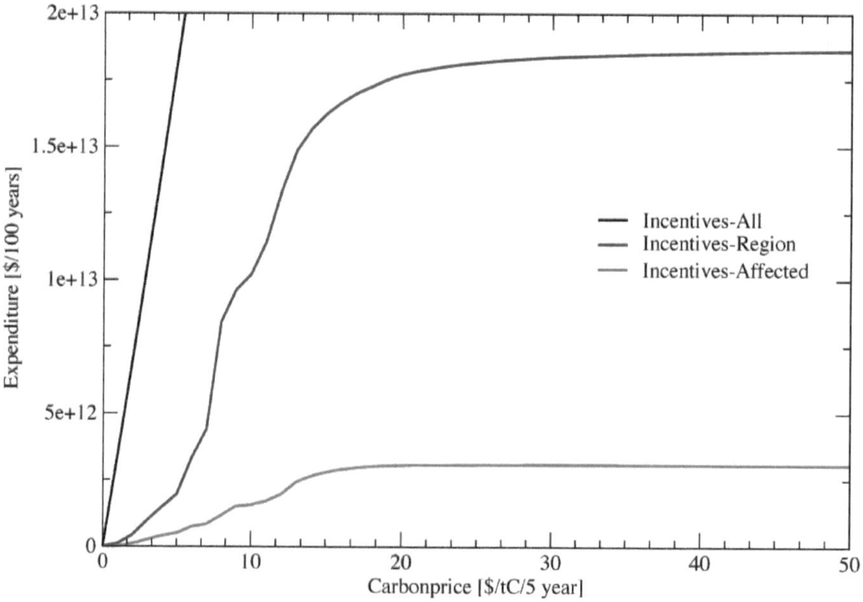

Figure 16. Carbon in forest biomass. Regions with no carbon in forests are gray. Regions with high values of carbon in forests are black.

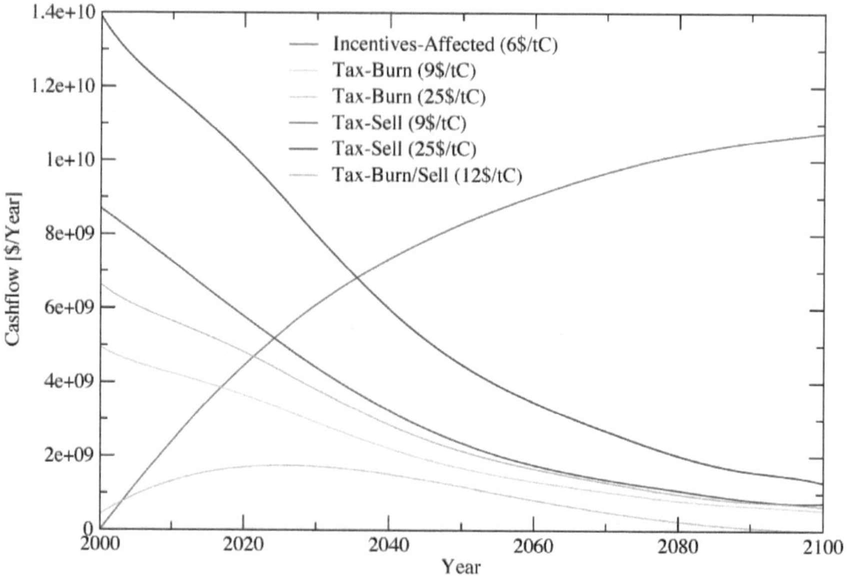

Figure 17. Purchasing power parity (PPP).

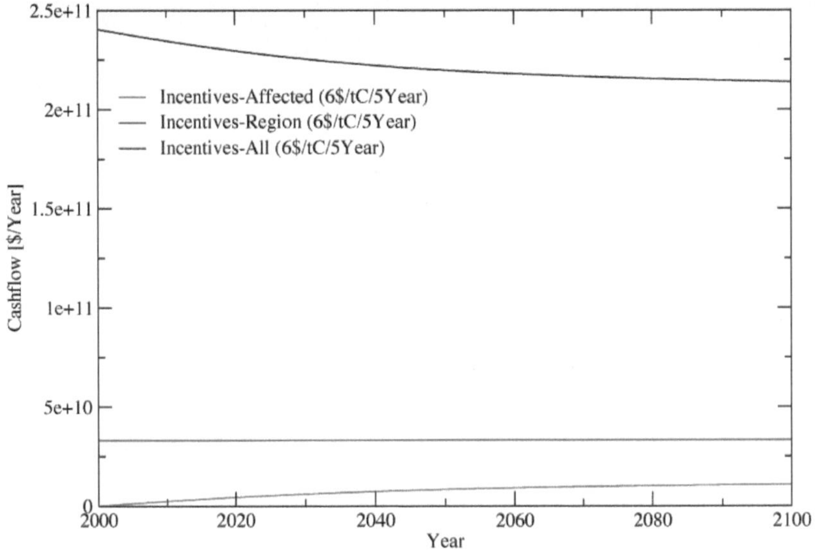

Figure 18. Discount rate.

Regional Effects of Carbon Prices on Deforestation

Sources of deforestation in the model are expansion of agriculture and buildup areas as well as from unsustainable timber harvesting operations impairing sufficient reforestation. Deforestation results from many pressures, both local, and international. While the more direct causes are rather well established as being agricultural expansion, infrastructure extension and wood extraction, indirect drivers of deforestation are made up of a complex web of interlinked and place-specific factors. There is large spatially differentiated heterogeneity of deforestation pressures. Within a forest-agriculture mosaic, forests are under high deforestation pressure unless they are on sites which are less suitable for agriculture (swamp, slope, altitude). Closed forests at the frontier to agriculture land are also under a high deforestation pressure while forest beyond this frontier are under low pressure as long as they are badly attainable. The model was build to capture such heterogeneity in deforestation pressures.

Figure 19 shows that the model predicts deforestation to continue at the frontier to agricultural land and in areas which are easily accessible. Trans-frontier forests are also predicted to be deforested due to their relative accessibility and agricultural suitability. Forests in mosaic lands continue to be under strong pressure. Figure 20 illustrates the geography of carbon saved at a carbon tax of 12 US$/tC compared to biomass lost through deforestation. Under this scenario deforestation is mainly occurring in clusters, which are sometimes surrounded by forests (e.g., Central Africa) or are concentrated along a line (Amazon). The geography of the remaining deforestation pattern indicates that large areas are prevented from deforestation at the frontier by the 12 US$/tC tax. The remaining emissions from deforestation are explained mainly by their accessibility and favorable agricultural suitability.

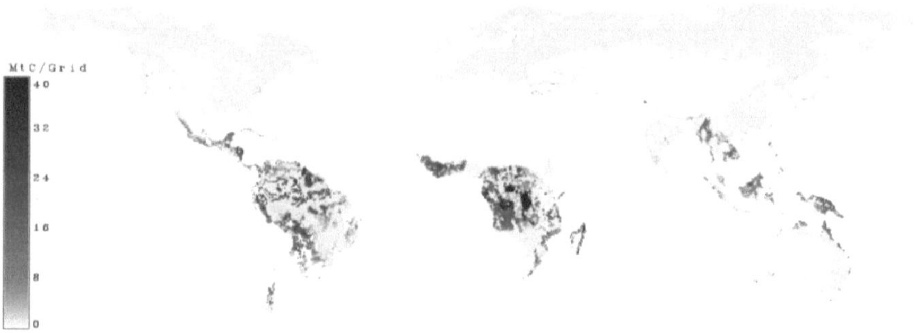

Figure 19. Effectiveness (Corruption). Countries with high values of corruption are marked in red, moderate countries in yellow and low values in green.

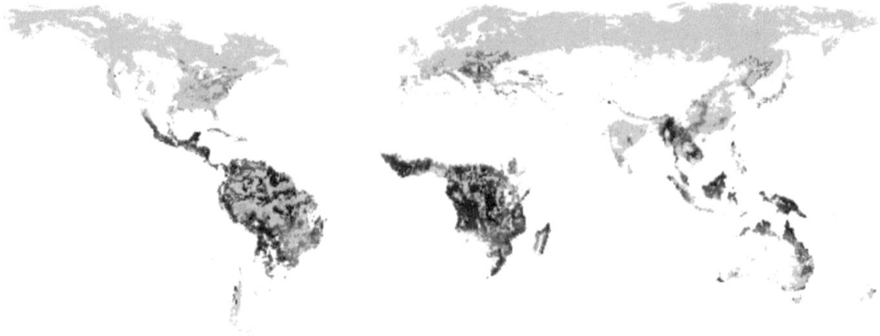

Figure 20. Share of long living products. Countries which use their wood mainly for fuelwood are marked in blue, those who use it for sawn-wood are in green.

CONCLUSIONS

Avoiding deforestation requires financial mechanisms that make retention of forests economically competitive with the currently often preferred option to seek profits from other land uses. According to the model calculations, even relatively low carbon incentives of around 6 \$/tC/5 year, paid for forest carbon stock retention or carbon taxes of 12 \$/tC would suffice to effectively cut emissions from deforestation by half. Taxes revenues would bring about annual income of US\$6 billion in 2005 to US\$0.7 billion in 2100. The financial means required for incentives are estimated to range from US\$3 billion to US\$ 200 billion per year, depending on the design of the avoided deforestation policy. Our scenario, where incentives are payed in regions where deforestation will appear and the payment has an effect, estimates the necessary funds to cut emissions from deforestation in half in the magnitude of some US\$ 33 billion per year, without including costs for transaction, observation, and illegal logging protection. Increasing the value of forest land and thereby make it less easily prone to deforestation would act as a strong incentive to increase productivity of agricultural and fuelwood production.

LIST OF SYMBOLS

α_i: Importance of agriculture
γ_i: Importance of population
ν_i: Land price level = minimum land price of reference country × price index (px_i) [$/ha]
ω_i : Carbon uptake per year [tC/year/ha]
θ_i : Fraction of carbon benefits in products [1]
A_i: Net present value of agriculture [$/ha]
AgSi: Agricultural suitability [0–1]
b_i: Baseline, how much carbon uptake will be if there is no forest, e.g., 0.1 [1]
BMP_i: Biomass in Products [tC/ha]
BM_i: Aboveground living wood biomass [tC/ha]
B_i: Present value of carbon benefits [$/ha]
Bul: Share of buildup land [1]
C2W: Conversion factor form 1t Carbon to $1m^3$ wood [m^3/tC]
cp_i: Planting costs [$/ha]
cp_{ref}: Planting costs reference country [$/ha]
CU: Carbon uptake, share of NPP stored in wood [1]
Crl: Share of crop land [1]
dec_{llp}: Decay rate of long living products e.g., 0.03 [1]
dec_{slp}: Decay rate of short living products e.g., 0.5 [1]
DV_i: Deforestation value [$/ha]
epc_i: Effectiv carbon price [$/tC]
f_i: Net present value of forestry for one rotation period [$/ha]
F_i: Net present value of forestry [$/ha]
F_s: Actual share of forest [0–1]
F_{dec}: Decrease of the forest share
fr_i: Frequency of incentives money payment [Years]
$frac_{llp}$: Fraction of long living products e.g., 0.5 [0–1]
$frac_{sb}$: Fraction of slash burned area e.g., 0.9 [0–1]
$frac_{slp}$: Fraction of short living products e.g., 0.5 [0–1]
F_s: Forest area share [0–1]
Fs_{year}: Forest share of a certain year [1]
fsx_{year}: Theoretical forest share of a certain year [1]
Ft_{dec} : Theoretical decrease of the forest share
GDP: Gross domestic product [$1995/Person]
H_i: Hurdle e.g., 1.5 [1]
HL_i: Harvesting losses e.g., 0.2 [1]
i: Grid number

$leak_i$: Factor of money which will in real reach the forest [1]
IP_i: Incentive payment [\$/ha]
MAI_i: Mean annual wood volume increment [m_3/ha]
NPP_i: Net primary production [tC/ha/year]
pc_i: Carbon price [\$/tC]
pca_i: Incentives carbon price [\$/tC/fr_i]
Pd_i: Population density [People/km^2]
PL_{max}: Maximal land price of reference country × price index (px_i) [\$/ha]
PL_{min}: Minimal land price of reference country × price index (px_i) [\$/ha]
PPP_i: Purchasing power parity [\$]
PPP_{ref}: Purchasing power parity of reference country [\$]
pr_i: Ratio of area planted [0–1]
pw_i: Stumpage wood price [\$/m^3]
pw_{max}: Maximum revenue of wood, e.g., 35\$/fm [\$/fm]
Pw_{min}: Minimum revenue of wood, e.g., 5\$/fm [\$/fm]
px_i: Price index [1]
r: Discount rate [e.g., 0.05]
R_i: Rotation interval length [years]
SAgSi: Standardized agricultural suitability [1–10]
SFs: Standardized not forest area share [1–10]
SPd: Standardized population density [1–10]
V_i: Harvest wood volume [m3]
x_i: Theoretical decrease of the forest share if $Fs_i > 0 \circ AgSi > 0$

KEYWORDS

- **Biomass**
- **Carbon sequestration**
- **Deforestation**
- **Forest degradation**
- **Greenhouse gas**

AUTHORS' CONTRIBUTIONS

Georg Kindermann has developed the deforestation rate model, implemented the whole model, collected some data sources, and organized them to be used for the implementation, runs the simulations, created figures and tables, and wrote a first draft of the chapter.

Michael Obersteiner developed the core model describing the forest value, agricultural value, and decision of deforestation, worked on the chapter, introduced the maximum tax income and contributed to the payment possibilities.

Ewald Rametsteiner contributed to the carbon price and incentives model and their practical implementation, worked on the chapter and brought in many background informations.

Ian McCallum collected and organized the data source and produced some figures of the chapter.

ACKNOWLEDGMENTS

We acknowledge the support by the Greenhouse Gas Initiative (GGI) project, an institute-wide collaborative effort within IIASA. The interdisciplinary research effort within GGI links all the major research programs of IIASA that deal with research areas related to climate change, including population, energy, technology, and forestry, as well as LUCs and agriculture. GGI's research includes both basic and applied, policy-relevant research that aims to assess conditions, uncertainties, impacts, and policy frameworks for addressing climate stabilization, from both near-term and long-term perspectives. Support from the EU FP 6 Project Integrated Sink Enhancement Assessment (INSEA, SSPI-CT-2003/503614 with DG RTD) is gratefully acknowledged. We graceful thank the reviewers for their reports.

COMPETING INTERESTS

The author(s) declare that they have no competing interests.

Chapter 8

Volcanic Hazards and Evacuation Procedures

D. K. Bird, G. Gisladottir, and D. Dominey-Howes

INTRODUCTION

Katla volcano, located beneath the Myrdalsjökull ice cap in southern Iceland, is capable of producing catastrophic jökulhlaup. The Icelandic Civil Protection (ICP), in conjunction with scientists, local police, and emergency managers, developed mitigation strategies for possible jökulhlaup produced during future Katla eruptions. These strategies were tested during a full-scale evacuation exercise in March 2006. A positive public response during a volcanic crisis not only depends upon the public's knowledge of the evacuation plan but also their knowledge and perception of the possible hazards. To improve the effectiveness of residents' compliance with warning and evacuation messages, it is important that emergency management officials understand how the public interpret their situation in relation to volcanic hazards and their potential response during a crisis and apply this information to the ongoing development of risk mitigation strategies. We adopted a mixed methods approach in order to gain a broad understanding of residents' knowledge and perception of the Katla volcano in general, jökulhlaup hazards specifically and the regional emergency evacuation plan. This entailed field observations during the major evacuation exercise, interviews with key emergency management officials and questionnaire survey interviews with local residents. Our survey shows that despite living within the hazard zone, many residents do not perceive that their homes could be affected by a jökulhlaup, and many participants who perceive that their homes are safe, stated that they would not evacuate if an evacuation warning was issued. Alarmingly, most participants did not receive an evacuation message during the exercise. However, the majority of participants who took part in the exercise were positive about its implementation. This assessment of resident knowledge and perception of volcanic hazards and the evacuation plan is the first of its kind in this region. Our data can be used as a baseline by the ICP for more detailed studies in Iceland's volcanic regions.

The Icelandic term "jökulhlaup" is defined as a sudden burst of meltwater from a glacier and may occur for a period of several minutes to several weeks (Björnsson, 2002). All confirmed historic eruptions of Katla, the volcano underlying the Myrdalsjökull ice cap in southern Iceland (Figure 1), have produced jökulhlaup (Thordarson and Larsen, 2007). A Katla eruption can melt through the ~400 m of ice covering the Katla caldera in 1–2 hr, producing a catastrophic jökulhlaup with a peak discharge of 100,000–300,000 $m^3 s^{-1}$ (Björnsson, 2002).

Transporting volcanic debris and large ice blocks, jökulhlaup have been the most serious hazard during historic Katla eruptions but not the only hazard. Local communities 30 km from the eruption site have been subjected to heavy tephra fallout and lightning strikes (Larsen, 2000) while jökulhlaup have triggered small tsunami during past volcanic events (Guomundsson et al., 2008). Earthquakes, felt by local communities, signify the start of an eruption. They are not however, of sufficient magnitude to cause major damage (Guomundsson et al., 2008). Furthermore, not all Katla eruptions have been subglacial. Lava covered ~780 km^2 of land during the 934–938 AD Eldgjá flood lava eruption which occurred along a 75 km discontinuous and predominately sub-aerial volcanic fissure extending from the Katla caldera (Thordarson and Larsen, 2007).

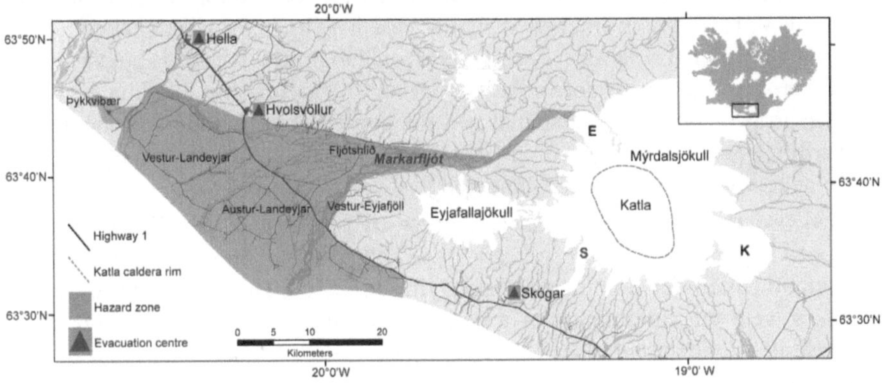

Figure 1. The jökulhlaup hazard zone of Rangávallasysla. The hazard zone is determined to be the maximum flood area for a catastrophic jökulhlaup. Communities located within the hazard zone are Vestur-Eyjafjöll, Fljótshlíd, Austur and Vestur-Landeyjar, and Pykkvibær. Evacuation centers are located in Hella, Hvolsvöllur, and Skygar. The three catchment areas of Myrdalsjökull: Entujökull, Sylheimajökull, and Kötlujökull are represented by E, S and K respectively.

Since settlement in the 9th century Katla has erupted approximately 1–3 times per century (Thordarson and Larsen, 2007). At least 21 eruptions have occurred during this time with the last confirmed eruption in 1918 AD (Larsen, 2000). All historic jökulhlaup have emanated from the catchment areas of Kötlujökull and Sólheimajökull while none have come from the Entujökull catchment. Unconfirmed volcanic activity may have created the jökulhlaup which occurred in 1955 AD and 1999 AD from the Kötlujökull and Sólheimajökull catchments, respectively (Björnsson et al., 2000; Guomundsson, 2005; Russell et al., 2000).

The Markarfljót valley was subjected to volcanic jökulhlaup emanating from the Entujökull catchment prior to settlement. A series of large, valley-ficatch prehistoric jökulhlaup were identified by Smith (2004) and Larsen et al (2005) from sedimentary deposits within the Markarfljót valley. Further, Smith and Haraldsson (2005) determined that the last volcanic jökulhlaup on the Markarfljót occurred 1,200 years before present. Other types of jökulhlaup have flooded the Markarfljót in more recent times. In 1967 AD, a rock/ice avalanche caused an outburst flood from the proglacial lake of Steinsholtsjökull on the northern flank of Eyjafjallajökull. This flood transported boulders measuring up to 80 m^3 5 km from the rockslide scar (Kjartansson, 1967). Lastly, geothermal meltwater drains from subglacial lakes in small, more frequent jökulhlaup from all three catchment areas (Björnsson et al., 2000).

Flood simulation models based on data from prehistoric jökulhlaup were used to identify peak discharge and temporal and spatial distribution of a possible catastrophic jökulhlaup flooding from the Entujökull catchment down the Markarfljót (Hólm and Kjaran, 2005). This populated farming region forms part of the Rangávallasysla municipality. The models show that a catastrophic jökulhlaup with a peak discharge of 300,000 $m^3 s^{-1}$ would reach its maximum within 2 hr, flooding to a depth of up to 15 m, at the uppermost farms in Fljótshlío and up to 10 m in Vestur-Eyjafjöll. However, many of the farmhouses in these communities are elevated above the floodplain. In contrast, the roads leading up to these farms parallel the Markarfljót and some sections of these roads are positioned at similar base heights to the river channel. Dykes approximately 2 m in height have been constructed to protect the roads but these fland mitigation structures were not built to withstand a catastrophic jökulhlaup. Within 3 hr Highway 1 would be inundated and the entire outwash plain surrounding the Markarfljót would be flooded within 10 hr. With a maximum flood depth of up to 2 m, low lying regions could remain submerged for over 24 hr.

In view of the potential future hazard presented by jökulhlaup, the ICPorganization developed regional evacuation strategies based on a worst case scenario as described in the report edited by Guomundsson and Gylfason (2005). This report and consequent strategies were the culmination of a multidisciplinary investigation into the physical threat of jökulhlaup produced from a Katla eruption. It did not however, include research from a societal aspect. Researchers argue that a collaboration between the physical and social sciences is a key step toward achieving a greater understanding of the consequences of volcanic hazards (e.g. Johnston et al., 1999). Following the investigation communication sessions were held with residents from communities located within the hazard zone in Rangávallasysla: Vestur-Eyjafjöll, Fljótshlíd, Landeyjar, and Pykkvibær. These consisted of information meetings in 2005 and 2006 regarding the possibility of a future Katla eruption and the proposed evacuation plan for a jökulhlaup hazard. During these meetings residents were informed that they could collect an evacuation and hazard information sign from local police (Figure 2) (K. Porkelsson, personal communication, 2006).

If an eruption is imminent residents would be notified via a text message to their mobile phone. If residents do not have a registered mobile phone number a recorded message would call through to their landline. Upon receiving this message residents have 30 min to prepare to evacuate. However, if an eruption occurs without precursory activity, residents will be instructed to evacuate immediately. Before leaving, they are required to hang the evacuation sign outside their house to indicate that they have left. Certain residents in each region have volunteered to "sweep" their local area to ensure their neighbors have left for the evacuation centers (EC) located in Hella, Hvolsvöllur, and Skógar. In order to reach these centers, some residents must evacuate via the roads that parallel the Markarfljót and along Highway 1.

To test the proposed evacuation plan the ICP conducted a full scale evacuation exercise on March 26, 2006 in Rangávallasysla. Approximately 1,200 residents live within the hazard zone (K. Porkelsson, personal communication, 2006) and for the purpose of fully testing the evacuation plan residents were not informed of the timing of the eruption scenario. Instead residents were instructed to go about their business as usual until they received an evacuation message (R. Ólafsson, personal communication, 2006). The mock eruption began at 10:55 local time (LT) and the first evacuation message was communicated to residents at 10:59 LT. Residents then had 30 min to complete the instructions on the hazard sign (Figure 2) before evacuating their homes to their designated center.

To improve the effectiveness of residents' compliance with warning and evacuation messages it is important that emergency management officials understand how the public interpret their situation in relation to volcanic hazards and their potential response during a crisis (Bird and Dominey-Howes, 2006, 2008; Dominey-Howes and Minos-Minopoulos, 2004; Gregg et al., 2004; Haynes et al., 2008; Paton et al., 2008; Ronan et al., 2000). Therefore, this study (1) investigates resident's knowledge and perception of Katla, jökulhlaup hazard and their views of the evacuation plan and exercise, and (2) reports the findings to help the ICP improve mitigation strategies. To achieve this, field observations were made during the evacuation exercise, semi-structured interviews with key emergency management officials were held after the evacuation exercise, and questionnaire survey interviews were conducted with local residents. The rationale for using this sequential mixed methods approach is to better understand the evacuation procedure from both a management and public perspective and to develop and implement a questionnaire survey interview to further explore participant views and knowledge. Before addressing the aim of our research we will describe the methods used to conduct the analysis.

MATERIAL AND METHODS

A mixed methods approach, drawing from both qualitative and quantitative data collection practices was used to obtain public perception data. We were invited to observe the evacuation exercise from within the emergency headquarters (EH) in Hella in addition to monitoring the proceedings at the EC in Hvolsvöllur and Hella. Following the exercise, we conducted semi-structured interviews with emergency management officials and face-to-face questionnaire survey interviews with local residents living within the hazard zone. Public perception research based solely on data generated from questionnaire surveys is unable to capture the complexity of a hazard in a societal context whereas a mixed-methods approach, employing both qualitative and quantitative techniques, provides the researcher with the opportunity to acquire a variety of information on the same topic allowing for a more accurate interpretation of the issues at hand (Haynes et al., 2007; Horlick-Jones et al., 2003). In this section, we describe the methods employed for field observations and interviews followed by those adopted to construct and deliver the questionnaire survey.

Observing the Evacuation Exercise

Located within the main EH, we (Bird and Gisladottir) observed and documented the development and management of the evacuation exercise. We were at the EH during the most critical stages of the eruption scenario. As the eruption developed we visited the EC in Hella and Hvolsvöllur to observe the emergency management proceedings of the Red Cross and to witness how the public behaved and responded to the evacuation. Some informal discussions were held with evacuees and Red Cross personnel at both centers. During our observations we made written notes to ensure the most significant points were recorded.

Interviewing Emergency Management Officials

Follow-up interviews were conducted with the project manager of ICP, the Chief of Police in Rangávallasysla, the president of the Icelandic Association for Search and Rescue (ICE-SAR), a research scientist involved in the hazard assessment report and coordination of the eruption scenario for the evacuation exercise, a regional manager for the Red Cross, and the Director of Communication for the Red Cross. The format of the interview was semi-structured whereby specific questions were asked about their departments' role in an emergency situation, their role during the exercise, their perception of the response behavior of evacuees, and whether or not they viewed the exercise to be a success.

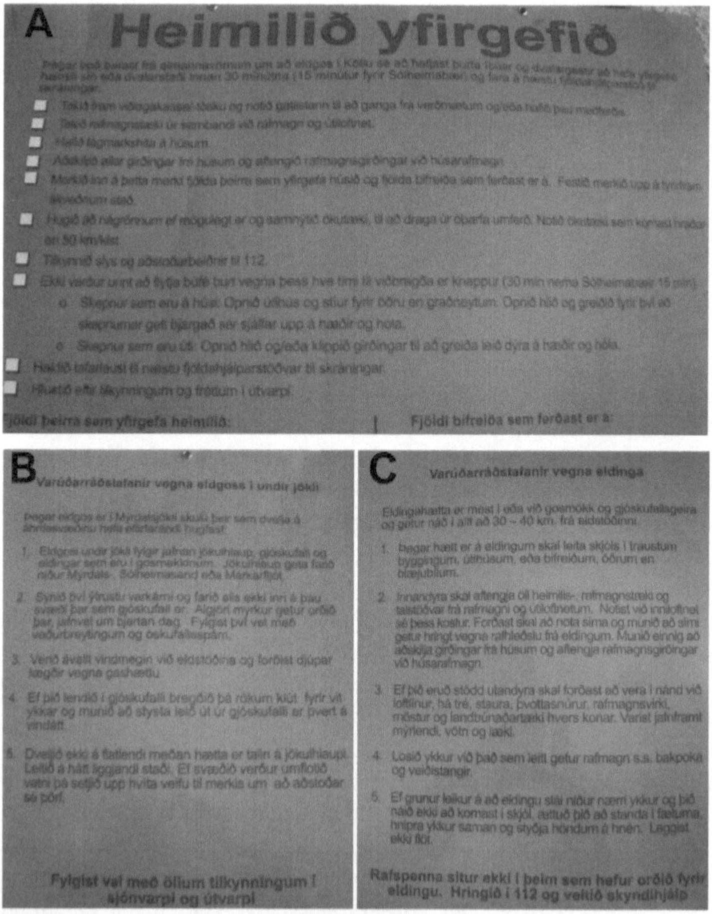

Figure 2. Evacuation and hazard information sign distributed to residents located in the volcanic hazard zone surrounding Katla. English translations follow.

A. House Evacuation

When a warning is given by the ICP that an eruption in Katla is starting residents and their guests must evacuate within 30 min (15 min for Sólheimar) to the nearest evacuation center.

- Get the first aid kit, follow this list and secure or collect the valuables you want to take with you.
- Unplug all electrical equipment as well as antennas.
- Set household heaters to a minimum temperature.
- Remove fencing from the house and unplug all electric fences from the house electricity.
- In the space provided indicate how many people have evacuated from this property and the number of vehicles used to evacuate. Fasten this sign on the predetermined spot.
- Check on neighbors if possible and share vehicles to avoid unnecessary traffic. Use vehicles that can drive faster than 50 km/hr.
- Call 112 if there has been an accident or if you need help.
- It is not possible to move animals due to short evacuation time (30 min, except for Sólheimar 15 min).

For animals that are housed, open the house and pen for all animals except bulls. Open gates and ensure that they can flee to higher ground.

For animals that are outside, open gate and/or cut fences so that they can flee to higher ground.

- Go straight to the nearest evacuation center and register.
- Listen to announcements and news on radio.

Number of people evacuated from house: Number of vehicles used for evacuation:

B. Precautions due to subglacial eruptions (back left hand side)

During an eruption in Myrdalsjökull those staying in the hazard area should think of the following:

1. Jökulhlaup, tephra fall, and lightning within the plume usually follow a subglacial eruption. Jökulhlaup can go down Myrdalssandur, Sólheimasandur, or the Markarfljót.
2. You should be very careful not enter areas of tephra fall as it can be completely dark even during the day. You should be observant of weather changes and forecast of tephra fall.
3. Always stay on the side of the volcano in the direction of the wind. Avoid deep topographical depressions due to the accumulation of poisonous gases.
4. If you happen to be in tephra fall use a moist cloth to cover your mouth and nose. Remember that the shortest distance from the ash plume is transverse to the wind direction.

5. Do not stay on flat land while the risk from jökulhlaup is predicted. Go to higher areas. If you are in an area that is flis pr by water use a white flag to signal for assistance. Follow all announcements on TV and radio.

C. *Precautions due to lightning (back right hand side)*
The risk for lightning is greatest in or close to the plume and can reach to a distance of 30–40 km from the volcano itself.

1. When there is the risk of lightning you should seek shelter in secure buildings, out-houses or cars (not convertibles).
2. Unplug all equipment from electricity inside the house and from outdoor antennas including electrical equipment, radio transmitters. Use indoor antennas if possible. Avoid using the telephone and remember that a phone may ring due to electricity from the lightning. Disconnect all fences from the house and unplug electrical fences from the house electricity.
3. If you are outdoors you should avoid being close to high lines, high trees, poles, laundry lines, electrical poles, masts, and agricultural equipment of any kind. Try to avoid wetlands, water, and rivers.
4. Unload things that can attract electricity such as rucksacks and fishing rods.
5. If you think that lightning will hit close to you and you cannot find shelter, stay on your feet and crouch down with your hands on your knees. Do not lay flat.

Electricity does not remain in someone who has been hit by lightning. Call 112 and administer first aid.

In addition to reviewing their perception of the evacuation exercise, the contents of the resident questionnaire were discussed with each person. A tape recorder was used for interviews when permission was granted. Written notes were taken during all interviews and these were transcribed into Microsoft Word® directly after each interview.

Conducting Questionnaire Survey Interviews

Our questionnaire was constructed using a format developed and tested by Bird and Dominey-Howes (2008) and adapted to the geographic and hazard focus of Katla. Further questions were developed based on residents' experience and discussion during the evacuation exercise. The final structure of the specific questions we included were discussed and negotiated with regional emergency personnel to ensure that the survey generated data of value to them in reviewing and improving their emergency management strategies. Therefore, it was important to pre-test our new questionnaire in order to highlight any errors or inconsistencies and to assess whether or not it would generate valuable data which are conducive to the goals of the project (Bird and Dominey-Howes, 2008; McGuirk and O'Neill, 2005; Parfitt, 2005). The pilot phase was carried out with local residents in April 2006. A few minor problems arose with respect to wording and sequencing of two questions. These issues were addressed prior to the main study.

Each questionnaire was printed in English with Icelandic translations. Translations were undertaken by a bilingual translator and then sent to another bilingual translator for verification. Participants were given the choice of conducting the interview in either English or Icelandic. To avoid misinterpretations and miscommunications translations were conducted during the interview and only one translator was used during the course of the study. Special and concise training of translators is critical to ensure that questions are asked exactly as intended and that participant responses are translated fully and completely (Patton, 1990). Our translator received thorough training prior to the study.

Face-to-face questionnaire survey interviews were conducted with local residents in the hazard zone of Rangávallasysla from May to October 2006. Since this was the first time an evacuation plan had been introduced to these communities and this study was the first of its kind to be held in this region, face-to-face interviews were deemed to be the most effective method for data collection. This is because it allows the interviewer to probe for more detailed responses when required as well as providing clarification if necessary (McGuirk and O'Neill, 2005; Parfitt, 2005).

Participants were recruited using two non-probability qualitative sampling methods. Firstly, a purposive sampling technique was used to target residents living within the hazard zone (i.e. residents registered in each community within the hazard zone were directly contacted). Purposive sampling is used to deliberately select subjects who are thought to be relevant to the research topic (Sarantakos, 1998). Secondly, a snow-ball sampling technique was employed whereby the first recruitment of participants suggested other residents who might be available to participate during the research period (Sarantakos, 1998). Despite apparent biases with both these sampling techniques, each was deemed appropriate to the study as we were actively seeking knowledge and perception data from residents from each community in the hazard zone. Furthermore, it is not our intention to generalize our results from this sample to the population as a whole, but rather provide a more descriptive preliminary investigation of public perception in this region.

All residents were initially contacted by telephone and interviews were arranged at a time convenient to them. Residents over 18 years of age were targeted and all participants were guaranteed anonymity. Prior to the interview each participant was informed about the purpose of the questionnaire and the proposed use of the data. They were also told that they were free to withdraw from the survey at any given time without consequence. Participants were required to sign Human Ethics forms to indicate that they agreed with the terms of the survey interview.

The questionnaire was divided into three sections. The first section gathered classification data about the participant. The second section gathered information about their knowledge and perception of Katla, jökulhlaup hazards, and emergency procedures. While the third section gathered information about their attendance at, and their perception of, the information meetings on Katla, the evacuation plan and exercise and their use of hazard information available through various media sources. Each section contained both open (free answer) and closed (check-list) questions. In total, the questionnaire contained 52 questions and took approximately 45 min to complete. However, participants were given as much time as needed to complete the interview. All data were analyzed within SPSS® 15.0 (Statistical Package for Social Science) and Microsoft Word®.

It is beyond the scope of this chapter to present data generated from all 52 questions. The questions we present here were selected on the basis of the information they provide (i.e. we believe they have generated significant data which may be useful to emergency managers charged with the responsibility of the ongoing development of risk mitigation procedures). An electronic copy of the questionnaire is available at http://www.nat-hazards-earth-syst-sci. net/9/251/2009/nhess-9-251-2009-supplement..pdf or from the corresponding author.

DISCUSSION

A unique opportunity was presented during and after the evacuation exercise to assess resident knowledge, behavior and perception of Katla, jökulhlaup hazard, and the evacuation plan—a task which had never been done for volcanic hazards in Iceland. A short time window was offered to capture residents' views of the exercise before they forgot this practical experience of risk mitigation. Our small sample size reflects this brief window of opportunity but the data collected provide an in-depth account stemming from a mixed methods approach which incorporated field observations, semi-structured interviews with emergency management officials, and questionnaire survey interviews with residents.

The problem of poor communication became evident through our field observations at the EC and was later reiterated during interviews with emergency management officials and residents. The issue of communication between scientists, emergency management officials, and the public can inhibit a successful response to evacuation orders (Chester et al., 2002). During the exercise, communication of the evacuation warning was not adequate and some residents were unaware the drill had commenced. This was confirmed in a post-exercise assessment report, where it was stated that the evacuation warning was not communicated effectively to residents (Almannavarnir, 2006). Effective communication not only refers to broadcasting hazard information but also the public and media's ability to understand the nature, meaning, and intent of the warning (Dominey-Howes et al., 2007). Communication strategies should be developed with respect to the intended audience and in consideration of social psychological factors which may influence whether or not people assimilate this information and respond accordingly (Paton and Johnston, 2001).

The particular role of communication was noted by the president of the International Union of Geodesy and Geophysics (IUGG) during the 2008 International Association of Volcanology and Chemistry of the Earth's Interior (IAVCEI) conference held in Iceland. He emphasized the need for successful communication in volcanic crises and questioned the reliance on modern technology to relay hazard information. This strong dependence on modern technology created problems during the evacuation exercise. To exacerbate this situation, approximately half the farmers in this region stated they do not carry a mobile phone with them at all times and it is these residents who are most likely to be away from a landline. It is therefore critical they receive an evacuation message through an alternative mode. The sweepers in some regions were able to notify those residents who were unaware that the evacuation had commenced. However, through our interviews we were able to ascertain that certain residents were not contacted by phone or sweeper.

Residents were concerned about their own personal safety due to the time it would take them to release livestock from the enclosures. Other residents were concerned about the safety of their animals after being released. They believe it would be safer to leave them inside especially with respect to tephra fall out. Time was a recurring issue as people were confused about the time allocated for them to evacuate particularly with residents located 30 min from the EC.

Empowerment is described by Paton et al. (2008) as an individual's capacity to have control over their personal affairs and confront hazard issues while receiving the necessary support from emergency management officials. Some residents described a loss of empowerment as they were not involved in the development of the evacuation plan and they were told they had to follow the plan (or be arrested) contrary to their own knowledge and perception. Furthermore, during the interview period residents had not received any feedback regarding the success of the exercise. Despite these shortcomings all the emergency management officials interviewed in this study deemed the evacuation exercise a success. This notion was enforced by the majority of our participants who took part in the exercise.

The questionnaire survey interviews revealed that even though most participants were able to demonstrate an accurate understanding of the eruptive history of Katla and nearly all participants correctly defined jökulhlaup, many (32%) think their area of residence would not be affected by a jökulhlaup. Alarmingly, 80% of participants from Vestur-Eyjafjöll share this view even though 93% of them live within 2 km of the river. However, these participants clarified their beliefs by stating their homes, like others in this community, are located approximately 30–40 m above the river bed. Considering that the hazard assessment and consequent hazard map modeled a catastrophic jökulhlaup reaching a maximum flood depth of at least 15 m upstream of these houses it is understandable that many participants feel it is safer to stay in their homes during a Katla eruption.

Notably, none of the participants from the 18–30 year age group and very few from the 31–50 year age group could correctly describe a brief volcanic history of Katla. An important element for community resilience is inherited memory of volcanic activity (Dominey-Howes and Minos-Minopoulos, 2004). Those residents whose parents experienced the 1918 Katla eruption displayed inherited memory of the eruption. However, this knowledge has not been passed down to the next generation.

Reassuringly, nearly all participants are aware of the emergency procedures they need to follow if an evacuation warning is issued even though some participants stated they would not evacuate. Again, Vestur-Eyjafjöll participant responses stood out from the group with 60% of them replying they would stay in their homes. In addition to their homes being located higher than the river, the evacuation route for this community travels alongside the Markarfljót. To further exacerbate their concerns residents feel that the evacuation route may place them in a vulnerable position to other hazards such as rock fall and tephra. However, non-hazard related factors may also influence residents' decision making process during a Katla eruption. It is possible that socioeconomic constraints such as personal and economic connection to livestock may influence residents' decision on whether or not to evacuate.

Regardless of the communication failures during the evacuation exercise most participants said they would call the emergency number 112 or the police to obtain information about a Katla eruption. However, telephone communication is likely to fail or yield busy signals for specific phone numbers if the network is oversaturated with calls. Exceeding the capacity of regional telecommunication systems complicates the task for emergency management officials and scientific agencies to gather and distribute hazard information by telephone (Gregg et al., 2004). It is therefore optimal for emergency management officials to promote public use of the media during a volcanic crisis. The media can provide an important source of volcanic hazard information for the public and attention should focus on increasing the planned use of this resource and ensuring that it provides consistently accurate information (Johnston et al., 1999). Risk mitigation strategies should include developing a mutually productive relationship between media organizations and emergency management officials in the form of a crisis communication plan to manage the media during a disaster (Hughes and White, 2006).

Participants demonstrated good knowledge of possible hazards that can occur during a future Katla eruption with jökulhlaup, tephra, and lightning cited as the most serious. Possessing knowledge of possible hazards ensures that the individual is better equipped to decide whether they should engage in personal preparedness measures and the most appropriate way to achieve this goal (Paton et al., 2008). Our participants' knowledge and concern of tephra was highlighted by one individual who stated that they had taken their own preparedness measures for tephra by purchasing gas masks.

Participant feedback on information provided at the town meetings held to discuss the possibility of a Katla eruption and the proposed evacuation plan was positive. Nearly all participants stated that the scientific information presented through talks, simulations and displays was very informative. A fundamental element of the pathway of information from scientists, emergency management officials, and the media is ensuring that it is delivered to the public in a form that represents community needs and functions (Gregg et al., 2004; Ronan et al., 2000;). Critical feedback relating to the lack of knowledge and experience of those presenting material at the meetings and technical difficulties should be addressed. Considering that the public are more than just passive receivers of hazard information (Horlick-Jones et al., 2003; Murdock et al., 2003), an integrated approach, that facilitates active participation from both residents and emergency management officials within a risk mitigation framework will help increase public trust, risk acceptance, and willingness to adopt personal preparedness measures (Paton et al, 2008).

Participation during the evacuation exercise was reasonably good with approximately 65% of residents taking part. Our sample group of residents reflected this rate with 68% stating that they took part. Apart from participating in order to improve personal safety and preparedness, many participants stated they took part in the exercise as they believed it was "their duty" to do so. Similarly, Haynes et al. (2008) reported that during an ongoing volcanic crisis on the Caribbean Island of Montserrat participants followed orders because it was the right thing to do.

Although an overwhelming majority of participants have followed media discussions concerning Katla most have not actively sourced hazard information available on the internet. Internet usage was quite low even though Bird et al. (2008) reported that 83% of Icelandic households have internet connection and 79% of internet users interact with public authorities. Despite this, it is important to utilize all forms of media as individuals prefer various means of acquiring information (Haynes et al., 2008). Furthermore, the perceived credibility and trust in hazard information can be compromised if forms of distribution are limited (e.g. just pamphlets and TV advertising) (Paton et al., 2008).

The precise location of a future eruption is uncertain therefore making it impossible to predict which direction the jökulhlaup will flow from the glacier margin (Sturkell et al., 2008). Furthermore, adequate preparation for all hazard consequences, such as lightning and tephra, is essential for all residents. The infrequent and complex nature of volcanic hazards increases the public's need to have easily accessible expert information in order to guide their risk management decisions (Paton et al., 2008).

In summary, the key outcomes of this research are:

- Improve the communication system.
- Emphasise the sweepers' role in supporting the dissemination of warning and evacuation information.
- Provide more detailed information on the effects of other volcanic hazards such as tephra, lightning, and rock fall and what preparedness measures can be applied to best protect person, property, and livestock.
- Ensure that all residents know exactly how much time they have to evacuate.
- Empower residents through involvement in risk mitigation planning.
- Provide feedback on proposed strategy outcomes within a reasonable timeframe (for example, within 3 months after completion).
- Continue to provide hazard information within an appropriate timeframe at town meetings with knowledgeable experts. The timeframe should be based on the level of alert (i.e. meetings should be more frequent when there is a higher risk of an eruption).
- Promote the use of all media sources for volcanic hazard information.

Further Developments and Future Research

Sturkell et al. (2008) report on seismic and geodetic measurements from around Katla between 1999 and 2005. Although increasing rates of crustal deformation and seismicity have lowered considerably, they believe that the volcano remains in an agitated state and an eruption in the near future should be expected. Therefore continued development of risk mitigation procedures is essential.

Improvements have been made to the communication system following the failures during the evacuation exercise and plans are underway to test the network (K. Porkelsson, personal communication, 2008). The ICP has confirmed the problem is being rectified and that the chief of police in Rangávallasysla is charged with the responsibility of testing the communication system during a follow-up exercise (R.

Ólafsson, personal communication, 2008). Town meetings were organized with local residents in Rangávallasysla during 2008. Residents were given the opportunity to voice their concerns with the evacuation plan (K. Þorkelsson, personal communication, 2008). In order to better suit community needs and expectations, information gathered during these meetings is being used to develop more appropriate evacuation procedures.

Our preliminary investigation entails a descriptive view of public knowledge and perception from a select group of residents living in each community in the Rangávallasysla hazard zone. As a result it is impossible to infer that results generated through our research apply to the population as a whole. In order to establish a clear idea of how the general public will respond during a future volcanic event and the complex range of natural and social phenomena that affect the decision making process, more detailed research needs to be conducted with a much larger sample group. Considering that the residents of Rangávallasysla are not the only ones located in the hazard zone this investigation has been expanded to include residents located in the hazards zones to the south and east of Myrdalsjökull. A parallel study is also being conducted with tourists and tourism employees within Þórsmörk, a popular tour destination located west of Myrdalsjökull. Following the recent meetings with residents and current progress toward developing more appropriate evacuation procedures further studies should investigate whether or not they suit community needs and expectations.

RESULTS

Our results are divided into three sections. Firstly, we report on our observations during the evacuation exercise on March 26, 2008. Secondly, information derived from the interviews with emergency management officials is documented. Thirdly, we present results generated from the questionnaire survey interviews with the residents. Comments recorded verbatim are presented in bullet form. In total, 60 individuals were interviewed; 6 emergency personnel, and 54 residents.

The Evacuation Exercise of March 26, 2008

All people involved in the evacuation exercise were instructed to treat it as a real volcanic emergency situation. Details on weather conditions were determined by ICP and emergency personnel were expected to consider wind speed and direction in relation to the development of the volcanic plume. Regular updates of the height and width of the plume were broadcast. Due to the possible hazard from tephra, helicopter pilots refused to fly until EH gave them a direct order. Following this, one helicopter was dispatched with a leading scientist to assess the eruption and another was on standby at a nearby airstrip.

All officials within EH held a round table meeting to discuss the progress of the eruption and evacuation every half hour. The Chief of Police of Rangávallasysla was in charge. Everybody reported to him and he delegated responsibilities as the day progressed. He enforced the need to stay in constant contact with all personnel out in the field. To test the emergency teams for different situations actors were employed to role play residents who refused to evacuate, residents who required medical assistance, people located in a high risk area and in need of helicopter evacuation, and tourists travelling within the hazard zone. The police were instructed to arrest residents if they refused to evacuate (this did not actually occur but residents who were refusing to evacuate were told that they would be arrested in a real evacuation).

The main problem brought to the attention of the Red Cross at the EC was the failure in communication–many residents did not receive the evacuation message and during the evacuation, the EH did not receive this message from the EC. Despite this, approximately 65% of the population located within the hazard zone of Rangávallasysla registered at the ECs. Talk amongst the residents at the EC included the communication failure while many voiced their concerns about leaving their animals. Another problem witnessed at the EC was the time it took to manually register residents.

Several instances occurred where residents had not received an evacuation warning but were asked to leave by the sweepers and one family was rescued by the emergency helicopter. Four elderly men arrived at the EC 3 hr after receiving the initial evacuation message. They were surprised that no one had come to check on them. They were not aware they were allocated 30 min for preparation before evacuating. Red Cross personnel reported a misunderstanding about the time allocation for evacuation. Some people were anxious to get to the EC within 30 min while others thought they had a lot longer. Furthermore, the EC in Hvolsvöllur was not well signposted and some people (including the present authors) could not easily find it.

Regardless of the problems that arose during the evacuation exercise, the general mood at each center was good-humored. Residents joked about the fact that the communication system did not work as planned. Some participants light-heartedly explained that they would have been inundated by flood water due to the fact that they had not received any evacuation message (these residents went to the evacuation center on their own accord since they knew the exercise was taking place). Resident behavior and comments indicated that many of them were there for the social aspect of the day.

As a result of our observations during the exercise, specific questions were developed for the questionnaire survey to investigate the failure in communicating the evacuation message, the time allocated to residents to evacuate and whether residents would refuse to evacuate during a real situation.

Interviews with Emergency Management Officials

All emergency management officials gave a clear description of their departments' role and their own personal role during an emergency situation. Each person that was in direct contact with the evacuees reported an overall positive public response. Comments in relation to this included:

- Approximately 65% of residents took part in the exercise which suggests that people are probably taking this seriously.
- Almost everyone was positive about the evacuation. Some who did not receive the evacuation message were mixed. Those who were not positive did not bother coming.

- The evacuees were extremely positive about the exercise. People were willing to participate probably due to the major earthquakes that occurred in 2000.

The evacuation was viewed as a success by all emergency management officials. The main negative comments that arose were attributable to the problem with the communication system. Comments in relation to this included:

- The information that is given to the people is crucial. They need to know how long they have before the flood comes. Also timing of the warnings should allow time for the rescue teams to help the evacuees if the weather conditions are bad. The sweepers can play this role.
- Phone calls and sms (text messages) were not good. People joked about this at the time but once they went home they were probably more concerned that they could have been stuck in a real flood.
- It is always the communication that breaks down and therefore the sweeper's role should be more concentrated on (providing warning and evacuation information to people). Technology can break down especially in a volcanic disaster. It must be organized as a door-to-door operation.
- We have broadcast advertisements asking people to report if they did not receive a message during the exercise. We have asked them to give their details to the local police and ICP directly so we can try to sort out this problem.

Questionnaire Survey Interviews with Residents

This section is divided as per the three sections of the questionnaire. The first section describes participant demographics based on their responses to classification questions. Participants' responses to both open and closed knowledge and perception questions of Katla, jökulhlaup hazards, and emergency procedures are presented in the second section. The third section reports participants' responses to questions relating to their attendance at and their perception of the information meetings on Katla, the evacuation plan and exercise, and their use of hazard information available through various media sources. The sequence of questions presented here is the same sequence as that within the questionnaire. Quick-look summary tables have been provided in each section for specific closed questions.

Participant Demographic

A total of 54 participants were recruited from 67 residents who were approached to take part in the questionnaire survey interviews, providing a response rate of 81%. Our sample included 19% of participants from Vestur-Eyjafjöll, 26% of participants from Fljótshlío, 15% of participants from Vestur-Landeyjar, and 20% of participants from each Austur-Landeyjar and Pykkvibær (Table 1). The majority (57%) of participants were 51 years of age or over and 57% of participants lived within 2 km of either the river Markarfljót or Pverá. Nearly all participants (98%) had lived in Iceland most of their lives. Education qualifications of our participants was quite diverse; 28% held a trade certificate or diploma, 15% had a university degree or higher, and a further 13% stated an education qualification from another source. Fifty percent of participants were full-time farmers while another 9% were part-time farmers.

Residents' Knowledge and Perception of Katla, Jukulhlaup Hazard, and Emergency Procedures

Participants were asked if they could give a brief eruptive history of Katla and a definition of jökulhlaup. In order to be counted as correct for the history of Katla, participants were expected to mention: The last confirmed eruption in 1918; or, the possible eruptions in 1955 and/or 1999; and, the frequency of Katla eruptions as 1, 2, or 3 times per century. However, some participants were counted as correct if they mentioned just one of the above in addition to detailed information about other aspects of Katla. Based on this, a correct response was given by 63% of participants, 7% were incorrect while a further 30% stated they did not know (Table 2). None of the participants in the 18–30 year age group gave a correct answer while only 27% of the correct answers came from the 31–50 year age group. A correct response for jökulhlaup was credited to answers that defined a flood of water from a glacier. Nearly all participants (94%) gave a correct response. Only 6% stated they did not know.

Sixty-seven percent of participants perceive that their region could be affected while 32% of participants stated no they do not perceive the hazard could affect their region. Eighty percent of participants from the community of Vestur-Eyjafjöll do not perceive the threat to their area and 93% of these people live within 2 km of the Markarfljót.

When the participants were asked if they are aware of the emergency procedures they need to follow if a jökulhlaup warning is issued 89% responded "yes". Seventy-one percent of participants correctly described the evacuation procedure, 19% stated that they would stay in their homes while the remaining 10% said that it would depend on:

- If it was occurring right away we would stay. If we had a few hours we might go to Hvolsvöllur;
- I would go to higher ground if at night or during bad weather. If the weather is good and it is daylight I would follow the evacuation procedure and go to Hvolsvöllur; and
- I would follow the plan to some extent but I would use commonsense especially if they tell me to do something that I know is wrong or dangerous.

Of the participants that live in Vestur-Eyjafjöll 60% of them said they would stay in their homes. Reasons given to clarify their response were:

- We consider ourselves safe where we live and therefore we will not evacuate. Also, for health reasons I feel better about staying at home;
- All farms in this community are 30–40 m higher than the river bed;
- I would not evacuate as I feel safe and comfortable in my own home. I am concerned about driving along the road which in my opinion is very dangerous as the road is in the lowland area and close to the river. After 30 min we will spend much time in the danger zone driving out of this area; and
- We would not evacuate. We would stay here on the farm. It is safer here than on the road. Tephra may block the road and rock fall may occur due to seismic activity.

If a Katla eruption commenced prior to the ICP issuing a warning 55% of participants stated that they would call 112 or the police (the most popular response) for information while a further 28% would seek information from the radio, television, or internet. Sixty-two percent of participants considered jökulhlaup as the most serious hazard in their area if Katla were to erupt while tephra was deemed most serious by 26% (Table 2). We then allocated scores to the rankings (i.e. the most serious hazard was allocated a score of 8; the second most serious was allocated a score of 7 and so on). A nil score was allocated if no ranking was given. Each hazard was ranked at least once (Figure 3) with jökulhlaup and tephra scoring the highest respectively.

Table 1. Participant responses from Sect 1: Classification questions. All data are given as a percentage. Some sections do not equal 100% due to rounding.

In what region of Rangavallasysla do you live?	Vestur-Eyjafjoll	Flj6tshlfo	Vestur-Landeyjar	Austur-Landeyja	pykkvibaer	
	19	26	15	20	20	
What is your age group?	18- 30 years old	31- 50 years old	51+ years old			
	7	35	57			
How far from the river do you live?	0<2km	2<5km	5<10km	10+ km		
	57	33	7	2		
In which country have you lived the longest?	Iceland	Other				
	98	2				
What is the highest level of education you have completed?	Some schooling	Educated	Educated	Trade certificate/ Diploma	University degree or higher	Other
	9	20	15	28	15	13
What is your occupation?	Full-time fa rmer	Part-time farmer	Other			
	50	9	41			

Table 2. Participant responses from Sect 2: Questions on Katla, jökulhlaup hazards, and the warning system. All data are given as a percentage. The second question does not equal 100% due to rounding. The last question totals more than 100% as participants were allowed to rank several hazards as the most serious.

	Correct	Incorrect	Don't know
Can you tell me a brief eruptive history of Katla?	63	7	30
How would you define jökulhlaup?	94	0	6
Do you think the region where you live could be affected by a jökulhlaup?	Yes 67	No 32	Don't know 2
Are you aware of the emergency procedures you need to follow if a jökulhlaup warning is issued?	Yes 89	No 11	
What would you define as the most serious hazard in your area if Katla were to erupt?	Jökulhlaup	62	
	Ice blocks	11	
	Lightning	9	
	Tephra	26	
	Poisonous gases	2	
	Lava	0	
	Tsunami	0	
	Earthquake	4	

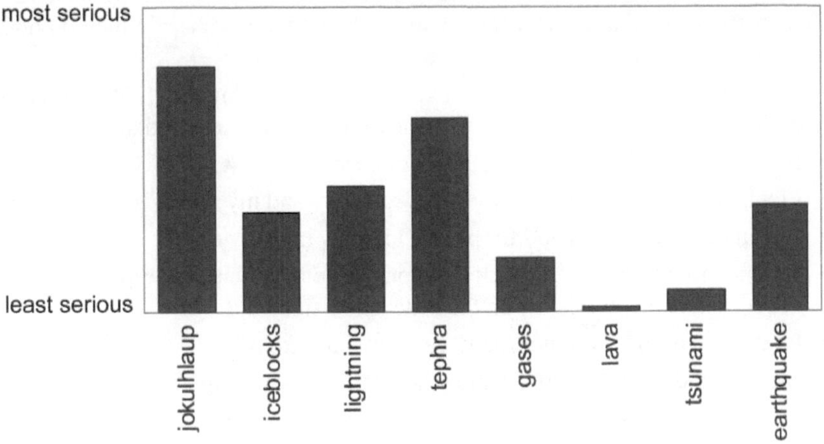

Figure 3. Participants' perception of the most serious hazards produced during a Katla eruption.

Residents' Knowledge and Perception of the Information Meetings on Katla, Evacuation Plan and Exercise, and Hazard Information in the Media

More than half the participants did not attend information meetings on Katla and the proposed evacuation plan and exercise. Reasons stated for not attending included:

- Could not attend due to health reasons;
- At work;
- Too busy when they were on; and
- Not interested.

Other people stated they did not attend but others within their household did. For those that did attend, we enquired whether they found them informative. Only 5% of participants did not find them informative. Participant perceptions of the meetings included:

- The simulation and displays were very informative but the sound system was very bad and therefore I could not hear the talks so well.
- It is good to talk about this and make people aware.
- I found the meetings very informative and now there is direct information on what to do if something happens. They educated people and now the local people should not be as afraid as they know what to do.
- I found the meeting informative but they needed more preparation. The people in charge lacked knowledge and those presenting the meetings were not the most experienced. There was no geologist at the last meeting.
- Most of it was nonsense. In the Westman Islands in 1973 everyone had to save themselves and it worked. Here will be the same.

Sixty nine percent of participants did not receive any evacuation message during the exercise (Table 3) and of these, 49% did not receive a message to their landline. When asked if they always carried their mobile phone 68% of participants responded "yes". However, only 52% of farmers carry their mobile phone with them at all times. Of those participants that always carry their mobile phone, 34% said they do not always have an active connection in their area.

Participation during the evacuation exercise was rather high with 68% of participants stating they did take part. Their reasons for participation included:

- It is part of my duties as an Icelandic citizen;
- I took part in the evacuation for my own safety and my family's;
- I thought it would be good for people to know how to act;
- I wanted to participate to check how long it would take us to prepare but we did not complete the whole list on the evacuation sign; and
- I did take part but I did not really gain anything from it.

Those who did not take part clarified their actions by stating:

- Too tired and sick;
- I was at work but everyone else in the house took part;
- We would have participated if we had received the evacuation message; and
- Not interested as I do not perceive that I will be in danger.

Despite some people's negativity toward the evacuation exercise of those who did participate 82% of them were positive about the exercise.

Thirty minutes was deemed enough time to complete the list as described on the evacuation sign (Figure 2) before evacuating their property by 52% of participants. Of the 48% that stated no or do not know they responded with:

- It is not enough time if you have to let the animals out (as per the instructions);
- 30 min is not enough time for farmers;
- 30 min may not be enough depending where I am on the farm; and
- It depends if the kids are at home from school and if I am at work in Hvolsvöllur then I would have to drive back to the house to collect them.

Table 3. Participant responses from Sect 3: Questions on Katla information meetings, evacuation plan, evacuation exercise, and hazard information available in the media. All data are given as a percentage.

	None	One	Two	Three
How many Katla information meetings did you attend?	55	29	4	12
How many evacuation messages did you receive on the March 26, 2006?	69	19	6	6
	Yes	No		
If you did not receive any messages did you receive a phone call to the landline or your mobile phone?	51	49		
Do you always carry your mobile phone with you?	68	32		

	Yes	No	Don't know
Do you always have service coverage to your mobile phone around your area?	64	34	
Did you take part in the evacuation exercise?	68	32	
	Positive	Negative	Mixed
If you did take part in the exercise on March 26, 2006 how did you feel about it?	82	8	10
	Yes	No	Don't know
Do you think 30 min is enough time to complete the list (on the evacuation sign) and evacuate?	52	42	6
Would you follow this procedure if there was a real evacuation?	74	18	8
	Yes	No	
Have you looked up the ICP website and familiarised yourself with information on the possible natural hazards connected to a Katla eruption?	19	81	
Have you ever used the Skja.Jftavefsja!IMO website for hazard information?	26	74	
Have you followed discussions in the media about natural hazards connected to a Katla eruption?	89	11	

With these comments in mind it is not surprising that 64% of farmers do not believe that 30 min is enough time. Furthermore, several participants were under the impression that they had 30 min to complete the list and get to the evacuation center. These people expressed great concern about this because for some of them it takes 30 min to drive to the closest evacuation center. These residents were located in Austur and Vestur-Landeyjar (Figure 1).

Only 19% of participants had accessed hazard information related to a Katla eruption from the ICP website (www.almannavarnir.is) while 26% of participants had accessed hazard information from the Skjálftavefsjá (earthquake web-viewer) website (drifandi.vedur.is/), and the Icelandic Meteorological Office (IMO) website (www.vedur.is). Media discussions about natural hazards connected to a Katla eruption were followed by 89% of participants and they sourced this information from television (88%), radio (82%), newspaper (72%), information brochures (54%), books (40%), and the internet (20%).

Once the questionnaire had been completed the participants were given the opportunity to engage in open discussion. Many participants stated their reluctance to leave their animals and some believe that due to this many farmers may choose to stay at home during an actual evacuation. Some participants would like to see the hazard zone reclassified in order to rank the areas according to the level of risk. These participants felt that people may be complacent as they do not recognizes, they are actually living in a high risk area and therefore they may prefer to stay at home with their animals during a Katla eruption. Furthermore, many people expressed concern about completing all the instructions on the evacuation list and of particular concern was the instruction to release animals from their enclosures.

Another important message communicated during the discussions was the great concern for tephra fallout. Participants not only feared personal health risks—one participant stated "we have bought ourselves gas masks in case of tephra"—but also related risks associated with the complete darkness that can be experienced during the middle of the day, the threat to agricultural land and the threat to car engines. However, one of the most important statements that arose during these discussions was regarding residents' involvement in the development of the evacuation plan. Several residents objected that they had no say in how the evacuation should be implemented within their communities and following the exercise they were not informed about how successful the drill had been.

CONCLUSIONS

The evacuation plan is the first to be developed and implemented in the municipality of Rangávallasysla and the ICP, scientists, local police, and rescue teams should be commended for their efforts. However, more work needs to be done to reduce the impact of a future Katla eruption. This can be achieved by addressing some of the main issues raised by our participants. The data provides an insight into how residents interpret their situation in relation to Katla, its associated hazards and their potential response during a crisis. This information highlights the importance of integrating the physical characteristics of Katla's volcanic hazards within context of the communities at risk. Our participants are aware of jökulhlaup, tephra, lightning, and rock fall hazards but they have not been provided with enough information to enable them to make an informed decision on whether to evacuate or take shelter in place and how to best protect their livestock. Comparatively, from the information provided, residents in Vestur-Eyjafjöll have been able to conclude that their homes will not be directly affected by jökulhlaup and therefore they are not willing to evacuate. However, non-hazard related factors such as not wanting to leave animals unattended may also influence their decision to evacuate. Furthermore, residents' participation in the evacuation exercise does not necessarily reflect their willingness to evacuate. These examples underline the complex range of natural and social phenomena that affect the individual's decision making process and as a result may inhibit a successful evacuation.

Results from our study highlighted problems associated with communication during the evacuation exercise and the possible need to find alternative modes which do not rely so heavily on technology. In light of this, scientists and emergency management officials should collaborate with media agencies and the public in order to promote the use of media resources and, to ensure hazard information is accurately distributed in an understandable form. Furthermore, the importance of the sweepers' role during an evacuation should be emphasized as they may provide the only communication link between emergency management and farming communities. Recent public meetings which involved residents in risk mitigation efforts are a positive step toward empowering residents with evacuation procedures and preparedness strategies.

This chapter presents the first results on residents' knowledge and perception of Katla, jökulhlaup hazard, and their views of the evacuation plan and exercise in Rangávallasysla. The key outcomes, as summarized above, should help provide considerable value to the ongoing development of an effective response capability. Considering this research is the first of its kind in this region the results can be used as a baseline by the ICP for more robust surveys in Iceland's volcanic regions.

KEYWORDS

- **Evacuation centers**
- **Evacuation exercise**
- **Hazard assessment report**
- **Icelandic Civil Protection**
- **Volcanic hazards**

ACKNOWLEDGMENTS

All participants are graciously thanked for their willingness to participate in this investigation. Gratitude is expressed to Árni Valur Kristinsson for Icelandic translations and Helga Birna Pétursdóttir for field assistance. Funding has been provided by the Department of Environment and Geography and the International Office at Macquarie University, Australia; Rannís—the Icelandic Centre for Research (Research Grant #081260008); Vegagerbin, Iceland (The Icelandic Road Administration); and Landsvirkjun, Iceland. This chapter benefited from insightful comments from Damian Gore, Chris Gregg, Katharine Haynes, Douglas Paton, and an undisclosed reviewer. Edited by: Giovanni Macedonio Reviewed by: C. Gregg, D. Paton, K. Haynes, and another anonymous referee.

Chapter 9

Identification of Earthquake Induced Damage Areas

Elif Sertel

INTRODUCTION

A devastating earthquake with a magnitude of Mw 7.4 occurred on the North Anatolian Fault Zone (NAFZ) of Turkey on August 17, 1999 at 00:01:39 UTC (3:01 am local time). The aim of this study is to propose a new approach to automatically identify earthquake induced damage areas which can provide valuable information to support emergency response and recovery assessment procedures. This research was conducted in the Adapazari inner city, covering a 3 × 3 km area, where 11,373 buildings collapsed as a result of the earthquake. The Satellite Pour l'Observation de la Terre (SPOT) high resolution visible infrared (HRVIR) Pan images obtained before (June 25, 1999) and after (October 4, 1999) the earthquake were used in the study. Five steps were employed to conduct the research and these are: (i) geometric and radiometric correction of satellite images, (ii) Fast Fourier Transform (FFT) of pre- and post-earthquake images and filtering the images in frequency domain, (iii) generating difference image using Inverse Fast Fourier Transform (IFFT) pre- and post-earthquake images, (iv) application of level slicing to difference image to identify the earthquake induced damages, (v) accuracy assessment of the method using ground truth obtained from a 1/5,000 scale damage map. The total accuracy obtained in the research is 80.19%, illustrating that the proposed method can be successfully used to automatically identify earthquake-induced damage areas.

Turkey is one of the most seismically active regions on the earth. Different fault systems in Anatolia and the surrounding regions were created due to the complex plate interactions among Arabia, Eurasia, and Africa (Tan et al., 2008). The NAFZ and the East Anatolian Fault Zone (EAFS) are the main strike-slip fault belts in Turkey where several earthquakes have occurred, resulting in huge numbers of fatalities over the past several hundred years (Table 1).

Table 1. Historical earthquakes in Turkey (USGS, 2008).

Date	Location	Magnitude (Mw)	Fatalities
August 17, 1668	Anatolia	8.0	8,000
August 9, 1912	Murefte (Ottomon Empire)	7.8	2,800
October 3, 1914	Burdur (Ottoman Empire)	7.0	4,000
December 26, 1939	Erzincan	7.8	32,700
November 26, 1943	Ladik, Samsun	7.6	4,000
March 18, 1953	Gonen	7.3	1,073
August 19, 1966	Varto, Mus	6.8	2,529
March 28, 1970	Gediz	6.9	1,086
October 30, 1983	Erzurum	6.9	1,342
August 17, 1999	Izmit	7.2	11,718

Earthquakes and other natural hazards can cause disasters of uncontrollable magnitude when they hit large urban areas. Emergency response and early recovery assessment in earthquakes require rapid and reliable damage assessment and loss estimation. In the case of suddenly occurring earthquakes, remote sensing data can be reliably used to create fast draft damage maps of the affected urban areas which provide valuable information to support emergency response teams and decision making during the recovery process. Remotely sensed images ranging from very high resolution to medium resolution have been widely used to derive information and estimation for damage assessment (Bendimerad, 2001; Kaya et al., 2005; Pal et al., 2006; Sertel et al., 2007; Stramondo et al., 2006; Turker and San, 2003). Moreover, multitemporal remote sensing data can serve as a basic data set to support post-disaster planning.

Different remote sensing methods have been used by many scientists to identify earthquake-induced damage areas. Sertel et al. (2007) investigated the relationship between semi-variogram metrics and degree of earthquake damage using transects over an earthquake area. Turker and San (2003) used differences between merged pre- and post-event SPOT high resolution visible (HRV) data to reveal the location of earthquake-induced changes. Stramondo et al. (2006) used coherence and correlation maps from Advanced SAR (ASAR) and change maps from advanced spaceborne thermal emission and reflection radiometer (ASTER) to analyze the capabilities and limitations of satellite remote sensing to detect damage due to earthquakes. Kaya et al. (2005) used government statistics and SPOT HRV data to estimate the proportion of collapsed buildings in an earthquake area. Although there have been a lot of earthquake damage assessment studies using different remote sensing methods, there has not been that much research on the application of Fourier transform to satellite image for an earthquake case. This research focuses on integrated usage of Fourier transform and level slicing to identify earthquake induced damage areas, also detailed accuracy assessment of proposed method was conducted using 1/5,000 scale damage map data and error matrix analyses.

Fourier transforms have been applied to different remote sensing applications. Lillo-Saavedra et al. (2005) used Fourier transforms to fuse panchromatic and multispectral data obtained from Landsat ETM+ sensor. Westra and De Wulf (2007) used Fourier analysis of Moderate Resolution Image Spectrometer (MODIS) time series data to monitor the flooding extent. Pal et al. (2006) used FFT filter to extract linear and anomalous patterns. Their results showed that numerous lineaments and drainage patterns could be identified and demarcated by FFT filters.

In this study, the following steps were conducted to accurately identify the location and magnitude of earthquake induced damages in an urban area and to quantify the accuracy of the proposed method: (i) pre- and post-earthquake images of the region were geometrically and atmospherically corrected, (ii) The FFT was applied to pre- and post-earthquake images and images were filtered in the frequency domain, (iii) a difference image was generated using IFFT-pre- and post-earthquake data, (iv) level slicing method was applied to difference image to identify the earthquake-induced damages, (v) accuracy assessment was performed by comparing the results of the proposed method with the 1/5,000 scale damage map of the earthquake area.

THE STUDY AREA AND DATA

A devastating earthquake with a magnitude of Mw 7.4 occurred on the NAFZ of Turkey on August 17, 1999 at 00:01:39 UTC (3:01 am local time). The center of the earthquake was at 40.74 N, 29.86 E. The earthquake struck Kocaeli and surrounding cities, namely Adapazari, Golcuk, and Yalova, and brought about massive destruction to these cities and their surrounding rural areas. This was one of the most destructive earthquakes of the 20th century considering the amount of damage and number of casualties. At least 17,118 people were killed, approximately 50,000 injured and the estimated financial loss in Istanbul, Kocaeli, and Adapazari was 3–6.5 billion US dollars (Kaya et al., 2005; Sertel et al., 2007; USGS, 2008).

This research investigates the earthquake induced changes in the city center of Adapazari where the recorded number of collapsed buildings was 11,373 (Sertel et al., 2007). The district of Adapazari is located in the northeastern part of the Marmara Region, Turkey, covering 29°57′–30°53′ N and 40°17′–41°13′ E. The population of the Adapazari inner city was 169,099, 184,013, and 172,000 in 1990, 1997, and 2000, respectively (TSI, 2008).

The SPOT HRVIR panchromatic images obtained before (June 25, 1999) and after (October 4, 1999) the earthquake were used in the research. These images have 10 m spatial and 8 bit radiometric resolution. The 1/5,000 scale digital damage map illustrating the degree of earthquake damage was used to analyze the accuracy of the proposed method. This map was produced by local and federal authorities by conducting a field survey on the building base after the earthquake.

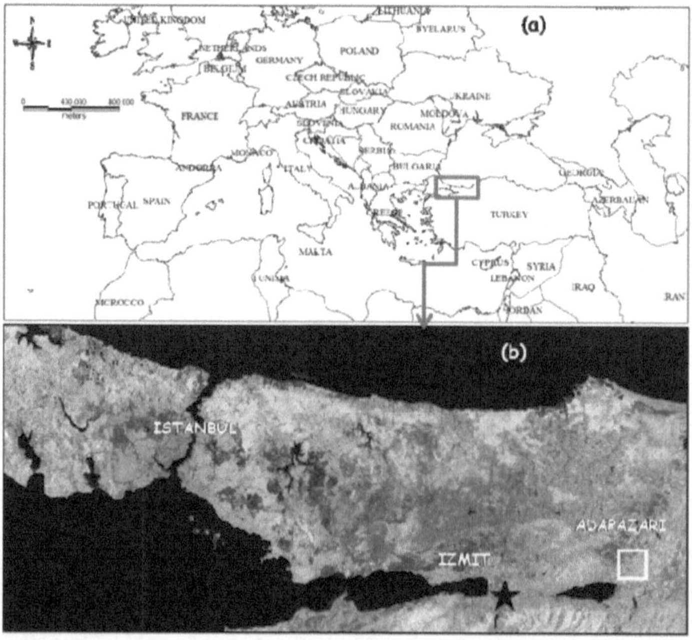

Figure 1. Location of the study area, (a) Turkey and surrounding countries, (b) Marmara Region, star shows the epicenter of the earthquake.

METHODOLOGY

Radiometric Normalization and Geometric Correction

Both images were first geometrically corrected into Universal Transverse Mercator (UTM) projection using first order polynomials and appropriate Ground Control Points (GCP) collected from topographic maps. Then, radiometric normalization was employed using a histogram matching algorithm.

Fourier Transform

Any one-dimensional function, *f(x)* (which might be a row or column of pixels), can be represented by a Fourier series composed of some sine and cosine terms and their associated coefficients combination. Different spatial frequencies over an image can be represented by many sine and cosine terms and with their associated coefficients. Fourier series are effective to identify and quantify spatial frequencies (Erdas Field Guide, 2005; Gonzalez and Wintz, 1977). Since an earthquake changes the spatial structure of a related area because of collapsed or damaged buildings, roads, and so on, Fourier series can be used to identify different spatial frequencies in images obtained before and after the earthquake which indeed lead information about the earthquake-induced damages.

The FFT calculation used in this research is shown in the equation 1 (Erdas Field Guide, 2005):

$$F(u,v) \leftarrow \sum_{x=0}^{M-1}\sum_{y=0}^{N-1}\left[f(x,y)e\frac{j2\pi ux}{M} - \frac{j2\pi vy}{N} \right] \tag{1}$$

where:

M = the number of pixels horizontally

N = the number of pixels vertically

u,v = spatial frequency variables

e = 2.71828, the natural logarithm base

j = the imaginary component of a complex number

Once the FFT is applied, a raster image from the spatial domain is converted into a frequency domain image. The Fourier image can be edited (mainly using filters) to reduce noise, to identify specific features or to remove periodic features. After editing the Fourier image, it is transformed back into spatial domain using IFFT equation (equation 2) (Erdas Field Guide, 2005):

$$f(x,y) \leftarrow \frac{1}{N_1 N_2}\sum_{u=0}^{M-1}\sum_{v=0}^{N-1}\left[F(u,v)e\frac{j2\pi ux}{M} + \frac{j2\pi vy}{N} \right] \tag{2}$$
$$0 \le x \le M-1, 0 \le y \le N-1$$

Difference Image and Level Slicing

A difference image was calculated by subtracting the inverse Fourier transformed post- and pre-earthquake images. The difference image then divided into slices based on the number of bins (10 for this research) using the following equations:

$$x = \frac{DN_{\max} - DN_{\min}}{numberofbins} \tag{3}$$

$$DN_{out} = \frac{DN_{in} - DN_{\min}}{x} \tag{4}$$

where:

DN_{max} = maximum value of digital numbers

DN_{min} = minimum value of digital numbers

DN_{in} = input digital number

DN_{out} = output digital number after level slicing

The DN_{out} values obtained after the level slicing were categorized as either damaged or non-damaged based on their values. Lower DN_{out} values represent the non-damaged areas whereas higher values represent damaged areas.

Accuracy Assessment

The inner city of Adapazari was divided into 99 blocks of 300 m × 300 m size and these blocks were used for the detailed accuracy assessment procedure. The results obtained after the level slicing of the difference image (calculated from inverse Fourier transformed images) illustrates the damaged and non-damaged areas. These areas were compared with the 1/5,000 scale damage map for each block on a parcel basis to investigate the applicability of this method to automatically identify earthquake-induced damage. The number of parcels in each block was evaluated individually and error matrix for each block was created as shown in Figure 2.

Data created from the proposed method	Data from the damage map (reference data)	
	Damaged	Non-damaged
Damaged	Cell1	Cell2
Non-damaged	Cell3	Cell4

Figure 2. Accuracy assessment of each block.

Each block has four values corresponding to parcels identified as damaged both in the damage map and with the proposed method (cell 1), parcels identified as non-damaged both in the damage map and with the proposed method (cell 4), parcels identified as damaged in the damage map but non-damaged with the proposed method (cell 3) and parcels identified as non-damaged in the damage map but damaged with the proposed method (cell 2). Overall accuracy of each block was calculated by summing diagonal elements and dividing them to total number of parcels within that block. The equation of the overall accuracy for a block based on the values described in Figure 2 is as follows:

$$Overall\ accuracy_{BLOCK(n)} = \frac{CELL1_{BLOCK(n)} + CELL4_{BLOCK(n)}}{CELL1_{BLOCK(n)} + CELL2_{BLOCK(n)} + CELL3_{BLOCK(n)} + CELL4_{BLOCK(4)}} \quad (5)$$

After calculating the overall accuracy of each block, total accuracy of the proposed method was calculated by rationing the sum of diagonals of all blocks to total number of parcels of all blocks. The performance of the proposed method was evaluated considering the total accuracy. Figure 3 shows the steps conducted during the study and it is also a summary of the methodology section.

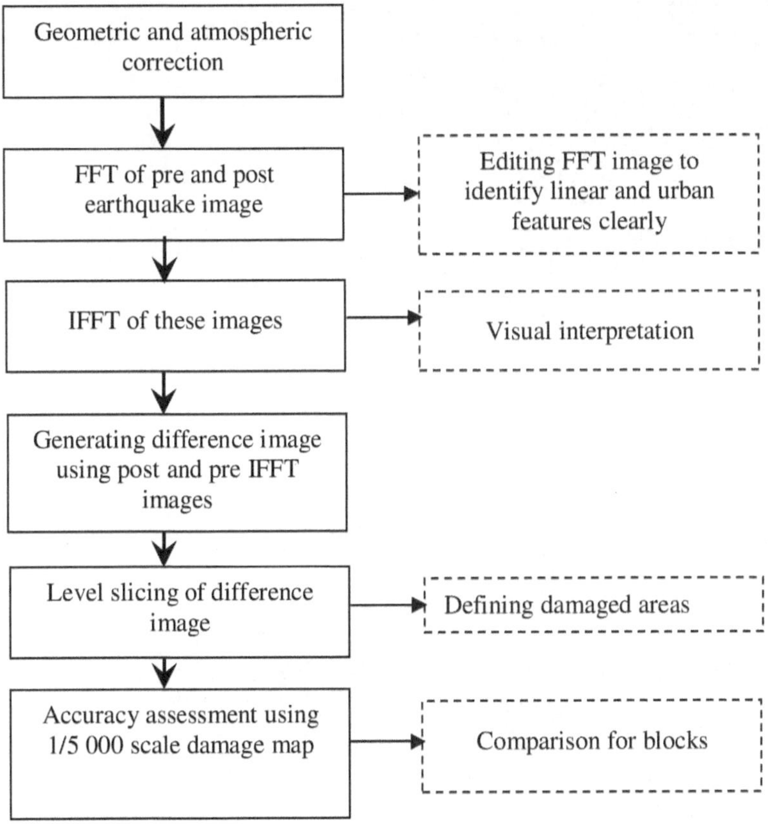

Figure 3. Procedures conducted in the methodology.

RESULTS

The original pre- and post-earthquake images are shown in Figures 4a and b and the FFT images generated from these data are illustrated in Figures 4c and d. As a result of collapsed buildings and roads, spatial structure and texture of the post-earthquake image had changed. This caused differences in spatial frequency which can be determined via Fourier transform. The differences in spatial frequency for pre- and post-earthquake data can be identified from Figures 4c and d.

The low frequencies are plotted near the origin (center) while the higher frequencies are plotted further out. Generally, the majority of the information in an image is in the low frequencies indicated by the bright areas at the center of the Figures 4c and d.

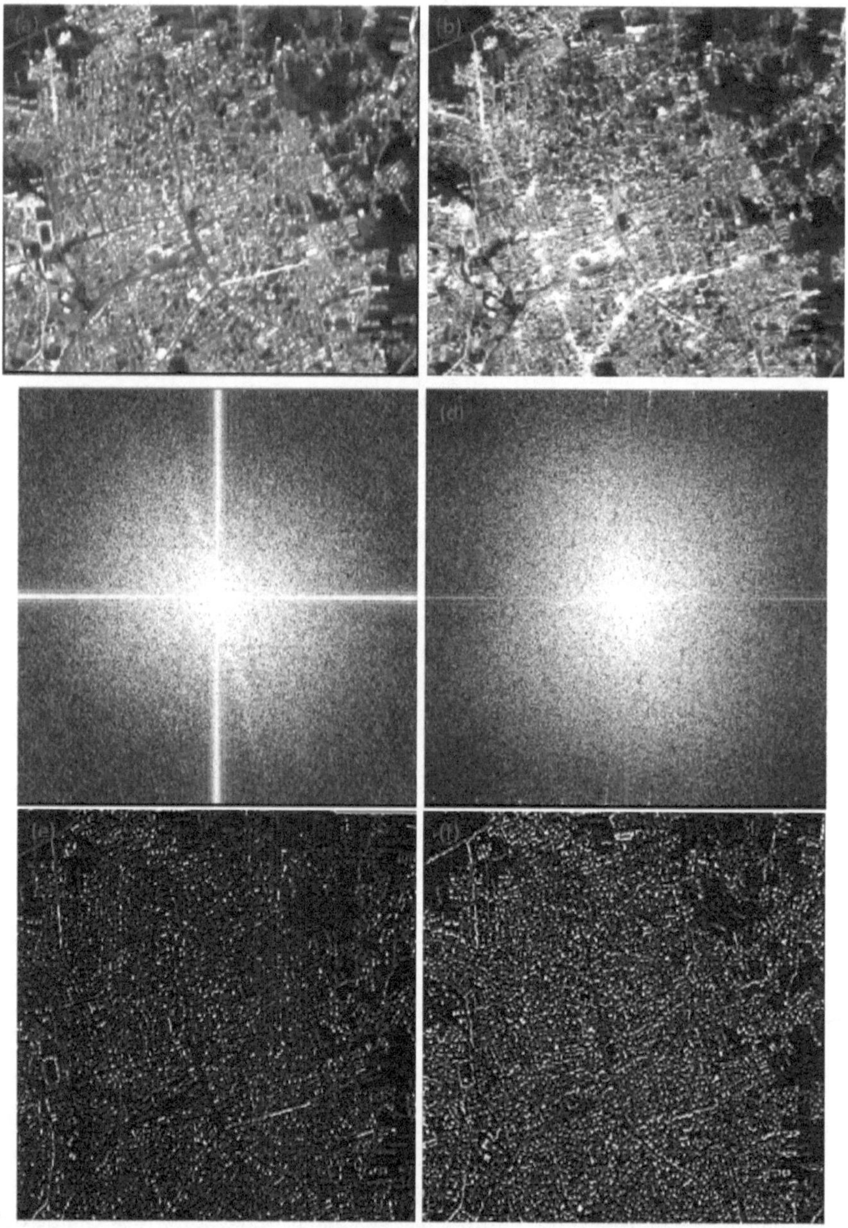

Figure 4. (a) Original pre-earthquake data, (b) original post-earthquake data, (c) FFT pre-earthquake image, (d) FFT post-earthquake image, (e) IFFT pre-earthquake image, (f) IFFT post-earthquake image.

High pass filter was applied to satellite images in frequency domain to delineate the border of linear objects like roads and buildings precisely. Filters were applied to the low frequencies which are around the center for both pre- and post-earthquake data. After the filtering, IFFT was applied and edited Fourier images were converted back into the spatial domain.

After the visual interpretation of IFFT images, a difference image was generated by subtracting post-IFFT-image and pre-IFFT-image. Level slicing method with 10 slices was conducted to identify the damaged and non-damaged areas. Different numbers of levels were tried to find out the optimum number of slices and the analyses shows that having a slice number higher than 10 did not contribute significant information since only a few number of pixels was assigned to a slice. The histogram of the difference image was investigated to see the general distribution of data and to determine a threshold value for damaged and non-damaged regions. Further analyses were conducted with different threshold values to find out the most appropriate value for the study to identify the changes. Standard deviation (σ) and mean (μ) values obtained from the difference image were used for the analysis. 3σ, 2.5σ, 2σ, 1.8σ and 1.6σ, 1.5σ, and1.4σ were tried and 1.4σ was found as the best threshold value to determine changes. Using this threshold value slices including data between $\mu-1.4\sigma$ and $\mu+1.4\sigma$ were assigned as non-damaged areas whereas slices outside this range were assigned as damaged areas. Figure 5 shows the result of level sliced-difference image and blocks overlaid on this image with parcel boundaries.

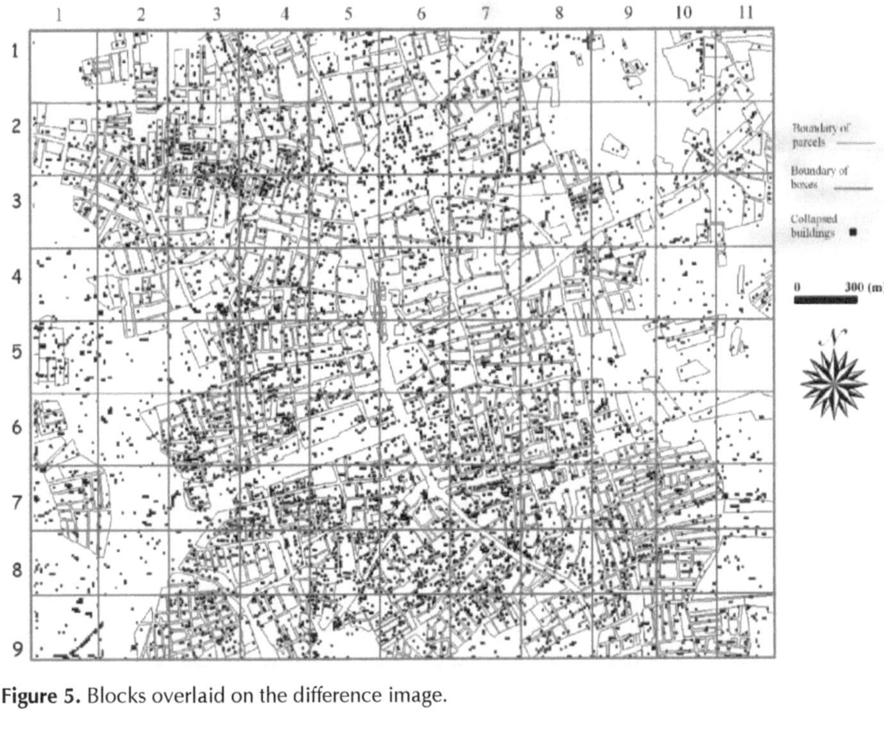

Figure 5. Blocks overlaid on the difference image.

Damaged areas obtained from remote sensing methods were compared with the 1/5,000 scale damage map to quantify the accuracy of the proposed method. Figure 6 shows the damage map and blocks where the detailed accuracy assessment was conducted and error matrixes created.

Comparisons were conducted for each block in parcel base. Number of damaged or non-damaged parcels within a block were calculated from difference image (Figure 5) and damage map (Figure 6). Table 2 includes each block with box number (BOX NO), box number are called based on their row and column location. For example, BOX NO 1–2 is corresponded to the box at row 1 and column 2 in Figure 5 and 6. Each block has four values corresponding to status of parcels. The N/A is corresponded to not available meaning that there is either no parcel or damage data in those regions. Based on Table 2, the corresponding values of each cell for BOX 1–2 will be as following:

- Cell 1: parcels identified as damaged both in damage map and with the proposed method, this value is 3 for block 1–2.
- Cell 2: parcels identified as non-damaged in damage map but damaged with the proposed method, 4 for block 1–2.
- Cell 3: parcels identified as damaged in damage map but non-damaged with the proposed method, 1 for block 1–2.
- Cell 4: parcels identified as non-damaged both in damage map and with the proposed method, 6 for block 1–2.

Table 2. Accuracy assessment of the proposed method using error matrixes.

BOX NO	PARCELS		BOX NO	PARCELS		BOX NO	PARCELS		BOX NO	PARCELS		BOX NO	PARCELS	
1-1	2	0	3-1	2	2	5-1	5	0	7-1	6	0	9-1	N/A	N/A
	0	1		0	3		0	1		1	5		N/A	N/A
1-2	3	4	3-2	8	1	5-2	N/A	N/A	7-2	N/A	N/A	9-2	N/A	N/A
	1	6		1	3		N/A	N/A		N/A	N/A		N/A	N/A
1-3	4	2	3-3	10	1	5-3	4	1	7-3	8	1	9-3	3	0
	1	5		1	1		1	1		2	1		2	5
1-4	7	1	3-4	6	1	5-4	8	1	7-4	13	1	9-4	6	0
	0	3		4	6		3	3		1	2		1	3
1-5	4	1	3-5	10	1	5-5	6	2	7-5	8	1	9-5	9	0
	1	2		0	2		2	4		1	3		1	2
1-6	5	3	3-6	5	1	5-6	3	0	7-6	7	0	9-6	4	1
	0	1		1	2		1	8		1	6		2	4
1-7	3	1	3-7	7	0	5-7	5	2	7-7	12	1	9-7	7	0
	2	2		1	1		1	4		1	3		2	3
1-8	2	1	3-8	5	1	5-8	9	4	7-8	8	0	9-8	4	0
	2	2		1	3		1	5		1	7		2	3
1-9	2	0	3-9	6	1	5-9	1		7-9	6		9-9	5	0
	3	2		1	2		0			2			1	4

Table 2. *(Continued)*

BOX NO	PARCELS		BOX NO	PARCELS		BOX NO	PARCELS		BOX NO	PARCELS		BOX NO	PARCELS	
1-10	1	2	3-10	4	0	5-10	3		7-10	4		9-10	4	0
	1	2		1	1		0			1			1	6
1-11	4	0	3-11	5	1	5-11	N/A	N/A	7-11	3	1	9-11	3	1
	2	3		1	2		N/A	N/A		0	3		1	6
2-1	6	1	4-1	N/A	N/A	6-1	3	1	8-1	N/A	N/A			
	1	0		N/A	N/A		0	3		N/A	N/A			
2-2	6	0	4-2	3	2	6-2	N/A	N/A	8-2	N/A	N/A			
	2	3		1	4		N/A	N/A		N/A	N/A			
2-3	6	0	4-3	4	2	6-3	10		8-3	5				
	1	3		1	1		3			1				
2-4	8	0	4-4	9	0	6-4	10		8-4	8				
	2	2		1	3		3			1				
2-5	6	0	4-5	7	2	6-5	6		8-5	5				
	1	2		0	2		0			7				
2-6	6	1	4-6	4	1	6-6	8		8-6	6				
	2	3		1	8		0			2				
2-7	4	0	4-7	2	1	6-7	10		8-7	5				
	2	2		2	4		3			2				
2-8	3	0	4-8	1	1	6-8	8		8-8	7				
	0	4		3	5		0			1				
2-9	2	0	4-9	5	0	6-9	5		8-9	4				
	0	0		3	7		2			3				
2-10	2	1	4-10	4	0	6-10	2	1	8-10	6				
	0	1		4	9		0	3		2				
2-11	4	0	4-11	N/A	N/A	6-11	N/A	N/A	8-11	N/A	N/A			
	2	1		N/A	N/A		N/A	N/A		N/A	N/A			

In most cases parcels identified as damaged in the damage map but non-damaged with the proposed method (cell 3) occurred because there were only one or two collapsed buildings within a parcel which could not be identified using SPOT images. Block 1–2, 1–3, 1–10, 4–5 are examples of this situation. On the other hand, most of the parcels having more than three collapsed buildings were easily identified using the proposed method. Higher resolution satellite images should be used to identify collapsed buildings individually.

The overall accuracy of each block ranges from 50 to 100% (Figure 7). Most of the blocks have accuracy value higher than 75%. The minimum accuracy was obtained for Block 1–10, because there are three damaged parcels and two of these parcels include only one collapsed building which is hard to identify with current spatial resolution. Most of the parcels which have plenty of collapsed buildings were easily identified with 75% or higher accuracy.

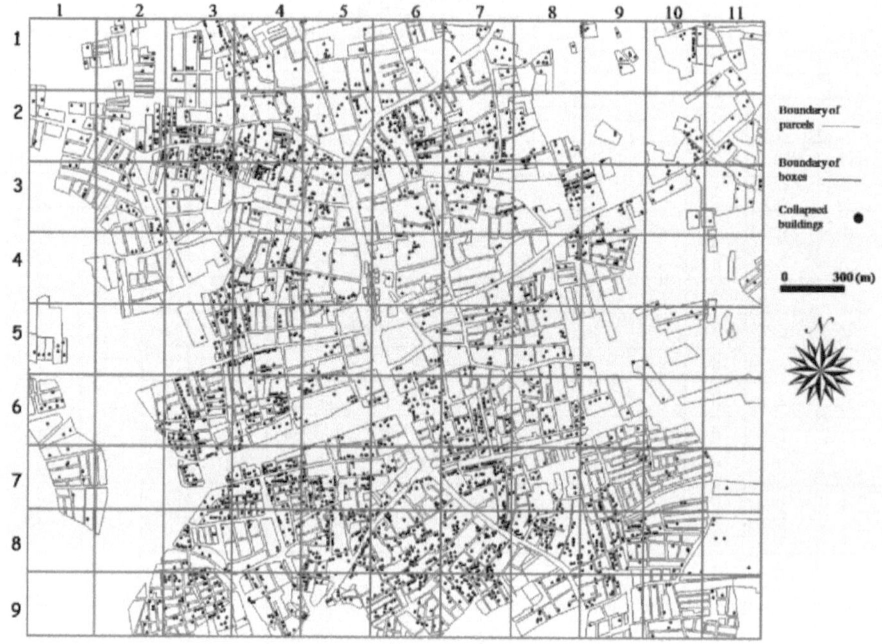

Figure 6. Blocks overlaid on the damage map.

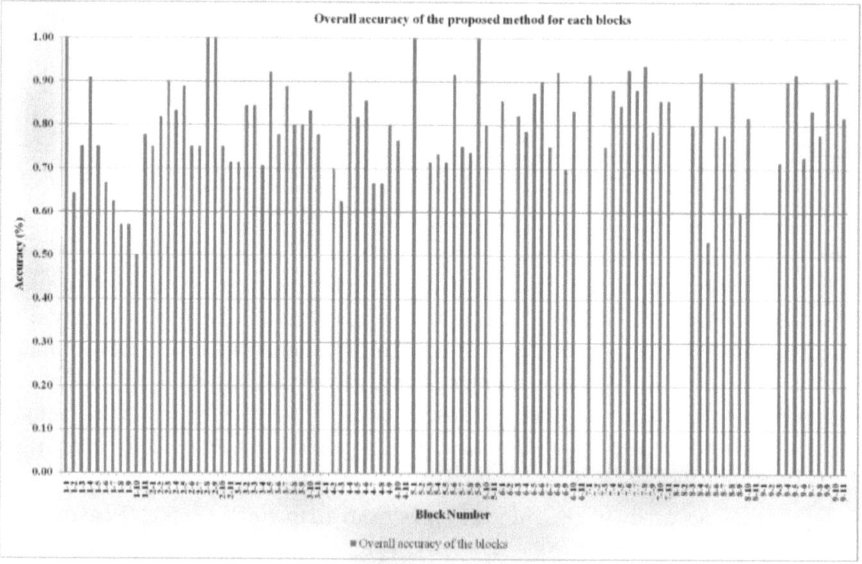

Figure 7. Overall accuracy the proposed method for each block.

Total accuracy of the proposed method was calculated by rationing the sum of diagonals of all blocks to total number of parcels. The total accuracy obtained from this ratio is 80.19%. The results illustrated that the proposed method can be successfully used to identify earthquake-induced damage areas automatically. Considering the spatial resolution of the satellite image used in this study, this accuracy value is reasonable. The results derived from this research can provide important information to many decision-makers and local authorities to determine location and magnitude of destructions and conduct emergency operations. However, depending on the end-user needs, if higher accuracy value is desired, high resolution satellite image should be used.

CONCLUSIONS

Remotely sensed data are crucial for disaster management, and rapid and reliable information extraction from these data is an important source for decision-making. Quick identification of heavily damaged areas in a disaster provides key information on potential damage and losses to buildings, transportation systems, industrial facilities, and critical emergency facilities. These data can be used by urban planners and emergency managers to manage vulnerabilities of a region and develop risk mitigation plans.

This research proposed a new approach which is the integration of FFT and level slicing to accurately identify the location of the damaged areas caused by an earthquake. A difference image obtained from IFFT-pre- and post-earthquake data can provide significant information about the location and degree of earthquake induced damage areas. Most of the parcels which include more than three collapsed buildings were easily determined using the proposed method. The total accuracy of the method is calculated for pre-defined blocks in parcel basis and found to be 80.19%. Higher resolution satellite images should be used to identify collapsed buildings individually. However, in some cases high resolution data of the region is not available and available satellite images of the region obtained before and after the disaster have to be used. Therefore, it is important to process the available data set rapidly and accurately and this research proposed an approach to fulfill this aim.

With the use of the proposed method, emergency managers, risk managers, and public policy/decision makers can understand the impact of earthquakes, identify the heavily damage areas to direct the emergency-response teams and incorporate the results into preparedness programs and early warning systems.

KEYWORDS

- **Fourier transform**
- **High resolution visible infrared**
- **Inverse fast Fourier transform**
- **Moderate resolution image spectrometer**
- **North Anatolian Fault Zone**

Chapter 10

Ionospheric Quasi-static Electric Field Anomalies During Seismic Activity

M. Gousheva, D. Danov, P. Hristov, and M. Matova

INTRODUCTION

The chapter proposes new results, analyses, and information for the plate tectonic situation in the processing of INTERCOSMOS-BULGARIA-1300 satellite data about anomalies of the quasi-static electric field in the upper ionosphere over activated earthquake source regions at different latitudes. The earthquake catalog is made on the basis of information from the United State Geological Survey (USGS) website. The disturbances in ionospheric quasi-static electric fields are recorded by IESP-1 instrument aboard the INTERCOSMOS-BULGARIA-1300 satellite and they are compared with significant seismic events from the period August 14–September 20, 1981 in magnetically very quiet, quiet, and medium quiet days. The main tectonic characteristics of the seismically activated territories are also taken in account. The main goal of the above research work is to enlarge the research of possible connections between anomalous vertical electric field penetrations into the ionosphere and the earthquake manifestations, also to propose tectonic arguments for the observed phenomena. The studies are represented in four main blocks: (i) previous studies of similar problems, (ii) selection of satellite, seismic and plate tectonic data, (iii) data processing with new specialized software and observations of the quasi-static electric field, and (iv) summary, comparison of new with previous results in our studies and conclusion. We establish the high informativity of the vertical component *Ez* of the quasi-static electric field in the upper ionosphere according observations by INTERCOSMOS-BULGARIA-1300 that are placed above considerably activated earthquake sources. This component shows an increase of about 2–10 mV/m above sources, situated on mobile structures of the plates. The chapter discusses the observed effects. It is represented also a statistical study of ionospheric effects 5–15 days before and 5–15 days after the earthquakes with magnitude M 4.8–7.9.

The study is devoted to statistical data of possible connections between anomalous ionospheric vertical quasi-static fields and the seismic activity. A 3–7 mV/m penetration in the vertical component in the upper ionospheric quasi-static electric fields from an electrostatic source were observed at first by Chmyrev et al. (1989) on the board INTERCOSMOS-BULGARIA-1300 satellite 15 min before an earthquake with M = 4.8. Several observations of ionosphere precursors for earthquakes, including those connected with a perturbation of an electrostatic field in the atmosphere and ionosphere was studied by Alperovich et al. (1999) and Kim et al. (1999). One of the most developed interpretations of these phenomena is based on the electrodynamic model about ionospheric precursors of earthquakes (Sorokin and Chmyrev, 2002; Sorokin et al., 2001a). This model provides an explanation of some electromagnetic and plasma phenomena connected to the effects of amplification of the DC electric field in the ionosphere. A theoretical model of the electric field disturbances caused by the conductivity currents in the atmosphere and the ionosphere initiated by external electric current was proposed by Sorokin and Yaschenko (2000) and Sorokin et al. (2001b). According to this model, the external current arises as a result of emanation of charged aerosols transported into the atmosphere by soil gases and the subsequent processes of upward transfer, gravitational sedimentation and charge relaxation. Further development of this model includes a new method for computation of the electric field in the atmosphere and the ionosphere over active faults for arbitrary spatial distribution of external current in oblique magnetic field (Sorokin et al., 2005, 2006). In our previous chapters (Gousheva et al., 2005a, b; 2006a, b, 2007, 2008a, b) on the basis of INTERCOSMOS-BULGARIA-1300 satellite data we found arguments for seismically-induced increases in the vertical components of the quasi-static electric field up to 2–10–15 mV/m in the near equatorial, low, middle, and high latitude ionosphere. In this chapter we present supplementary data about quasi-static electric field anomalies according INTERCOSMOS-BULGARIA-1300 satellite information over the Southern Atlantic Ocean, Tonga-New Hebrides region, Northern Islands of New Zealand, Indonesian region, Eastern Canada, Labrador Sea, Caribbean region, Central America, Western coast of South America, South-West Pacific Ocean, Kuril Islands, Aleutian Islands, Southern Pacific Ocean, Southern Iran during seismic events in August–September 1981.

SATELLITE AND SEISMIC DATA SELECTION

The INTERCOSMOS-BULGARIA-1300 satellite is launched on August 7, 1981. It has a perigee of 825 km, an apogee of 906 km and an orbital inclination of 81.2°. The INTERCOSMOS-BULGARIA-1300 satellite operates during 2½ years. The registrations of the quasi-static electric field are carried out by the IESP-1 instrument. The IESP-1 instrument measures the electric field using the Langmuir double probe floating potential method, identical with a voltmeter. There is a potential difference between two top-hat probes (Pedersen et al., 1984, 1998). Two sensors are applied to obtain the values of the both horizontal and one vertical component. The basis for the X and Y components is 7.5 m and for the Z component—1.8 m. The dynamical range is ±300 mV/m for the X component, ±600 mV/m–for the Y one and ±90 mV/m–for the Z one. The sensitivity is 1 mV/m for each component. The *Ex* is the horizontal component almost parallel to the magnetic field line; *Ey* is the horizontal component perpendicular to the magnetic field line; *Ez* is the vertical component to the Earth surface. It is a difficult task to study the relations between the quasi-static electric field data and the seismic activity because the passes of satellite over the epicenter zones were rarely. The first task is to select the satellite data about ionospheric anomalies recorded over seismic zones with information for earthquakes in the time period August 14–September 20, 1981. Twenty six orbits are chosen above sources of 73 earthquakes complying with the following conditions:

- observations for satellite orbits over earthquake sources at different latitudes;
- observations for satellite orbits above areas with light, moderate, strong, and major
- earthquakes in different tectonic structures of the planet;
- angular distances of $\Delta\lambda \leq 25^\circ$ between the earthquake epicenter and the closest point of the satellite orbit;
- seismic events in magnetically quiet days (the average geomagnetic activity index $K_p \leq 5$);
- exclusion of events at the beginning or in the end of the orbit when there are calibrations and other operations that made the data unreliable;
- elimination of intervals with clear instrumental effects; and
- exclusion of orbits that do not contain terminator crossing.

The second task is to select the seismic data. We consider that it is necessary to take in account the seismic situation 15 days before and after the monitoring for regions with strong seismic activity and 5 days before and after the monitoring for relatively quiet seismic regions. The earthquake data and related details for the same period are based on USGS website.

DATA PROCESSING AND OBSERVATIONS OF THE QUASI-STATIC ELECTRIC FIELD

New specialized software is used for the processing of experimental data about quasi-static electric fields. We present our last results from observations of the quasi-static electric field on board INTERCOSMOS-BULGARIA-1300 satellite in the upper ionosphere (h = 800–900 km) above earthquake sources in the Southern and Northern Hemisphere at different latitudes. In the topic, the arrows indicate the moments when the satellite passed at the closest distance $\Delta\lambda$ from the earthquake epicenter. The data are presented as a function of the Universal time (UT), satellite altitude (ALT), the geographic latitude and longitude (LAT, LONG), and the invariable latitude (Inv LAT).

South Atlantic Ocean

Five days before and after the passing of satellite (orbits 184, 198) and 5 days before and after the passing of satellite (orbits 305, 347, 348, 403) the seismic manifestations in the region is shown in Figure 1a. Two events EQ_1 and EQ_{46} (see Table 1) occurred near the Scotia Sea in the southern Atlantic Ocean. The territory is a complex area of marginal basins bordered by the Southern America and the Antarctic Plates. The boundary motion between the cited two large plates represents predominately left lateral strike-slip. The small Scotia microplate is twisted between the both plates as a result of their horizontal movements (Earthquake Summary Poster, 2006). Two earthquakes EQ_{49} and EQ_{60} with magnitudes M 5.1 occurred in the area of South Mid-Atlantic Ridge (see Table 1). The ridge represents an oceanic rift that separates the Southern American Plate from the African one in the South Atlantic Ocean. According to plate tectonics, this ridge runs along a divergent boundary (The Mid-Atlantic Ridge, 2008). On the background of the trend in the quasi-statistic electric field we observe an increase in the E_z component of about 10 mV/m, 12 hr before EQ_{46} and 22 hr before EQ49. The ionosphere disturbance zone (Figure 1b—orbit 347) is concentrated around the earthquake epicenter. Figure 1b shows possible seismic pre-effects of about 2–3 mV/m, 78 hr before EQ_{60}. The obtained results of events EQ_1 (orbits 184 and 198), EQ_{46} (orbits 305, 348, and 403) and EQ_{60} (orbit 403) are summarized in the Table 1.

Table 1. Parameters of 73 earthquakes selected from USGS website and disturbances in the vertical component of the quasi-static electric fields observed by INTERCOSMOS-BULGARIA-1300 during limited period of time.

Earthquake Catalogue							Orbits	Date	Δt,h	1ndexof geomag-netic activity Kp	Disturbances in E;: componentof the quasi-static electric fields, mV/m; Foreshocks and aftershocks	Distance from satellite to epicentrer, km t1'-,deg
No	D , M,Y	Time, UTC, hhmmss. mm	Lat,	Long,	Depth, km	M						
EQ_1	198108 17	050444	-60.01	-25.89	33	5.0	184	20.08.1981	+80:30	2	5	1214 07.40
							198	21.08.1981	+104:15	4	5	1300 08.38
EQ_2	1981 08 17	170741	-25.45	-179.05	383	5.5	170	19.08.1981	+44: 15	3	2	1140 06.70
							213	22.08.1981	+96: 18	3	uncertain	1903 14.09
EQ_3	198108 17	191243	-30.14	-177.53	33	5.1	170	19.08.1981	+42:11	3	3, cumulative effect of EQ3,7	1067 05.70
							213	22.08.1981	+94:14	3	uncertain	1910 14.89
EQ_4	1981.08.18	031252	-4.03	127.32	41	5.0	215	22.08.1981	+113:27	3	foreshock of EQ13	1875 14.32
EQ_5	1981.08.18	052934	-4.09	127.33	33	5.1	215	22.08.1981	+108:27	3	8, cumulative ef-fect of EQs,33	1876 14.33
EQ_6	198108 18	132754	65.84	-89.89	18	5.1	193	21.08.1981	+62:28	2	5	1780 12.98
EQ_7	1981 08 19	014107	-33.46	179.66	184	5.1	170	19.08.1981	+11:43	3	3, cumulative effect of EQ3,7	1342 09.00
							213	22.08.1981	+60:44	3	absent	1655 12.00
EQ_8	1981 08 19	030107	-24. 12	180.00	507	4.8	170	19.08.1981	+10:21	3	absent	1207 07.50
							213	22.08.1981	+55:24	3	absent	1821 13.08
EQ_9	1981 08 19	060624	-22.98	170.50	25	5.6	170	19.08.1981	+07: 15	3	10	879 16.00
							213	22.08.1981	+87:53	3	5	974 04.57
EQ_{10}	198108 19	122533	-49.79	164.23	33	4.8	213	22.08.1981	+50:05	3	5	1102 06.20
EQ_{11}	19810820	021948	- 11.48	166.16	70	5.0	213	22.08.1981	+53:56	2	uncertain	933 01.03
EQ_{12}	1981 08 20	044847	-22.93	-70.55	23	4.9	250	25.08.1981	+120:25	3	7	1569 11.00
							251	25.08.1981	+125:36	3	7	1853 14.00
EQ_{13}	198108 20	151003	-27. 14	-179.15	346	4.9	170	22.08.1981	-25:47	3	absent	1308 08.60
							213	22.08.1981	+23:25	3	absent	1695 12.60
EQ_{14}	19810821	011508	-34. 11	-70.09	117	4.9	250	25.08.1981	+97:02	3	8	1392 09.58
							251	25.08.1981	+101:44	3	8	2196 17.35
EQ_{15}	1981 08 21	182414	-18.77	-174.51	33	4.8	170	19.08.1981	-54:03	3	absent absent	839 01.50
							213	22.08.1981	+20:04	3		1479 10.40
EQ_{16}	1981 08 21	225240	-26.50	-114.76	10	5.1	196	21.08.1981	-13:20	2	8	909 03.40
							244	24.08.1981	+68:46	4	absent	1394 09.23
EQ_{17}	1981 08 22	234741	-35.83	-103.30	10	5.2	196	21.08.1981	-14:12	2	10	1509 10.32
							244	24.08.1981	+43:49	4	absent	953 03.42
							251	25.08.1981	+55:13	3	8	2020 15.72

Table 1. *(Continued)*

EQ_{18}	1981 08 23	015950	-22.06	170.95	100	5.8	170 213	19.08.1981 22.08.1981	-84:21 -14:35	3 3	12 9-10	879 1022	16.00 05.10
EQ_{19}	1981 08 23	120026	48.71	157.39	40	6.0	283	27.08.1981	+91:57	5	10	1160	09.03
EQ_{20}	1981 08 23	180704	-35.67	178.25	164	4.8	170 213	19.08.1981 22.08.1981	+100:42 -27:39	3 3	absent absent	1507 1479	10.30 10.40
EQ_{21}	1981 08 23	195632	-17.07	120.52	33	4.8	215	22.08.1981	-26: I 0	3	8	1110	06.39
EQ_{22}	1981 08 23	234528	-63.57	-167.21	10	5.0	170 213	19.08.1981 22.08.1981	+106:15 -33:11	3 3	5 10	1045 1298	05.40 08.50
EQ_{23}	1981 08 24	112033	61.22	-59.01	10	4.8	193	21.08.1981	-79: 19	2	10	946	02.84
EQ_{24}	1981 08 24	154627	51.50	-178.35	56	5.2	240 283	24.08.1981 27.08.1981	-00:52 +72:21	3 5	10 uncertain	1128 2244	06.09 18.00
EQ_{25}	198108 25	015849	-23.51	-179.91	550	4.8	213	22.08.1981	-25:34	3	absent	1880	14.00
EQ_{26}	1981 08 25	052021	-34.61	-179.46	68	5.4	213	22.08.1981	-59:52	3	4	1724	12.90
EQ_{27}	19810825	071658	-22.89	-175.85	33	5.9	213	22.08.1981	-62:02	3	absent	2686	21.78
EQ_{28}	1981 08 25	072245	-22.90	-175.90	33	5.7	213	22.08.1981	-62:62	3	absent	2686	21.78
EQ_{29}	198108 25	165438	6.93	-76.59	33	5.2	250 251	25.08.1981 25.08.1981	-11:05 -10:06	5 5	5 4	1641 1788	12.00 13.00
EQ_{30}	198108 25	172907	7.01	-76.58	33	5.1	250 251	25.08.1981 25.08.1981	-12:23 -10:41	3 3	aftershock EQ29 aftershock EQ29	1639 1790	12.00 13.50
EQ_{31}	19810825	214025	-11.76	166.59	150	4.9	213	22.08.1981	-76:20	3	absent	933	01.03
EQ_{32}	1981 08 27	054509	44.99	146.09	197	4.8	283	27.08.1981	+07.11	5	absent	2331	18.39
EQ_{33}	1981 08 27	075455	-6.46	129.89	46	4.9	215	22.08.1981	-107:21	3	8, cumulative effect of EQs,33	2126	17.00
EQ_{34}	198108 27	135240	6.88	-76.72	33	4.8	250 251	25.08.1981 25.08.1981	+56:47 +55:04	3 3	aftershock EQ29 aftershock EQ29	1641 1788	12.00 13.00
EQ_{35}	19810828	123651	52.42	-169.28	39	5.1	240	24.08.1981	+96:00	3	uncertain	1114	05.94
EQ_{36}	1981 08 29	001549	19.28	-64.85	33	4.8	250 349	25.08.1981 01.09.1981	-85:13 +52:55	3 1	absent absent	836 2123	00.79 16.66
EQ_{37}	19810829	074151	-2.20	100.85	57	4.9	258	25.08.1981	-84:56	3	2-3	1386	09.50
EQ_{38}	198108 29	181747	12.89	-87.85	33	5.2	251 349	25.08.1981 01.09.1981	-101:57 -58:55	3 1	absent absent	889 1145	02.79 06.84
EQ_{39}	1981 08 30	205008	6.88	-76.58	33	4.9	250 251	25.08.1981 25.08.1981	-105:14 -106:54	3 2	aftershock EQzg aftershock EQ29	1641 1788	12.00 13.00
EQ_{40}	1981 09 01	072302	-15.13	-173.28	33	5.8	170 527	19.08. 1981 13.09.1981	~-3 06 ~+300	3 1	foreshock of EQ41 foreshock of EQ41	822 2519	00.28 19.61
EQ_{41}	1981 09 01	092931	-14.96	-173.08	25	7.9	170 527	19.08.1981 13.09. 1981	~-306 ~+300	3 1	18 12	822 2519	00.28 19.61
EQ_{42}	1981 09 01	095932	-15.15	-173.26	33	5.6	527	13.09.1981	~+300	1	aftershock of EQ41	2519	19.61
EQ_{43}	1981 09 01	105903	-15.01	-173.36	33	5.2	527	13.09.1981	~+300	1	aftershock of EQ41	2519	19.61
EQ_{44}	198109 01	123914	-15.05	-173.31	33	4.8	527	13.09.1981	~+300	1	aftershock of EQ41	2519	19.61
EQ_{45}	1981 09 01	152436	-15.23	-173.07	33	4.8	527	13.09.1981	~+300	1	aftershock of EQ41	2519	19.61

Table 1. *(Continued)*

EQ$_{46}$	1981 09 01	155557	-58.71	-25.36	115	4.8	305	29.08.1981	+61:06	3	absent	882	02.68
							347	01.09.1981	-12:47	0	10	1123	06.46
							348	01.09.1981	-12:00	4	8	1896	14.43
							403	05.09.1981	+81:19	4	8	1597	11.57
EQ$_{47}$	198109 01	183847	-15.31	-173.30	33	5.7	527	13.09. 1981	~+300	1	aftershock of EQ41	2519	19.61
EQ$_{48}$	1981 09 01	235545	-15.22	-173.17	33	5.6	527	13.09. 1981	~+300	1	aftershock of EQ41	2519	19.61
EQ$_{49}$	1981 09 02	002354	-55.31	-1.67	10	5.1	347	01.09.1981	-22:12	0	10	2034	15.75
EQ$_{50}$	1981 09 02	003427	-15.41	-172.86	33	4.8	527	13.09.1981	~+276	1	aftershock of EQ41	2519	19.61
EQ$_{51}$	198109 02	021025	-15.55	-172.59	33	4.8	527	13.09. 1981	~+276	1	aftershock of EQ41	2519	19.61
EQ$_{52}$	1981 09 02	062547	-15.08	-173.01	33	4.8	527	13.09.1981	~+276	1	aftershock of EQ41	2519	19.61
EQ$_{53}$	1981 09 02	084421	-15.47	-172.97	33	5.6	527	13.09.1981	~+276	1	aftershock of EQ41	2519	19.61
EQ$_{54}$	1981 09 02	103052	-14.91	-173.68	33	5.3	527	13.09. 1981	~+276	1	aftershock of EQ41	2519	19.61
EQ$_{55}$	1981 09 03	021255	-38.79	-92.40	10	4.8	349	01.09.1981	+47:56	1	8	2110	16.55
EQ$_{56}$	1981 09 03	053544	43.62	147.03	45	6.6	283	27.08.1981	- 153:38	5	15	2274	18.00
EQ$_{57}$	198109 03	062814	43.58	146.87	33	5.0	283	27.08.1981	-158:31	5	absent	2292	18.00
EQ$_{58}$	1981 09 05	080035	43.68	146.73	33	4.9	283	27.08.1981	-160:03	5	absent	2303	18.00
EQ$_{59}$	1981 09 03	080901	11.85	-87.46	48	5.1	349	01.09.1981	-55:24	1	absent	1290	08.00
							405	05.09.1981	+44:17	3	absent	1840	14.00
EQ$_{60}$	1981 09 04	082306	-15.69	-13.10	10	5.1	347	01.09.1981	-78:02	0	2-3	1847	14.13
							403	05.09.1981	+19:40	4	2	1295	08.00
EQ$_{61}$	1981 09 06	164319	-36.17	-100.70	10	5.4	482	10.09.1981	+95:05	3	3	2212	17.10
							483	10.09.1981	+96:49	3	2	1320	08.15
EQ$_{62}$	1981 09 10	143726	-22.67	-179.34	528	4.8	527	13.09. 1981	+73:41	1	absent	3072	24.93
EQ$_{63}$	1981 09 10	224300	-23.26	-177.11	33	5.2	527	13.09.1981	+69:36	1	absent	2819	22.64
EQ$_{64}$	19810911	120704	-15.03	-173.61	33	4.9	527	13.09.1981	+56: 13	1	absent	2531	20.02
EQ$_{65}$	198109 12	022916	27.85	56.97	33	4.8	505	12.09.1981	+04:43	3	5	959	03.00
EQ$_{66}$	1981 09 15	224412	-27.66	-71.57	33	4.9	482	10.09.1981	-123:44	3	absent	1320	08.50
							546	15.09.1981	+18:16	2	absent	1534	10.00
							602	19.09.1981	+79:22	5	absent	1610	11.00
EQ$_{67}$	1981 09 16	022348	-8.85	-109.13	10	4.9	505	12.09.1981	-20:40	3	5	1068	05.96
							547	15.09. 1981	-20:47	2	10	833	01.81
							581	17.09.1981	+40:34	1	3-4	1596	11.00
EQ$_{68}$	1981 09 17	150929	-15.20	-173.11	33	4.8	527	13.09.1981	-96:5 1	1	absent	2475	19.51
EQ$_{69}$	1981 09 18	102409	-24.67	-71.86	33	4.9	546	15.09.1981	-78:25	2	absent	1616	11.00
							602	19.09.1981	+12:47	5	absent	1109	06.00
EQ$_{70}$	1981 09 18	141732	-35.24	- 110.34	10	5.0	547	15.09. 1981	-80:57	2	10	871	02.35
							581	17.09.1981	-22:18	1	4-5	1596	11.00
EQ$_{71}$	1981 09 19	124910	10.42	-62.81	63	5.1	602	19.09.1981	+9:54	5	absent	2316	18.00
EQ$_{72}$	1981 09 19	114056	-39.08	-74.80	30	5.6	546	15.09.1981	-76:20	2	10	1280	07.50
							602	19.09.1981	-08:32	5	10	852	01.30
EQ$_{73}$	198109 20	104820	-23.08	-66.63	234	5.1	546	15.09.1981	-76:20	2	10	2183	17.00
							602	19.09.1981	-25:43	5	10	852	01.30

Tonga-new Hebrides region and North Islands of New Zealand

We receive satellite information for the seismic manifestations in the region in a relatively large time period August 17–September 17, 1981 when the major earthquake EQ_{41} with a magnitude Mw 7.9 (see Table 1) occurred along the Tonga trench on Tuesday, September 1, 1981 at 09:29:31.

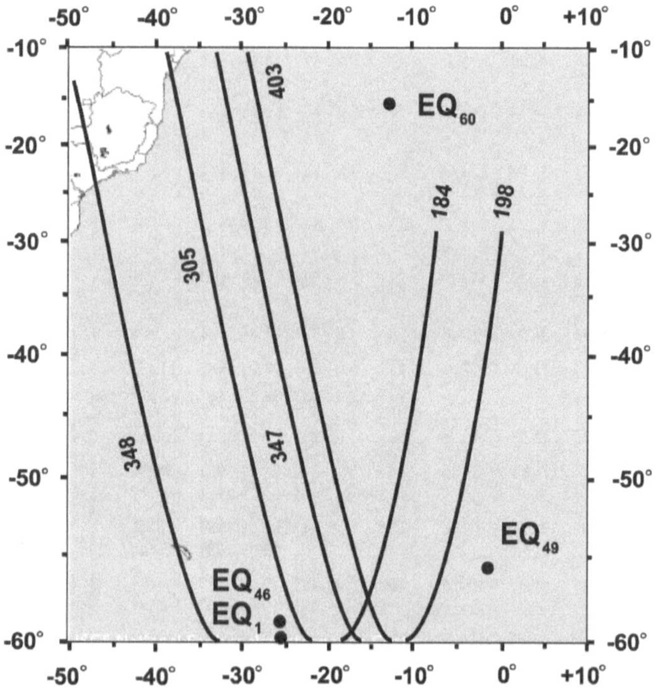

Figure 1a. Satellite orbits 184, 198, 305, 347, 348, 403, and positions of the earthquake epicenters for periods of seismic activity in August 13–31, 1981 and August 28–September 9, 1981.

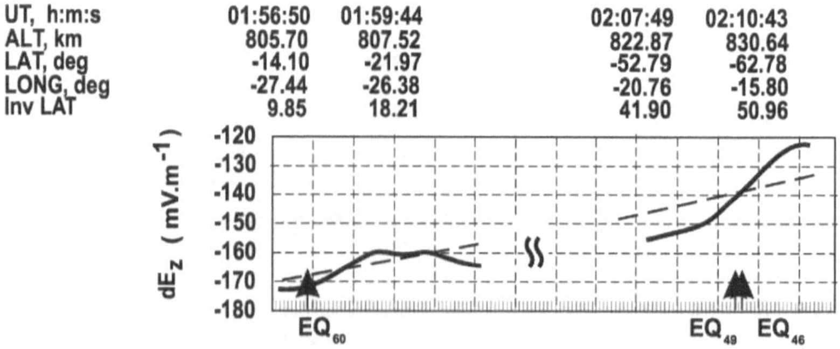

Figure 1b. Disturbances in *Ez* component of the quasi-static field, orbit 347.

In a geophysical sense, New Zealand sits in a precarious position because it gets astride the boundary between the Pacific and Australian Plates. There are other two potential sources of large seismic activity: the Tonga-Kermadec trench and the New Hebrides one. The Tonga trench extends from the southern periphery of the Samoa Islands up to the connection with the Kermadec trench. The Kermadec trench connects the Tonga Trench to the north with the Northern Island of New Zealand to the south. The Kermadec and the Tonga trenches can be considered as one representative structural unit of Tonga-Kermadec Trench (Goff et al., 2006; Walters et al., 2006). The pass of INTERCOSMOS-BULGARIA1300 for orbits 170, 213, and 527 is shown in Figure 2a. The earthquake epicenters take place in a relatively long and wide area. The analysis of the Figure 2b indicates that the ionospheric anomalous disturbance zone follows the same wide latitudinal interval of earthquake epicenters. The first changes in E_z component of the quasi-statistic electric field of about 3 mV/m are due to the cumulative seismic effect 11–42 hr after EQ_3 and EQ_7. The second changes, showing an increase in E_z component about 2 mV/m, we observe at the beginning of the big bulge. The established increase could be interpreted like post-seismic effects of EQ_2 (44 hr after the earthquake). We observe a wide latitudinal interval with big changes and an increase in E_z component about 10–12–18 mV/m. In our opinion, the amplitudes of this unusually disturbance zone are formed by the cumulative just-post seismic effects (7 hr) of EQ_9 and forthcoming ones (EQ_{18}, EQ_{41}, and its foreshock EQ_{40}). Summarizing, the indicated big increase of about 18 mV/m might be connected with the major earthquake EQ_{41} with magnitude M_w 7.9 which was happened 13 days later. Any disturbances from EQ_{15} and $EQ_{8,13,20}$ (with hypocenters on the depths 164–507 km and magnitudes M 4.8; 4.8; 4.9; 4.8) are not noted. In these cases the energetic levels are very low. The mentioned observations take place in a quiet day $K_p = 3$.

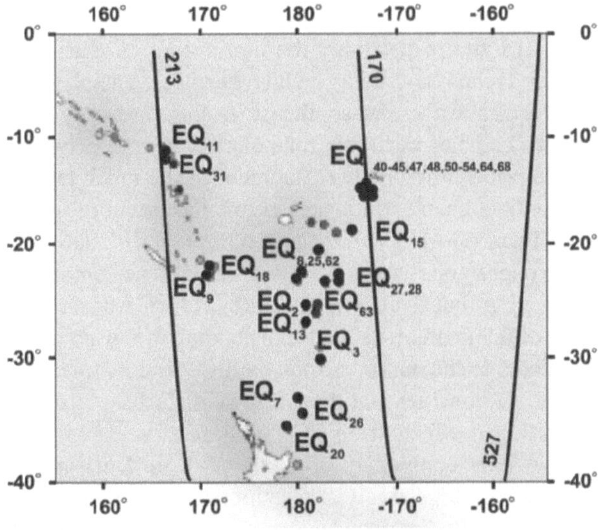

Figure 2a. Satellite orbits 170, 213, 527, and positions of the earth-quake epicenters for period of seismic activity in August 17–September 17, 1981.

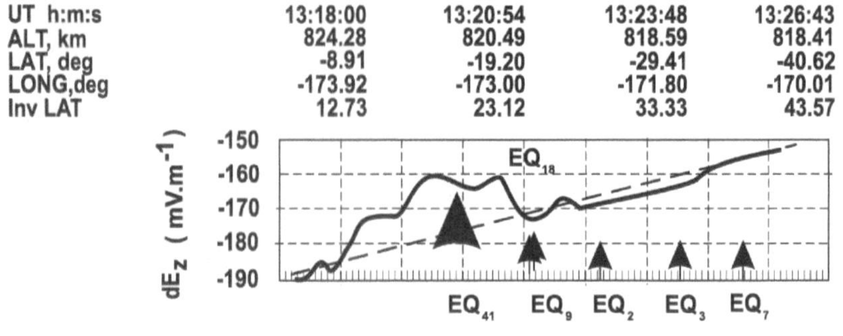

Figure 2b. Disturbances in E_z component of the quasi-static field, orbit 170.

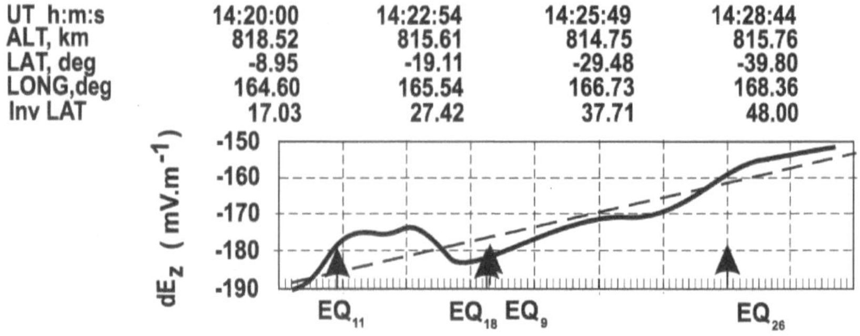

Figure 2c. Disturbances in E_z component of the quasi-static field, orbit 213.

Data from orbit 213 and respectively the monitoring of seismic situation in the area of New Hebrides Trench, is used as well (Figure 2c). The New Hebrides Trench extends from New Guinea to the east-southeast. In this case the zone of the shallow seismic manifestations coincides with the zone of convergence between the Australian plate and the New Hebrides microplate (a segment of the much larger Pacific plate). The occurred intermediate-deep earthquakes define the extent of the downgoing slab from the Australian Plate (Goff et al., 2006; Kolobov et al, 2006). It is of great interest, that the disturbances observed on Figure 2c propose some information about earthquakes in the region, but 3 days later (orbit 213 on August 22, 1981). Figure 2c shows new form of the anomalous disturbance zone in a wide latitudinal interval. Any disturbances related to the major earthquake EQ_{41} and its foreshock EQ_{40} are not observed ($\Delta\lambda > 20°$). We consider that the width of disturbance zone is probably determined by time-shift post effects from EQ_{11} and pre-effect (59 hr) of EQ_{26}. We can also observe increases in E_z component of about 4 mV/m, 5 mV/m, and 9–10 mV/m from EQ_{26}, EQ_9, and EQ_{18}, respectively. These increases are shifted to the north from the earthquake epicenter. Any disturbances from $EQ_{8,13,15,20,25}$ are not observed because these earthquakes are at a relatively low energetic level. There are difficulties in the determination of the influence of the events $EQ_{2,3}$ and EQ_7 (near to EQ_{26}). Any disturbances from pair $EQ_{27,28}$ are not observed ($\Delta\lambda > 20°$) as well. It should be also noted that these measurements are made in a quiet day ($K_p = 3$), so the observed anomalies are not caused by a solar-terrestrial disturbance. The satellite passes (orbit 527) at $\Delta\lambda = 19.61°E$ (Figure 2a) about 300 hr after the major earthquake EQ_{41}.

The disturbances in the E_z component of the quasi-static electric field: the first one to the north of the earthquake epicenters (projection over the equipotent magnetic field lines in the near-equatorial ionosphere at satellite altitude) and second one—in the magnetic conjugate region of about 12 mV/m for EQ_{41} are shown in Figure 2d. The major earthquake event is followed by two strong aftershocks $EQ_{42,43}$ the next 2 hr, also by a series of other ones in subsequent days ($EQ_{44,45,47,48,50,51,52,64,68}$) and two strong aftershocks $EQ_{53,54}$ (see Table 1). Thus major and great earthquakes occur frequently in this region. It is unlikely, that these post effects could provoke an increase in *Ez* component of about 12 mV/m. Any post effects (Figure 2e) from $EQ_{62,63}$ cannot be observed ($\lambda > 20°$). It should be noted that the above mentioned data are taken in a very quiet day ($K_p = 1$).

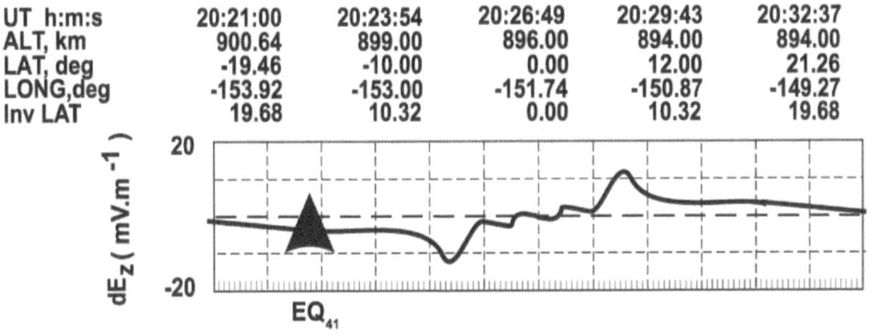

Figure 2d. Disturbances in E_z component of the quasi-static field, orbit 527.

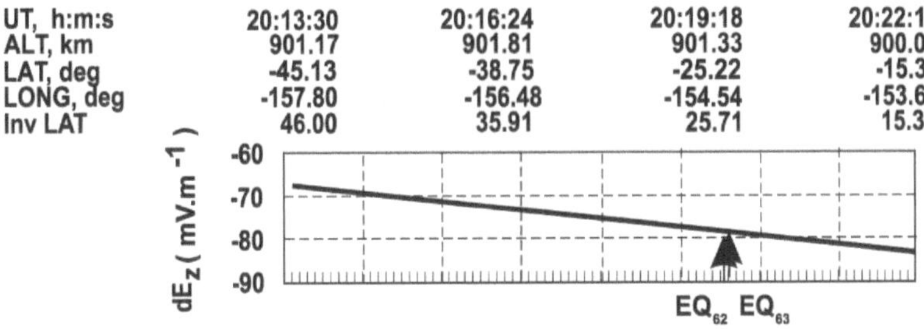

Figure 2e. Vertical component E_z of the quasi-static field, orbit 527.

Indonesian Region

The Indonesian region is one of the most seismically active zones on the Earth. It is an island-arc structure of about 17,000 islands. The islands of South-East Indonesia lie at the junction of the Eurasian, Australian, Pacific, and Philippine Sea Plates, resulting in rugged topography, frequent earthquakes, and intensive volcanism (Indonesia, 2008). The seismic manifestations of the region, bounded between latitudes 0–17°S and longitudes 100–130°E, 5 days before and after the passing of satellite (orbits 215, 258), are shown in Figure 3a. Five events ($EQ_{4,5,21,33,37}$) take place in the area and they are with magnitude M 4.8 during the studied time period (see Table 1). On the background of the trend, to the north of the earthquake epicenter, we observe an increase in the E_z component of about 8 mV/m, 26 hr before EQ_{21} (Figure 3b). Figure 3b illustrates also the next increase in the E_z component of about 8 mV/m to the north of EQ_5 and EQ_{33}. It is highly probable that the disturbances in the E_z component are due to the cumulative effect of occurred EQ_5 and forthcoming earthquake EQ33 (it is happened 107 hr later). The earthquake EQ37 of magnitude M 4.9 occurs on September 19, 1981 in the Sumatra region. The Sumatra region from the western part of Indonesia is considered as a part of the Sunda arc, which results from the convergence between the Indo-Australian and Eurasian Plates. As a product of the plate convergence, the Sumatra region is considered to be one of the most seismically active regions in Indonesia (Nanang and Gunawan, 2005).

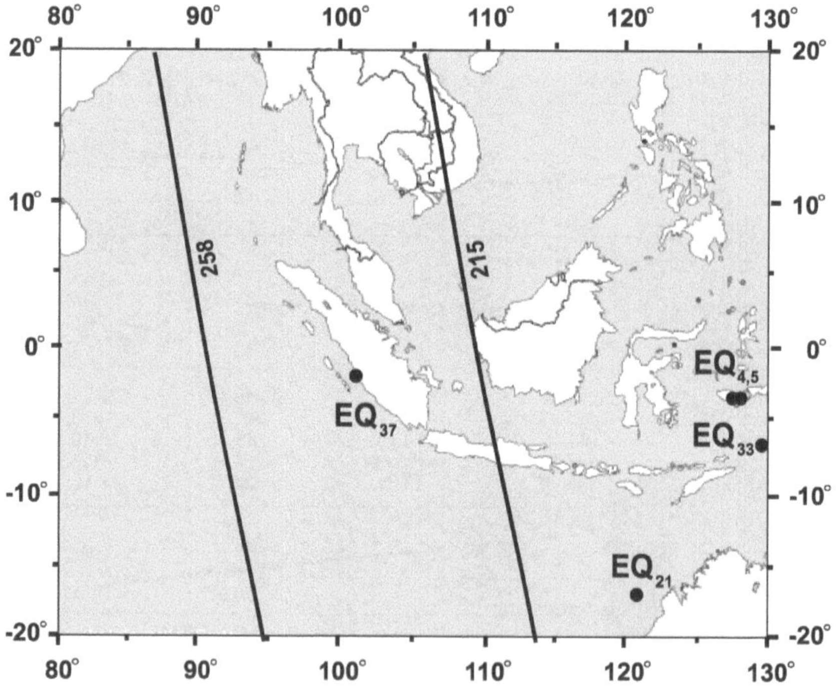

Figure 3a. Satellite orbits 215, 258, and positions of the earthquake epicenters for period of seismic activity in August 12–September 1, 1981.

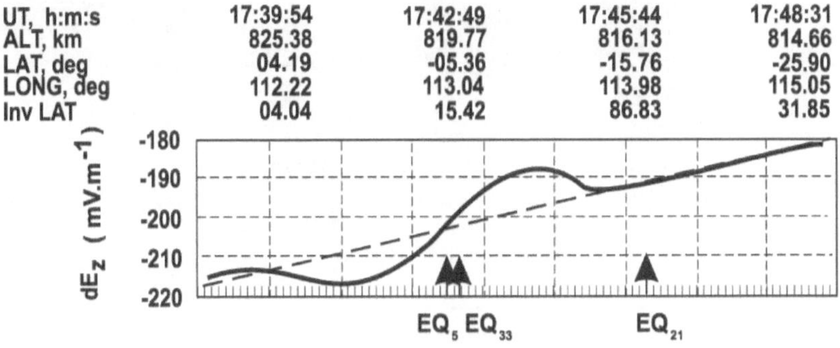

Figure 3b. Disturbances in *Ez* components of the quasi-static field, orbit 215.

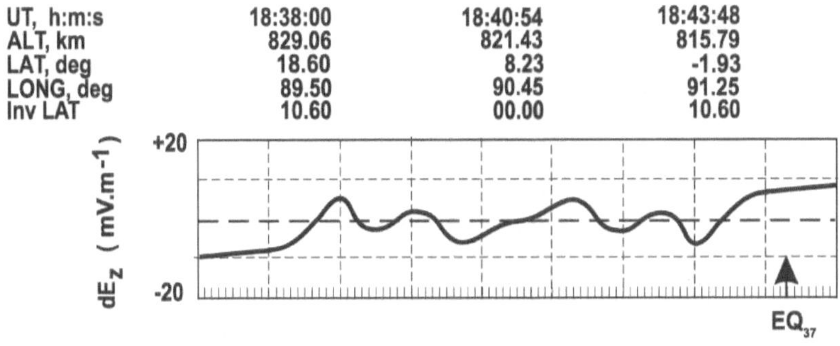

Figure 3c. Disturbances in *Ez* components of the quasi-static field, orbit 258.

The pass of INTERCOSMOS-BULGARIA-1300 (orbit 258) 84–85 hr before the seismic manifestations is shown in Figure 3c. The disturbances in the E_z component of the quasi-static electric field are the following ones: The first one to the north of the earthquake epicenters (projection over the equipotent magnetic field lines in the near-equatorial ionosphere at satellite altitude) and second one in the magnetic conjugate region of about 2–3 mV/m for EQ_{37}. The both of them are shown in Figure 3c. It should be noted that these measurements of orbits (215 and 258) are made in a quiet day ($K_p = 3$), so the observed anomalies could not be caused by a solar-terrestrial disturbance.

Eastern Canada and Labrador Sea

Two events (EQ_6, EQ_{23}) are recorded on August 18 and August 24, 1981 with magnitudes M 5.1 and 4.8 in the time period August 18–24, 1981, respectively (see Table 1). The EQ_6 occurred in the region of Eastern Canada. The causes of earthquakes in the Eastern Canada are not well understood.

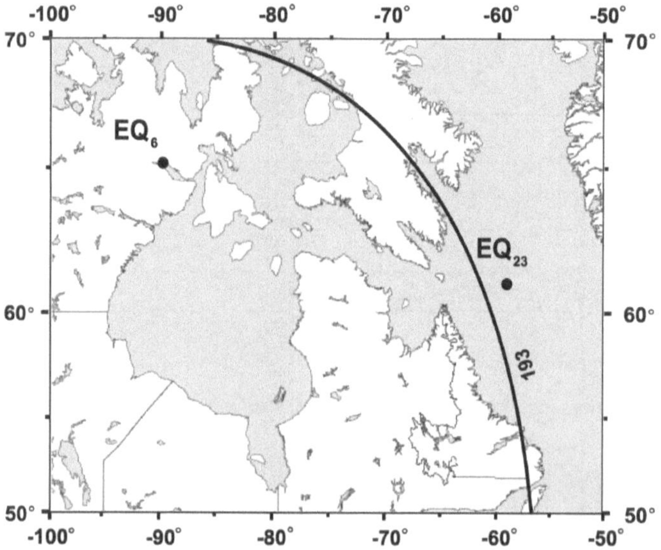

Figure 4a. Satellite orbits 193 and positions of the earthquake epicenters for period of seismic activity in August 18–24, 1981.

The Eastern Canada is a part of the relatively stable interior of the North American Plate, but not on a plate boundary. May be the North American Plate is in processes of transformation, fragmentation and division. The EQ_{23} (see Table 1) occurred in the Labrador seismic zone on August 24, 1981 (latitude 61.22°N, longitude 59.01°W, 11:20:33 UTC and depth 10 km) with M 4.8. The pass of INTERCOSMOSBULGARIA-1300 for orbits 193 is shown in Figure 4a. (Observations in) Figure 4b indicates two anomalous disturbance zones: the first one is to the south of EQ_6 with an increase in the *Ez* component of about 5 mV/m, 62 hr after this event and the second one to the south of EQ_{23} and it is of about 10 mV/m, 79 hr before this event. It should be noted that the observations are taken in a quiet day ($K_p = 2$) and the events EQ_6 and EQ_{23} occurred in isolated time-space domains.

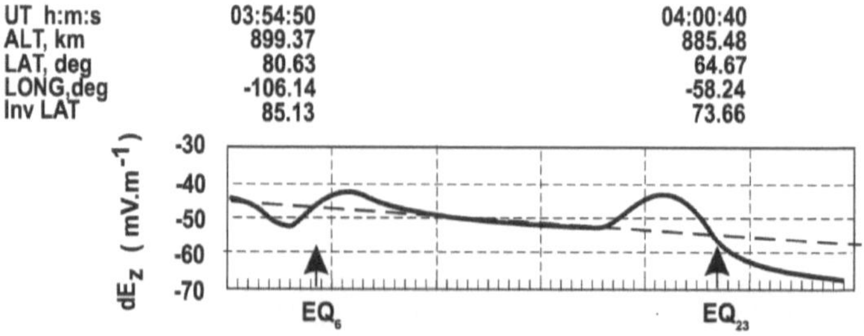

Figure 4b. Disturbances in *Ez* component of the quasi-static field, orbit 193.

Caribbean Region, Central America, West Coast of South America, and South–West Pacific Ocean

The pre- and post-seismic effects of the region bounded between latitudes 20°N–50°S and longitudes 60°W and 120°W are shown in Figure 5a. Four events $EQ_{36'38,59,71}$ with magnitudes M > 4.8 occurred in the region (Table 1) during the time period August 19–September 19, 1981.

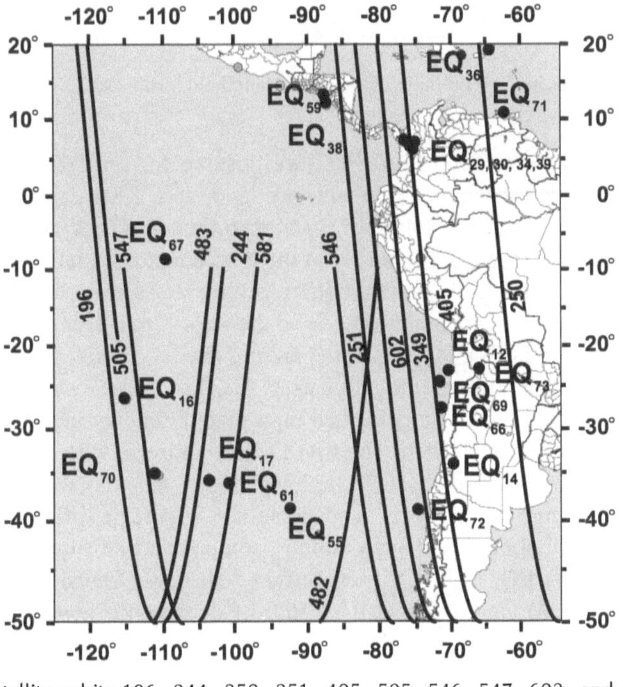

Figure 5a. Satellite orbits 196, 244, 250, 251, 405, 505, 546, 547, 602, and positions of the earthquake epicenters for period of seismic activity in August 5–September 29, 1981.

Their epicenters lie on the Caribbean plate (a region with relatively quiet seismicity). Any disturbances around 55–101 hr before to and 9–58 hr after these earthquakes are not marked (see Table 1). Now we do not explain these phenomena. We show also Figure 5b (orbit 250) when the satellite passes at $\Delta\lambda$ = 12°E–13°E, about 11 hr before earthquake EQ_{29}. The earthquake EQ_{29} occurred in the Andes Mountains region of Peru on August 25, 1981 (latitude 6.93°N, longitude 76.59°W, 16:54:38 UTC and depth 33 km) with M 5.2. Apart from the main shock, a number of earthquakes with comparatively smaller magnitudes happened in the same region like $EQ_{30,34,39}$ as well (see Table 1). Two disturbances in the E_z component of the quasi-static electric field are noted in this situation. The first one situated to the south of the earthquake epicenters (projection over the equipotential magnetic field lines in the low ionosphere at satellite altitude) and the second one—in the magnetic conjugate region of about 5 mV/m (for the main event) are shown in Figure 5b.

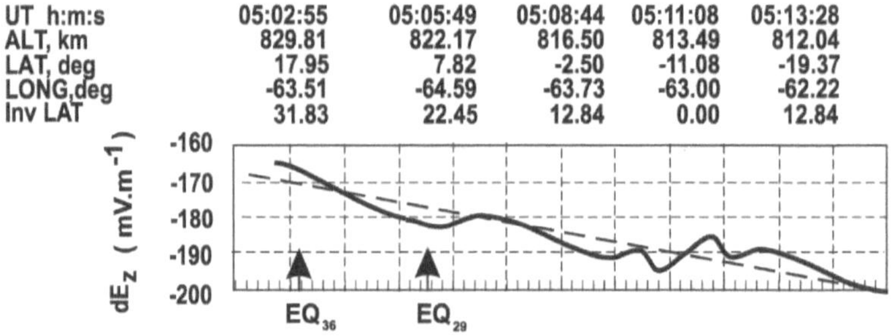

Figure 5b. Disturbances in E_z component of the quasi-static field, orbit 250.

A similar disturbance zone from EQ_{29} we observe for orbit 251 (see Table 1). It should be noted that these measurements are made in a quiet day ($K_p = 3$), so the anomalies are not caused by a solar-terrestrial disturbance. The West coast of South America is outlined by the eastern border of the Nazca tectonic plate and it is characterized by its extremely high seismicity. There is a narrow seismic belt (100–150 km wide) between the Andes Mountain Range and the Peru–Chile Trench (Gagnon et al., 2005). Five earthquakes $EQ_{12,14,66,69,72}$ in the area of the West coast of South America between 22 and 40°S latitudes are as large as M 4.8. The seismic events occur in and near the territory of Chile. Chile is located on a plate boundary and in a subduction zone called the Peru–Chile Trench. The first EQ_{12} is an event with magnitude M 4.9 and occurs in the Province of Antofagasta. Later in the same day another earthquake EQ_{14} with M 4.9 occurs in the region near Valparaiso. Figure 5c (orbit 250) illustrates a disturbance zone of about 7 and 8 mV/m in E_z component of the quasi-static electric field, 97–120 hr after EQ_{12} and EQ_{14}, respectively. The disturbance zone is in a wide latitudinal interval. We consider that the width of disturbance zone is probably determined by time-shift post effect of EQ_{12} and EQ_{14}. A similar disturbance zone from EQ12 and EQ_{14} we establish for orbit 251 (see Table 1). Anomalies of about 10 mV/m are marked in the E_z components of the quasi-static electric field 8–25 hr before EQ_{72} and EQ_{73} (see Figure 5d—orbit 602 and Table 1). Such pre-and post seismic influence of $EQ_{55,72,73}$ (orbits 349, 546) is observed in the E_z components of the quasi-static electric field and listed in Table 1.

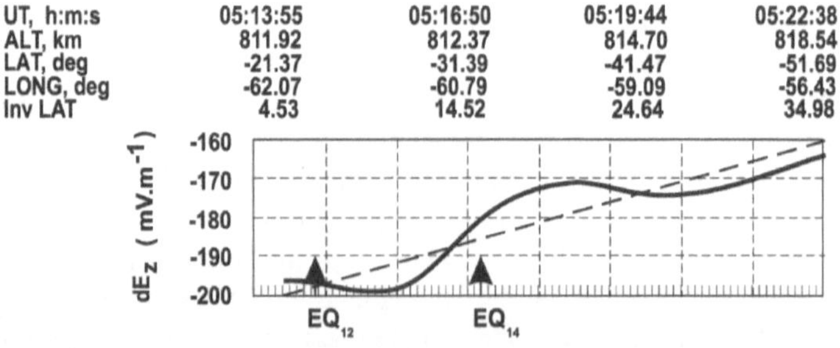

Figure 5c. Disturbances in E_z component of the quasi-static field, orbit 250.

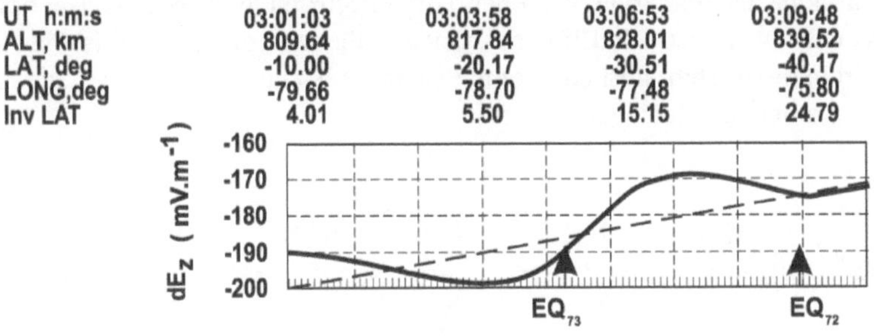

Figure 5d. Disturbances in *Ez* component of the quasi-static field, orbit 602.

Any disturbances (orbits 482, 546, and 602) from EQ_{66} to EQ_{69} (see Table 1) are not observed. Data from orbits 196 and 244 for observations about EQ_{16} and EQ_{17} is used as well. The satellite passes (orbit 196) about 13–14 hr before EQ_{16} and EQ_{17} in the region of South-West Pacific Ocean (Figure 5a). Figure 5e shows two disturbances in E_z component of the quasi-static electric field in relation with seismic activity. The first change is an increase in E_z component of about 8 mV/m, 13 hr before EQ_{16}. The second change represents another increase in *Ez* component of about 10 mV/m, 14 hr before EQ_{17}. These data is taken in a medium quiet day $K_p = 4$. We can also observe an increase in the E_z component of about 8 mV/m, 55 hr after EQ_{17}.

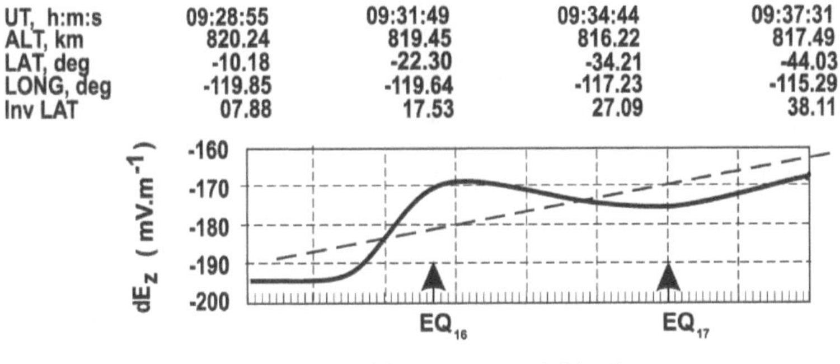

Figure 5e. Disturbances in E_z components of the quasi-static field, orbit 196.

Any disturbances (orbit 244) from EQ_{16}, EQ_{17} (see Table 1) are not noted. Similar anomalies in the quasi-static electric field are summarized as an increase in the vertical component of the quasi-static electric field of about 2–10 mV/m, around 20 hr before to and 96 hr after earthquakes EQ_{61} (orbits 482 and 483), EQ_{66} (orbits 482 and 546) and EQ_{67} (orbits 505, 547, and 581). All of them are listed in Table 1.

Kuril Island and Aleutian Islands

We take observations over seismic events in the region with the help of two satellite orbits (240 and 283) in the time period August 16–September 9, 1981 (Figure 6a). Two strong earthquakes in interval of about 10 days occurred in the region. The first event EQ_{19} (M 6.0) on August 23, 1981 took place in the area of the Kuril Islands that are located from the westernmost point of the Japanese Island of Hokkaido to the southern tip of the Kamchatka Peninsula.

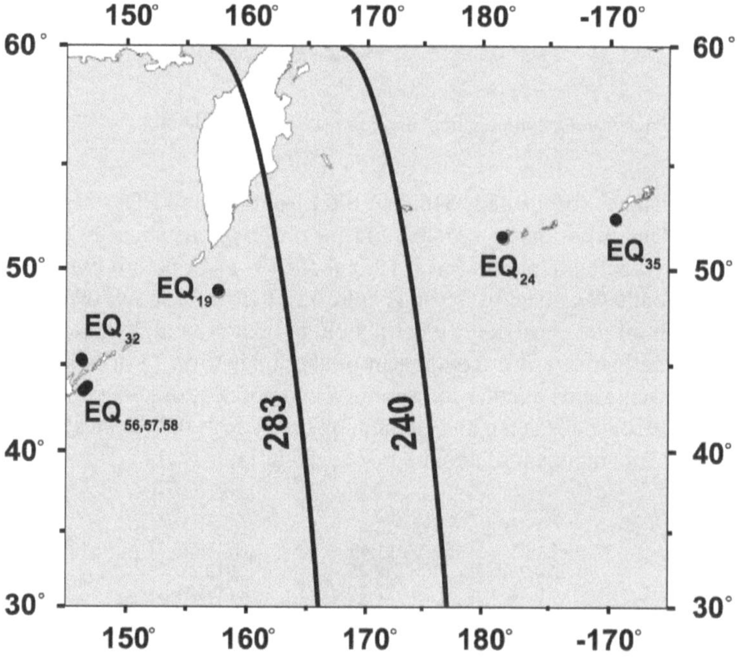

Figure 6a. Satellite orbits 240, 283 and positions of the earthquake epicenters for period of seismic activity in August 23 to September 5, 1981.

The Kuril Trench was formed by the subduction of the Pacific plate under the North American plate and extends from the central area of Kamchatka to Hokkaido. The Kuril Trench is one of the most active seismogenic regions due to the subduction of the Pacific Ocean Plate beneath Hokkaido. Figure 6b (orbit 283) shows an increase in the E_z component of about 10 mV/m 92 hr after the EQ_{19} event. The ionospheric disturbance zone shows concentration around the earthquake epicenter. Probably, it is not impossible that there is also influence of EQ_{24} (the satellite passed 72 hr after the earthquake EQ_{24}). A strong earthquake EQ_{56} (M 6.6) occurs in the region of Aleutian Islands, about 10 days later, on September 3, 1981. Another earthquake EQ_{56} occurs near Shikotan Island, in the South Kuril Island Group of Russian Far East. On the background of the trend in the quasi-static electric field, we establish an increase in the E_z component of about 15 mV/m (Figure 6b), 157 hr before this event. The ionospheric disturbance zone is shifted to the south of the earthquake epicenter. The information is taken in a medium quiet day $K_p = 5$. Any disturbances from aftershocks EQ_{57}, EQ_{58}, and EQ_{32}, events of relatively low energetic levels, are not marked. Several anomalies in vertical component E_z of the quasi-static electric field related to EQ_{24} are listed in Table 1. It is difficult to determine the influence of the event EQ_{35} that is in vicinity of EQ_{24}.

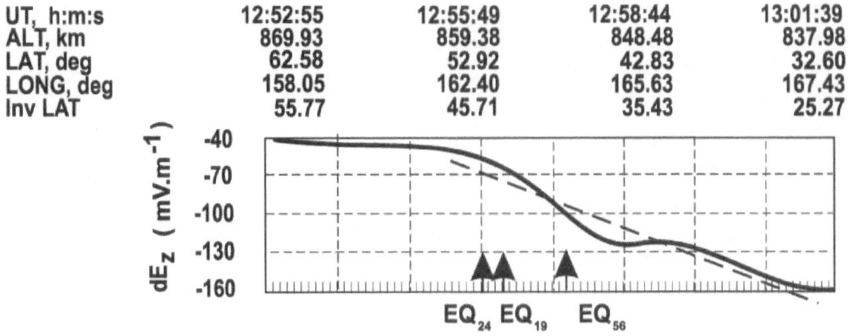

Figure 6b. Disturbances in *Ez* component of the quasi-static field, orbit 283.

The appearance of similar anomalies in vertical component E_z of the quasi-static electric field could be provoked by EQ_{24}, EQ_{35}, and EQ_{65} in the South Pacific and Southern Iran (see Table 1).

CONCLUSIONS

Quasi-static electric fields anomalies in the upper ionosphere associated with seismic activity during August–September 1981 are investigated by using the observation data of the INTERCOSMOS-BULGARIA-1300 satellite in conditions of magnetically very quiet, quiet, and medium quiet days. The observations suggest the presence of quasi-static electric field disturbances related to seismic activity above earthquake sources in the Southern and the Northern Hemispheres at different latitudes. The dates of seismic events, their origin time, locations of epicenter, magnitude, and depth in the observed period are obtained from USGS website.

The analyzed period is short and includes 38 days. New specialized software has been used for processing of ionospheric disturbance zones on the background of the trend for all orbits' tracing and of data for the represented plates of the Earth. The obtained results about 73 events selected from INTERCOSMOS-BULGARIA-1300 data-base are summarized in the Table 2. We exclude the possible effects from two foreshocks to 14 aftershocks. Twenty six orbits for observation of 92 main shocks are used. No disturbances from 35 events are observed. In the chapter 52 positive pre- and post-possible effects are shown as:

- 25 pre-effects of nighttime and two pre-effects of daytime observations;
- 21 post-effects of nighttime and four post-effects of daytime observations; and

- We focus our attention on the considerable rising of data only of nighttime observations for the ionospheric anomalies that are in possible associations with earthquakes.

Table 2. General results from observations.

Total number of earthquakes from USGS website	73
Total number of positive, uncertain and negative effects from main shocks for twenty six orbits	92
Total number of positive and negative effects	87
Number of possible positive effects	52
Number of pre effects	27
Number of nighttime pre effects for different latitudes	25

Near-equatorial latitudes		Low latitudes			Middle and hight latitudes		
2	2	2	6	1	2+1	7+1	1+0
-Δ*t*,*h*	-Δ*t*,*h*	-Δ*t*,*h*	-Δ*t*,*h*	-Δ*t*,*h*	-Δ*t*,*h*	-Δ*t*,*h*	-Δ*t*,*h*
≤12	≤11-20	≤12	≤13- 25	≤306	≤12	≤80	≤80
M Δ λ, deg	M Δ λ, deg	M Δ λ, deg	M Δ λ, deg	M Δ λ, deg	M Δ λ, deg	M Δ λ, deg	M Δ λ, deg
4.8-4.9≤1-9	5-6≤13	4.8-4.9≤1-9	5-6≤1-14	7-8≤1	4.8-4.9≤1-9	5-6≤1-18	6-7≤18

Number of nighttimepost effects for different latitudes

Low latitudes		Middle latitudes			Hight latitudes	
1	5	1	7	3	1	3
-Δ*t*,*h*	-Δ*t*,*h*	-Δ*t*,*h*	-Δ*t*,*h*	-Δ*t*,*h*	-Δ*t*,*h*	-Δ*t*,*h*
≤80	≤7-19	≤300	≤80	≤96	≤91	≤96
M Δ λ, deg	M Δ λ, deg	M Δ λ, deg	M Δ λ, deg	M Δ λ, deg	M Δ λ, deg	M Δ λ, deg
4-8-4.9 ≤ 1-14	5-6 ≤ 6-16	7-8≤20	4.8-4.9≤1-14	5-6≤1-14	6-7≤9	5-6≤1-14

Light Earthquakes

Over the earthquake source regions of earthquakes with magnitude M 4.8–4.9 at different latitudes, sometimes, but not always, disturbances (bulges) in the quasi-static electric field could be observed. Disturbances from light earthquakes with depth >117 km are not established because these seismic manifestations are at a relatively low energetic level and in a great depth. There are pre-effect disturbances of about 3–10 mV/m 12 hr before for all different latitudes and 80 hr after earthquakes for low and middle latitudes.

Moderate Earthquakes

Concerning Near Equatorial Latitudes

The anomalies in the quasi-static electric field at near-equatorial latitudes show mainly an increase in the vertical component E_z of about 5 mV/m during a period of 11–20 hr before moderate earthquakes in Northern and Southern Hemispheres. Similar effects are also observed in the magnetic conjugate region. Disturbances are observed for

$\Delta\lambda \leq 13°$. Over sources of moderate earthquakes (EQ_{29} and its aftershocks $EQ_{30,34,39}$) and light earthquake EQ_{37}, the electric field is perpendicular to the magnetic field line (or to the magnetic field) for orbits 250, 251, and 258. Ions and electrons are moving perpendicular to the plane determined by the electric and magnetic vectors. They cannot immediately compensate the charge that causes the electric field, so this field is expanded into a large area.

Concerning Low Latitudes

Numerous ionospheric anomalies in the quasi-static electric field at low latitudes indicate an increase in the vertical components E_z of about 10–8 mV/m (in the cases of $\Delta\lambda \leq$ 1–14°) to 10–2 mV/m (for $\Delta\lambda \leq$ 6–16°) around 13–25 hr to 7–19 hr respectively before and after moderate earthquakes which are related to seismic events in the Southern Hemisphere. There are not observations about analogous disturbances 100 hr before and 9 hr after moderate earthquakes in cases of seismic activity in the Northern Hemisphere.

Concerning High and Middle Latitudes

The observed anomalies in ionospheric zones, that have supposed relations with earthquakes in regions at high and middle latitudes of the Southern and Northern Hemisphere, are summarized as an increase in the vertical component E_z of about 10 mV/m (in the cases of $\Delta\lambda \leq$ 1–18°) to 2 mV/m (for $\Delta\lambda \leq$ 5–18°) around 80–96 hr before and after moderate seismic manifestations.

The recent results for high latitudes and the results from our previous chapter for Southern Ocean and Greenland Sea (Gousheva et al., 2008b) confirm the empirical models of Heppner (1977), Heppner and Maynard (1987), and the electric convection field model proposed by Heelis et al. (1982).

The obtained results strengthen our previous studies and conclusions for middle latitudes (Gousheva et al., 2008a). Often the intense disturbances in electric field components are especially observed at high latitudes under complicated conditions and these disturbances are marked usually in the open field lines that provoke rare penetration in geomagnetic mid-latitudes.

Strong and Major Earthquakes

Concerning Low Latitudes of the Southern Hemisphere

The observed anomalies in ionospheric zones, that have supposed relations with earthquakes in regions at low latitudes of the Southern Hemisphere, are summarized as an increase in the vertical component *Ez* of about 18 mV/m (for $\Delta\lambda \leq 1°$) to 12 mV/m (for $\Delta\lambda \leq 20°$) around 306–300 hr before and after major seismic manifestations.

Concerning Middle Latitudes of the Northern Hemisphere

The observed anomalies in ionospheric zones, that could be related to earthquakes in regions at middle latitudes of the Northern Hemisphere, are summarized as an increase in the vertical component E_z of about 15 mV/m (for $\Delta\lambda \leq 18°$) to 10 mV/m (for $\Delta\lambda \leq 9°$) around 153–91 hr before and after strong seismic manifestations.

The present statistical study of numerous ionospheric data permits to propose several conclusions:

- The study shows possible relations between the amplitude (or trend) of the disturbance zone, the magnitude M, depth D of the seismic event, and the distance from the satellite to the earthquake epicenter (r km and $\Delta\lambda$). Disturbances above 10 mV/m are only observed for strong and major earthquakes. No disturbances are observed for moderate earthquakes at $D > 234$ km and for light earthquakes at $D > 117$ km. Generally the disturbances are observed for $\Delta\lambda < 20°$.
- During the study two forms of quasi-static electric field disturbance zones in the upper ionosphere are recognized. The bulge with different amplitudes is observed in very narrow seismic belts where the earthquake manifestations are numerous. The wave discordance is established in regions with limited numbers of seismic events.
- The ionosphere disturbance zones are generated several days before the main shock. The ionosphere zones of electric field disturbances in cases of earthquakes in the Southern Hemisphere are shifted to the north from the earthquake epicenters. The same ionospheric zones (of electric field disturbances) related to earthquakes in the Northern Hemisphere take place to the south of the earthquake epicenters. This is connected with the electric field projection along magnetic field lines into the low ionosphere at satellite altitudes. The effects are also observed in the magnetic conjugate region as it has been already noted by Chmirev et al. (1989) and Gousheva et al. (2008b). Finally we suppose the presence of a source of quasi-static electric field of a seismic origin.
- At the moment of the main shock the disturbance zone is located above the epicenter and its amplitude increases.
- The disturbance zone changes its position with the time. It migrates back in l altitude several hours to 2–3 days later.
- The data about the seismic situation 15 days before and after the observations in regions with high seismic activity and 5 days before and after observations for regions of moderate and low seismicity gives us a possibility to separate fore- and after-shocks from the main shocks.
- Pre- and post-cumulative effects of several events are established very often. In several cases it is difficult to determine the predominant influence of one seismic phenomena.
- We found similar pre- and post effects during observations at magnetically very quiet, quiet, and medium quiet days for ionosphere zones of electric field disturbances in cases of earthquakes in the Northern and Southern Hemisphere.
- The ionospheric disturbance takes place above seismic active territories, mainly along the boundaries of the plates because these boundaries represent tectonic structures of the most considerable energy accumulation and liberation.
- The anomalous disturbance zone persists in a wide latitudinal interval and could represent one of numerous other indications for the activity of earthquake sources.

The obtained results about effects in the quasi-static field of the order 10 mV/m represent a confirmation of the new method for computation of the electric field in the atmosphere and the ionosphere over active faults proposed by Sorokin (2005, 2006).

The final results of the investigations indicate that the ionospheric anomalies, as phenomena accompanying the seismogenic processes, could be considered eventually as possible pre-, co-, and post-earthquake effects.

KEYWORDS

- **Indonesian region**
- **New Hebrides Trench**
- **North American Plate**
- **Quasi-static electric field**
- **United State Geological Survey**

Chapter 11

Ecosystem Changes in the Northern South China Sea

X. Ning, C. Lin, Q. Hao, C. Liu, F. Le, and J. Shi

INTRODUCTION

Physical and chemical oceanographic data were obtained by seasonal monitoring along a transect (Transect N) in the northern South China Sea (nSCS) during 1976–2004. Fluctuations of dissolved inorganic nitrogen (DIN), seawater temperature (SST and T_{av}–average temperature of the water column), N:P ratio and salinity (S_{av} and S_{200}—alinity at the 200 m layer) exhibited an increasing trend, while those of T_{200}, dissolved oxygen (DO), P, Si, Si:N, and SSS exhibited a decreasing trend. The annual rates of change in DIN, DO, T, and S revealed pronounced changes, and the climate trend coefficients, which was defined as the correlation coefficient between the time series of an environmental parameter and the nature number (namely 1,2,3, ...n), were 0.38–0.89 and significant ($p \leq 0.01$–0.05). Our results also showed that the ecosystem has obviously been influenced by the positive trends of both sea surface temperature (SST) and DIN, and negative trends of both DO and P. For example, before 1997, DIN concentrations in the upper layer were very low and N:P ratios were less than half of the Redfield ratio of 16, indicating potential N limitation. However after 1997, all Si:P ratios were >22 and the N_{av}:P_{av} was close to the Redfield ratio, indicating potential P limitation, and therefore N limitation has been reduced after 1997.

Ecological investigation shows that there have been some obvious responses of the ecosystems to the long-term environmental changes in the nSCS. Chlorophyll-a concentration, primary production (PP), phytoplankton abundance (PA), benthic biomass, cephalopod catch (CC), and demersal trawl catch (DTC) have increased. But phosphorus depletion in upper layer may be related to the shift in the dominant species from diatoms to dinoflagellates and cyanophytes. The ecosystem response was induced by not only anthropogenic activities, but also global climate change, for example ENSO. The effects of climate change on the nSCS were mainly through changes in the monsoon winds, and physical-biological oceanography coupling processes.

In this study physical–chemical parameters were systemic maintained, but the contemporaneous biological data were collected from various sources. Regional response to global climate change is clearly a complicated issue, which is far from well understood. This study was made an attempt to tackle this important issue. For the aim these data were valuable.

The South China Sea (SCS) is the largest semi-enclosed marginal sea in Southeast Asia with an area of about 3.5×10^6 km^2. Our study area is the nSCS, bounded by the mainland of China on the north and northwest sides, Taiwan Strait on the northeast, Taiwan Island and Bashi Strait on the east side, and the Hainan Island on the west side. The nSCS is connected to the East China Sea through Taiwan Strait, and it is connected to the open ocean through Luzon Strait, where a deep sill (>2,000 m) allows effective water exchange with the western Pacific. The topography of the area is characterized by the incline from the coast of mainland China towards the southeast, with a gradient from the coastal zone (<50 m), continental shelf (<200 m), the slope and open sea (>200 m), to the deep sea (>3,000 m) (Figure 1).

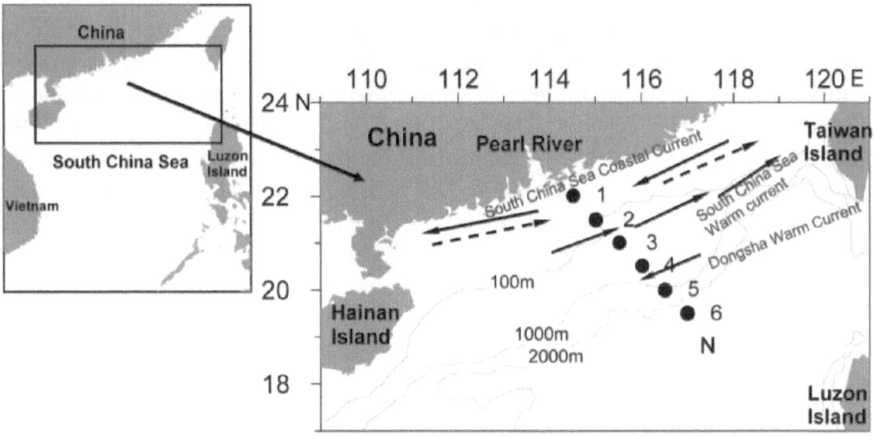

Figure 1. Geographical locations of the transect and stations and the circulation in the northern South China Sea (nSCS) (modified from Su, 1998; Xue et al., 2004). Transect N from the Pearl River Estuary towards the southeastern nSCS is the main transect with six stations (full circle).

The runoff from 29 rivers, with different sized input into the nSCS with total drainage area of 5.5×10^5 km^2, and an annual fresh water discharge of 3.8×10^{11} m (Han et al., 1998). Among them, the Pearl River is the largest with a drainage area of 4.3×10^5 km^2 and a discharge of 3.3×10^{11}my^{-1} (Han et al., 1998). It carries a large quantity of suspended solids (8.3×10^7 ty^{-1}, Han et al., 1998) and dissolved nutrients (N = 8.6×10^4 ty^{-1}; P = 1.2×10^4 ty^{-1}; Si = 184.3×10^4 ty^{-1}, before 1998, Wang and Peng, 1996; and N = 19.14×10^4 ty^{-1}, P = 0.8×10^4 ty^{-1} after 1998, SOAC, 2000, 2001a, 2002, 2003, 2004) into the nSCS. The Pearl River plume extends offshore to cover a large area of the nSCS (Yin et al., 2001). During the dry season in winter, the river plumes extend westward along the coast of Guangdong. Due to the strong northeast monsoon; during the flood season in summer, the river plume extends well into the nSCS, and its southeastward and southward tongue can reach up to 17°00'N, 112°E, about 5°, in latitude, away from the river mouth (Cai et al., 2007; Xue et al., 2001a, b).

The meteorological forcing over the nSCS is dominated by the East Asian Monsoon (Sadler et al., 1985). The upper ocean circulation follows closely the alternating monsoons (Wyrtki, 1961). During winter northeast monsoon, along the northern boundary, the warm and saline Kuroshio Current water with oligotrophic properties intrudes through Luzon Strait and flows westward along the continental margin of China to become the deep-water mass of the nSCS (Nitani, 1972; Shaw, 1991). The coastal water of the East China Sea flows southwestward through Taiwan Strait into the nSCS (Fang et al., 1998; Xue et al., 2004). On the contrary, during the summer southwest monsoon, the Guangdong Coastal Current flows eastward along the southern coast of mainland China, which eventually flows into the East China Sea through Taiwan Strait. The southwesterly winds also induce Ekman transport toward offshore and coastal upwelling. The deep water upwells and mixes with the upper water to form the SCS intermediate water, which flows out of the nSCS into the northwestern Pacific Ocean through Luzon Strait (Gong et al., 1992).

In the nSCS, the thermocline occurs all the year round, and the interannual change in its strength is pronounced (Shi et al., 2001; Yuan and Deng, 1997a, b). Previous studies have examined on variations in seawater temperature and salinity distributions (Yang and Liu, 1998; Yuan and Deng, 1998), DO distribution (Lin and Han, 1998), pollution status along the coast of the nSCS (Li and Chen, 1998) and the fisheries environment in the nSCS (Jia et al., 2005). Furthermore, it has been found that due to the combined effects of monsoons, topography, shape of the coastal line and the inertial effects, mesoscale eddies (Chen et al., 2005; Li et al., 2003; Xu et al., 2001; Zeng et al., 1989). Recent studies revealed that the effects of coupling between physical–chemical–biological oceanographic processes on phytoplankton biomass and production are important for understanding the influence on the long-term environmental changes and the ecosystem dynamics of the SCS (Liu et al., 2002, 2007; Ning et al., 2004).

However, the long-term changes in environmental conditions and the responses of the ecosystem in this region have not been well documented yet. The objective of this study was to analyze the 29 year time series of multidisciplinary observational data obtained during 1976–2004, aiming at understanding how the environment has changed and how the ecosystem and biological resources have responded to the environmental changes in the nSCS.

It must be pointed out that the data set we adopted in present analysis is large and from various sources (using data in this study were systemic maintained for the physical–chemical parameters, but for contemporaneous biological data were collected from various sources, and we have collected data from the same period of investigation as best as one can. Inevitably, there are some mismatches between the scales of physical and chemical parameters reflecting the processes of environmental change and the spatial and temporal dimensions of biological investigations, due to the less frequency for the latter). Although the quality of data might vary throughout the long period of observation, these data were valuable for the long-term changes in environmental conditions and the responses of the ecosystem in this region. Using these data, we can still find some disciplines about the response of the ecosystem to the environmental change in the SCS. Regional response to global climate change is clearly a complicated issue, which is far from well understood. This study made an attempt to tackle this important issue. And in order to improve our understanding, additional long-term study is mandatory.

MATERIALS AND METHODS

In this study, data were obtained from winter and summer monitoring along transect N (Figure 1, an observation transect, including six stations, crossing the nSCS, from the northwestern to southeastern), maintained by the survey team of the State Oceanic Administration (SOA), China during 1976–2004. These data include physical (seawater temperature (T) and salinity (S)) and chemical parameters (DO), phosphate (PO_4-P), silicate (SiO_3-Si), DIN (including NO_3-N, NO_2-N and NH_4-N)). The parameters of T, S, and DO data collection started from 1976; and the nutrients data (PO_4-P, SiO_3-Si, NO_3-N, NO_2-N, and NH_4-N) collection started from 1989. Seawater samples were collected using Nansen bottles from the surface, 5, 10, 15, 20, 25, 30, 35, 50, 75, 100, 150, and 200 m for T and S, and at the surface, 10, 20, 30, 50, 75, 100, 150, and 200 m for biogenic element determination. Seawater temperature was measured by using a reversing thermometer attached to the Nansen bottle, and salinity was measured using induction salinometer, according to SOAC (1975) and NBTS (1991). Nutrients (nitrate, phosphate, and silicate) were analyzed by standard spectrophotometric method, and DO was analyzed by the Winkler method (Strickland and Parsons, 1972). Photosynthetic pigments (Chl-*a*) were measured by the acetone extraction and fluorescence method (Holm-Hansen et al., 1965).

Annual mean values were the average for winter and summer which were derived from observations during February and August, respectively. The regional average was the average value for the all stations illustrated in Figure 1. First, we took the values at the sea surface (SS), the depth of 200 m and the average through the water column for 0–200 m (integrated) for each parameter for each station, since at the depth of 200 m concentrations of biogenic elements and other properties were relatively stable and much less influenced by the upper layers. Second, the regional means for each parameter on an annual scale were calculated. The average value for the water column was computed, according to the following equation:

$$X_{av} = \frac{1}{b}\int_0^b X(z)dz \tag{1}$$

Where X is an environmental parameter; b is the water depth (200 m, or 2 m above bottom if the water depth is shallower than 200 m) and z is the observation depth. In order to show the interannual changes in environmental parameters in the nSCS, the time series of various parameters was determined. The parameters include physical parameters, such as SST, T_{av}, T_{200} m, SSS, S_{av}, S_{200}, and chemical parameters, such as SSDO, DO_{av}, DO_{200}, SSP, P_{av}, P_{200}, SSSi, Si_{av}, Si_{200}, SSDIN, DIN_{av}, DIN_{200}, $SSNO_2$-N, NO_2N_{av}, NO_2-N_{200}, $SSNO_3$-N, NO_3-N_{av}, NO_3-N_{200}, $SSNH_4N$, NH_4-N_{av}, NH_4-N_{200}, and the ratios of the chemical parameters SSN:SSP, N_{av}:P_{av}, N_{200}:P_{200}, SSSi:SSN, Si_{av}:N_{av}, Si_{200}:N_{200}, where av = average for the whole water column and 200 = 200 m depth. Statistical test and linear regression analyses were conducted on time series (Chen and Ma, 1991) and climate trend coefficients (R_{xt}) were estimated. The R_{xt} was used to assess whether there was a significant linear climate-trend in a time series (Shi et al., 1995). This coefficient was defined as the correlation coefficient between the time series of an environmental parameter, $\{X_i\}$, and the nature number $\{i\}$, $i = 1, 2, 3..., n$. In this study, n is the total span of the years covered by the data. The coefficient was computed from the following equation:

$$R_{xt} = \frac{\sum_{i=1}^{n}(x_i - \bar{x})(i - \bar{t})}{\sqrt{\sum_{i=1}^{n}(x_i - \bar{x})^2 \sum_{i=1}^{n}(i - \bar{t})^2}} \tag{2}$$

where t = (n + 1)/2. Its significance level is determined from the Student t-test. A positive/negative value of R_{xt} indicates that the time series, $\{X_i\}$, has a linear positive/negative trend. In order to compare the environmental change rates between coastal/shelf waters with the slope/open sea, data for water depths <200 and >200 m, respectively, were analyzed.

Biological oceanography data, such as chlorophyll-a, PA, PP, zooplankton biomass (ZB), benthos biomass (BB), CC, etc. were obtained during the period by marine ecosystem surveys conducted by the South China Sea Fisheries Research Institute (SCSFRI), the SCS Institute of Oceanography, Chinese Academy of Sciences and the Second Institute of Oceanography, SOA. For the observational methods, Chl-a was determined by acetone extraction fluorescence method (Holm-Hansen et al., 1965) and often calibrated by spectrophotometry. Primary productivity was measured using the 14C tracer method established by Steemann-Nielsen (1952) and modified for scintillation counting by Wolfe and Schelske (1967). Phytoplankton samples were collected by vertical haul using a Judy net with a mesh size of 76 μm. The samples were preserved with Lugol's solution, and the species identification and cell counts were made using a microscope to get PA (Sournia, 1978). Zooplankton samples were collected by vertical haul using plankton net with a mesh size of 505 μm, and the samples were preserved with neutral formaldehyde solution (5%). The species identification and individual counts were made using a stereo microscope, and the wet weight biomass (ZB) was measured by an electronic balance after removing the body surface water in the lab, according to SOAC (1975) and NBTS (1991). Benthic macrofauna samples were collected by using a grab with a sampling area of 0.1 m^2, and the animals were sorted after removing the mud by elutriation. The species identification and individual counts were made using a stereo microscope, and the wet weight biomass (BB) was measured gravimetrically after removing the body surface water in the lab (NBTS, 1991; SOAC, 1975). The nekton samples were collected by using a cystoids net with a mesh size of 20 mm, towed by a pair of boats at a speed of 3–4 km for 1 hr at each station (NBTS, 1991; SOAC, 1975).

DISCUSSION

Increasing Trends and the Response of the Ecosystem

The positive increasing trends in SST and T_{av} in the nSCS during 1976–2004 are consistent with the increasing trends in the mean air temperature (AT) observed throughout the Northern Hemisphere (Fu et al., 2006; Houghton et al., 1996), South China (Chen et al., 1998, 1999; Zhai and Ren, 1997) and the annual means of AT and SST observed along the coast of the SCS (He et al., 2003; Martin and Arun, 2003). The increasing trends were also in phase with the changes in SST observed along the coast of the Yellow Sea and Bohai Sea (Lin et al., 2001, 2005). However, these annual rates of water temperature change were higher in the nSCS than in the Yellow Sea and the Bohai Sea (Lin et al., 2001, 2005).

The increase in DIN in the nSCS during 1989–2004 was consistent with the increase in DIN along the coast of the nSCS, such as Qinzhou Bay (Wei et al., 2002, 2003) and Daya Bay (Qiu et al., 2005). It also shows the same trend with the rise of DIN observed throughout the global marginal seas (Seitzinger et al., 2002). Along with the rapid economic development in China, DIN concentration in the Pearl River estuary and shelf of the China Sea has been dramatically increased, due to the increasing urbanization near the coastal areas, which resulted in more municipal sewage, agricultural fertilizer, mariculture waste, and so forth inputs (SOAC, 2001b). Through analysis based on DIN and PN models, combining with spatially explicit global databases, Seitzinger et al. (2002) showed that DIN input rates increased from approximately 21 Tg N y^{-1} in 1990 to 47Tg N y^{-1} by 2050. The largest increases are predicted for Southern and Eastern Asia, associated with predicted large increases in population, will increase fertilizer use, and increased industrialization. The DIN from the Pearl River discharge increased by three times in 2002 as compared to 1986 (He et al., 2004, Table 1), and NO_3-N input from the Pearl in River Estuary was 1.7 times in 1999 of that in 1987 (Guan et al., 2003; Wang and Peng, 1996, Table 1). The SSDIN should be influenced by the increase in DIN from the river discharges. In addition, significant inputs of DIN into the nSCS have also occurred through atmospheric dry and wet deposition (Zhang et al., 1999) and the upwelling of the deep waters (Zhao et al., 2005). The mitigation of N limitation in the upper layer since 1998 was clearly related to these DIN inputs (Figure 3). The increase in the annual rate of DIN was higher at the 200 m layer (DIN_{200}) than the water column average (DIN_{av}) and at the sea surface layer (SSDIN). This may be due to possible strengthening of the deep water upwelling. However, during 1989–1997, the DIN_{av} was low, the multi-annual mean was 4.2 μmol l^{-1} and the lowest value of DIN_{av} was only 1.7 μmol l^{-1} in 1989. Since 1998, the multi-annual mean of the water column average DIN concentration has reached 7.7 μmol l^{-1} and the maximal value of the annual mean even reached up to 10.8 μmol l^{-1} in 2001 (Figure 2b). And therefore, the previous status of N limitation in the nSCS was significantly alleviated. Moreover, the increases in the annual rates of DIN and NO_3-N were much higher in the nSCS than in the Yellow Sea (Lin et al., 2005).

The increase in the N:P ratio was due to an increase in DIN and a decrease in P concentration. Before 1997, the N:P ratios (SSN:SSP, DIN_{av}:P_{av}, and DIN_{200}:P_{200}) were lower than 10. Since 1998, these ratios have rapidly increased to 28, 18, and 16 in 2004, respectively (Figure 4). In 2004, the average values of DIN_{av}:P_{av} and DIN_{200}:P_{200} were close to the Redfield ratio (16:1), and therefore favorable to phytoplankton growth (Hu et al., 1989; Jiang and Yao, 1999; Richardson, 1997). The high value of SSN:SSP (28) in 2004 was probably due to dry and wet deposition. The peak value of SSN:SSP appeared in 2002 (up to 86, Figure 3a), due mainly to high rainfall as high Pearl River discharge, and this high ratio corresponded to the lowest surface salinity (Figure 3a). Moreover, the fact that the annual rates of DIN and DIN:P were higher in the shallow water area (<200 m) than in the deep water area (>200 m) (Table 2) suggested the influence of the anthropogenic activities on the ecosystems of the shallow or coastal waters. These results agree well with Xu et al. (2008) who also showed that

N input from the Pearl River has caused the nSCS to become P-limited. Phosphate decline may be ascribed to the implementation of phosphate detergent ban in the late 1990s, and phytoplankton great consumption.

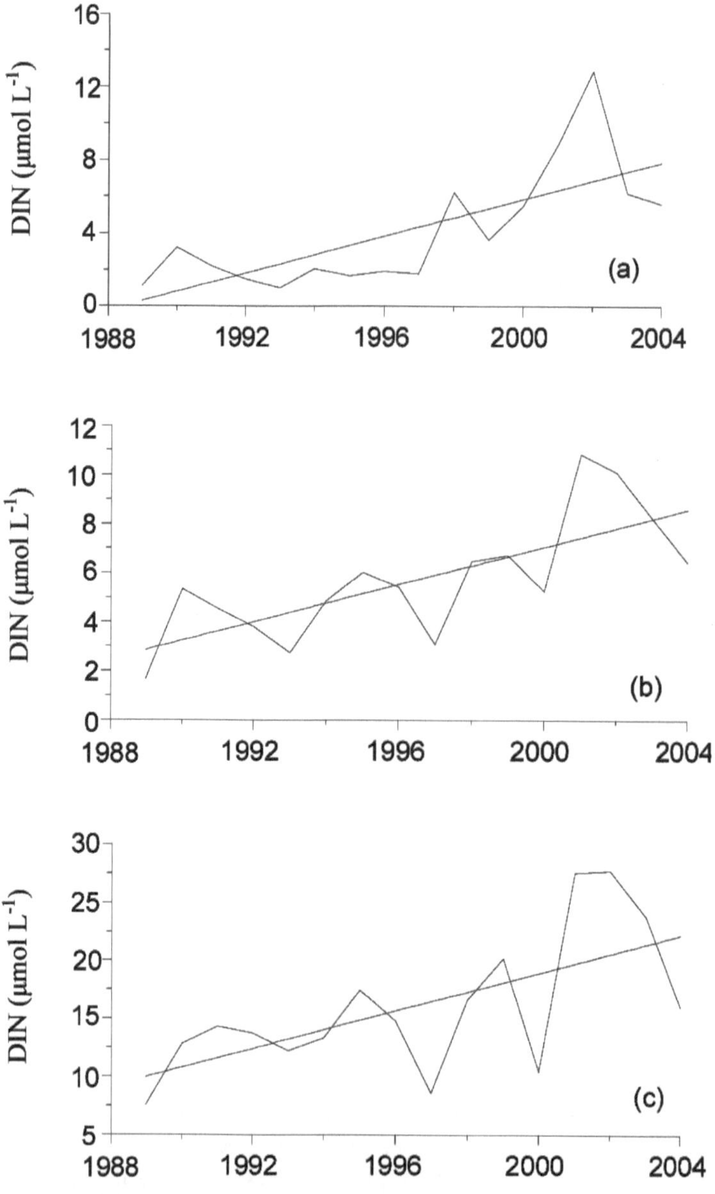

Figure 2. Variation trends in dissolved oxygen (DO) concentration in the nSCS during 1976–2004. (a), (b), and (c) show the annual means of sea surface DO (SSDO), water column average in the upper 200 m DO (DOav) and DO at the 200 m (DO200), respectively. The lines are linear regressions.

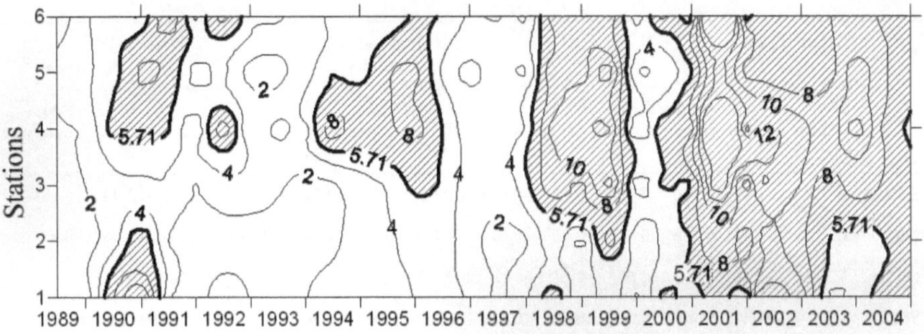

Figure 3. Variation trends in P concentrations in the nSCS during 1989–2004. (a), (b), and (c) show the annual means of sea surface P (SSP), water column average P in the upper 200 m (Pav) and P in the 200 m layer (P200), respectively. The lines are linear regressions.

Table 1. Changes in annually average concentration (μmol dm^{-3}) of NO_3-N and dissolved inorganic nitrogen (DIN), from discharges of the Pearl River and other rivers.

	Annual rate			R^a_{xt}			Amplitude of ftuctuationb			Mean± SD^c		
	Mean <200m	Mean >200m	Mean whole area	Mean <200m	Mean >200m	Mean whole area	Mean <200m	Mean >200m	Mean whole area	Mean <200m	Mean >200m	Mean whole area
SST	0.078	0.063	0.078	0.84^A	0.80^A	0.89^A	2.67	3.28	2.97	24.75±1.82	26.06±0.67	25.63±0.74
Tav	0.089	0.095	0.090	0.80^A	0.85^A	0.85^A	4.07	4.41	3.86	23.49± 0.97	22.42±0.95	22.29±0.89
T^d_{200}	0.045	-0.107		0.45^B	0.71^A		2.90	3.94		20.36±0.86	15.41±1.30	
SSS	- 0.007	-0.015	-0.022	- 0.19	- 0.35	-0.41^C	3.61	2.30	2.20	33.10±0.78	34.01±0.40	33.55±0.48
S_{av}	0.008	0.005	0.007	0.32	0.34	0.38^C	0.97	0.71	0.74	33.89±0.30	34.22± 0.19	34.05±0.21
S^d_{200}	0.005	0.005		0.30	0.44^B		0.42	0.48		34.34± 0.20	34.59± 0.10	
SSDO	-0.964	-0.927	-0.913	-0.46^B	-0.52^A	-0.48^A	80	85	83	426.5± 76.0	425.4±15.0	434.6±15.9
DO_{av}	-1.857	-2.037	-1.926	-0.67^A	-0.82^A	-0.81^A	102	92	94	422.6±33.0	384.9±21.0	402.3±20.3
DO^d_{200}	-2.230	-1.653		-0.70^A	-0.63^A		99	104		376.3±28.0	317.2±22.4	
SSP	- 0.011	- 0.004	-0.008	- 0.41	- 0.30	- 0.31	0.54	0.76	0.45	0.31±0.14	0.32±0.11	0.32±0.12
P_{av}	-0.008	-0.007	-0.009	-0.29	-0.35	-0.37	0.32	0.63	0.50	0.39±0.12	0.69±0.17	0.52± 0.13
P^d_{200}	- 0.006	- 0.017		- 0.22	-0.50^D		0.48	0.78		0.57±0.17	1.19±0.21	
SSSi	-0.642	-0.284	-0.071	-0.19	-0.30	-0.22	12.19	12.90	11.50	9.93±3.46	10.33± 4.81	10.20±3.71
Si_{av}	-0.284	-0.240	-0.161	- 0.24	- 0.19	- 0.16	17.84	21.46	17.86	11.51±6.05	15.15±7.40	12.64±7.17
Si^d_{200}	-0.038	-0.387		-0.06	-0.19		17.57	32.50		14.98±5.16	26.80±9.53	
SSDIN	0.613	0.391	0.504	0.70^A	0.63^A	0.74^A	17.27	9.69	11.85	4.44± 4.24	3.73± 3.00	4.09±3.29
DIN_{av}	0.416	0.343	0.381	0.75^A	0.61^A	0.73^A	9.68	10.53	10.53	4.71±2.63	6.33±2.65	6.71±2.60
DIN^d_{200}	0.721	0.813		0.73^A	0.64^A		15.71	20.13		7.61±4.68	16.08± 6.03	

Table 1. *(Continued)*

	Annual rate			R^{a}_{xt}			Amplitude of ftuctuationb			Mean± SD[c]		
	Mean <200m	Mean >200m	Mean whole area	Mean <200m	Mean >200m	Mean whole area	Mean <200m	Mean >200m	Mean whole area	Mean <200m	Mean >200m	Mean whole area
SSN03	0.457	0.342	0.456	0.70^{A}	0.54^{D}	0.66^{A}	12.05	9.56	9.74	2.62±3.09	2.45±2.99	2.55±2.85
NO_{3av}	0.345	0.324	0.335	0.71^{A}	0.58^{B}	0.67^{A}	8.46	9.85	8.13	2.98±2.31	5.09±2.68	4.01±2.37
NO^{d}_{3200}	0.582	0.711		0.68^{A}	0.55^{C}		13.57	20.55		5.57±4.05	14.53± 6.17	
SSDIN: SSP	3.61	1.89	2.75	0.55^{C}	0.54^{D}	0.63^{A}	126.55	52.65	84.02	20.5± 30.7	14.9±16.8	17.8±21.2
DIN_{av} : P_{av}	1.66	0.75	1.21	0.70^{A}	0.57^{B}	0.69^{A}	40.26	23.24	28.65	14.0± 11.3	10.7± 6.3	12.3±8.4
DIN_{200}: Pd_{200}	1.51	1.04		0.74^{A}	0.64^{A}		31.26	27.3		13.8±8.6	14.0±7.8	
SSSi: SSDIN	-0.52	-0.46	-0.44	-0.75^{A}	-0.39	-0.67^{A}	12.32	22.68	9.69	4.1±3.3	5.4±5.7	4.3±3.2
Si_{av}: DIN_{av}	-0.46	-0.15	-0.22	-0.71^{A}	- 0.46	-0.70^{A}	13.05	5.62	5.41	3.4±3.1	2.9±1.6	2.8±1.5
Si_{200}: DIN^{d}_{200}	-0.32	-0.10		-0.75^{A}	-0.65^{A}		7.29	2.8		2.87±2.02	1.85±0.8	

Table 2. The annual rate, climate trend coefficients (R_{xt}) and the amplitude of fluctuations of the environmental parameters in the nSCS during 1976–2004 (units: °Cy–1 for annual change in temperature, and μmol $l^{-1}y^{-1}$ for DO and nutrients).

Observation year	Location of the observation	Parameter, value	Data source
1980-1985	southern Hong Kong waters (outside Pearl River estuary)	$N0_3$-N, 1-3	Han et al. (1990)
1986	Pearl River estuary	DIN, 19.3	He et al. (2004)
1987	Pearl River estuary	$N0_3$-N, 28.0 DIN, 31.7	Wang and Peng (1996)
1990	Pearl River estuary	DIN, 34.6	He etal. (2004)
1995	Pearl River estuary	DIN, 36.4	He etal. (2004)
1996	Pearl River estuary	$N0_3$-N, 39.8	Cai (2002b)
1999 (summer)	Pearl River estuary	$N0_3$-N, 48.9 DIN, 51.9	Lin et al. (2004)
1999 (summer)	Pearl River estuary	DIN, 53.2	Guan et al. (2003)
2002	Pearl River estuary	DIN, 76.4	He et al. (2004)

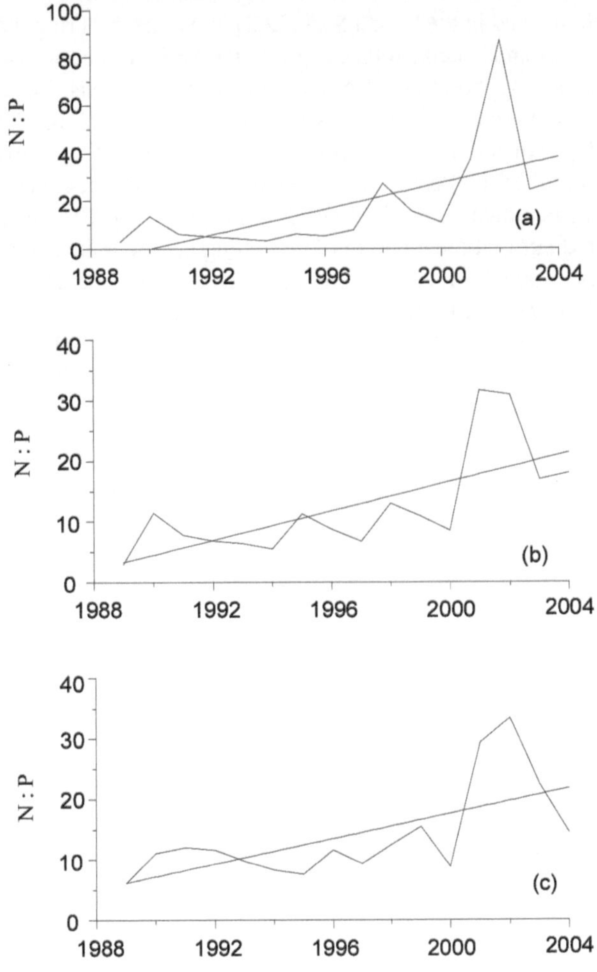

Figure 4. Variation trends in Si concentrations in the nSCS during 1989–2004. (a), (b), and (c) show the annual means of sea surface Si (SSSi), water column average Si in the upper 200 m (Siav) and Si in the 200 m layer (Si200), respectively. The lines are linear regressions.

Decreasing Trends and the Response of the Ecosystem

The decreasing trends in SSS may be related to the increase in the freshwater discharge into the nSCS since 1990 and less vertical mixing, because of the presence of a permanent thermocline (Shi et al., 2001; Yuan and Deng, 1997a, b). It has previously been observed that the occurrence of low SSS often corresponded to abnormally high discharge of the Pearl River (Xie and Zhang, 2003; Xu et al., 2008). During the period of 1990–2000, the Pearl River runoff increased by 22.5% in comparison with the mean discharge for 1960–1999 (Lei et al., 2003). In the 1990s (i.e., 1994, 1995, 1996, 1997, 1999) and 2002, SSS in the nSCS was especially low (Figure 5a), which was induced by the Pearl River floods (Lei et al., 2003). Particularly in 2002, the strongest typhoon rain storm resulted in the historically greatest flood of the Pearl River, and therefore, which induced in the lowest SSS (32.2) in the nSCS during that study period (Lei et al., 2003; Xie and Zhang, 2003). In 1999, the SSS was below normal due to the influence of frequent typhoon rain storms (total of 28 typhoons, including 7 landed ones). Furthermore, in the summer of 1999, the Dongsha upwelling (Station 3–6) was weak and the depth of the 20°C isopleth was deeper than 105 m. However in 1998, although large floods of the Pearl River occurred, the SSS did not decrease, due to the strong upwelling as indicated by shallow (84 m) depth of the 20°C isopleths. Chai et al. (2001) pointed out that the strength of upwelling can be indicated by the depth of 20°C temperature isopleth in the SCS. In this study, the range depth of 20°C isopleths is 84–150 m. However, the high SSS in 1993 was probably due to the reduced Pearl River discharge (Lei et al., 2003).

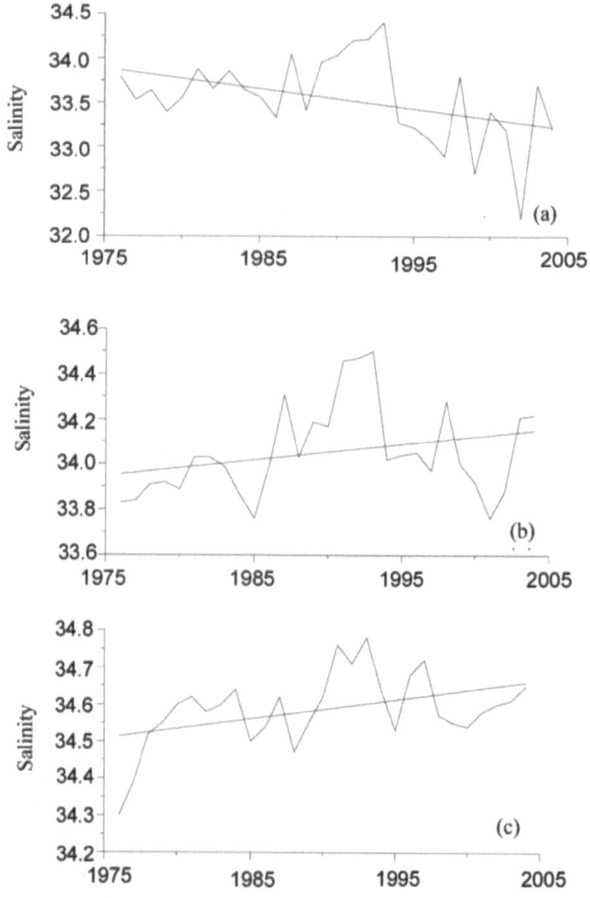

Figure 5. The spatial-temporal distributions of the DINav (mean concentrations in the water column). The shadow shows the area and period of time which exceeded the low limit of suitable N concentration for diatom growth (5.71 μmol l–1, Chu, 1949).

The decreasing trend in DO concentration (Figure 7) can probably be attributed to three reasons. First, a small decrease in DO solubility in the seawater can be induced by the increase in the seawater temperature (Figure 6). Second, an increase in DO consumption was an important reason. The DO consumption resulted from the decomposition of organic matter originating mainly from the Pearl River discharge and the decay of phytoplankton blooms with increasing frequencies in the coastal water was increasing (Peng, 1994; Tang et al., 2006). Before 1998, HABs (harmful algae blooms) occurred once or twice a year, but during 1998–2003, blooms increased to 10–20 a year in the nSCS (Tang et al., 2006), which was in phase with the dramatic decreasing trend in DO (Figure 7). Third, the mixing between the surface water and the deep layers was reduced by the stronger thermocline, due to the rapid rise in SST since 1995 (Figure 6a), which resulted in less transfer of oxygen from the atmosphere to the deeper waters. In addition in 1998, the lowest value of DO was probably attributed to the strongest upwelling, which occurred that year. According to the compute of this study, in 1998 the depth of the 20°C isopleths was 84 m, it show the strongest upwelling (see above paragraph and references in it).

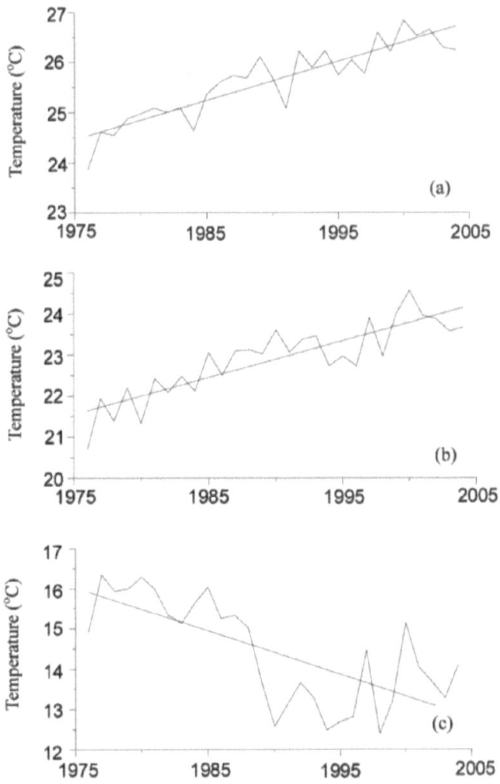

Figure 6. Variation trends in dissolved inorganic nitrogen (DIN) concentrations in the nSCS during 1989–2004. (a), (b), and (c) show the annual means of sea surface DIN (SSDIN), water column average DIN in the upper 200 m (DINav) and DIN at the 200 m layer (DIN200), respectively. The lines are linear regressions.

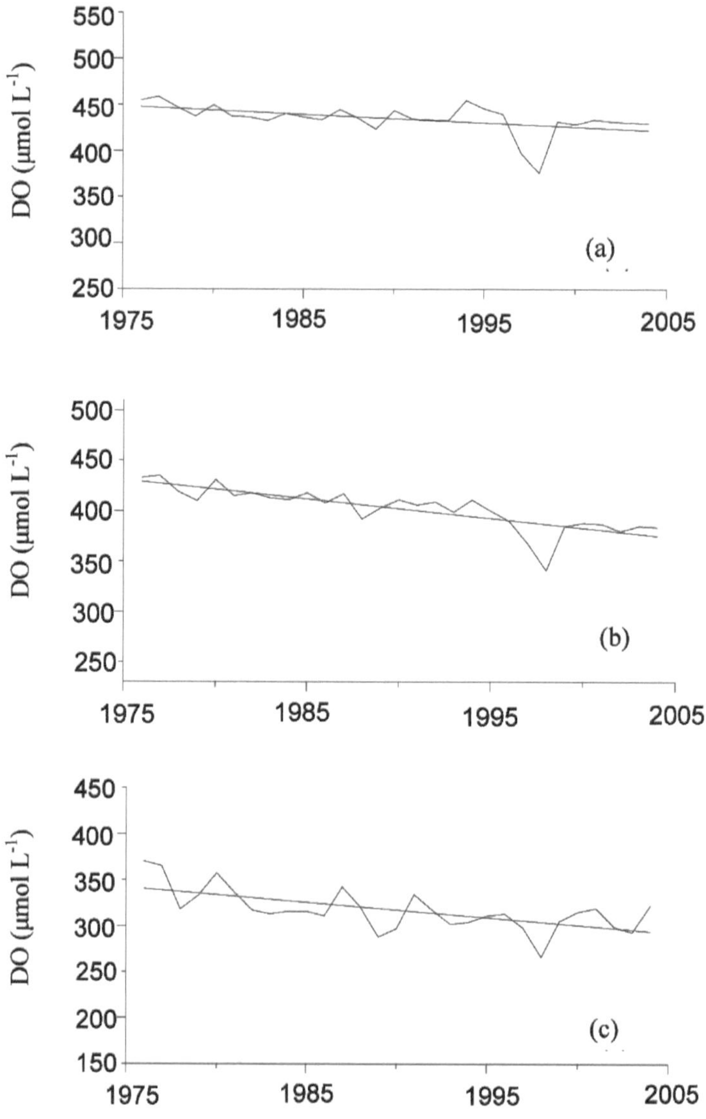

Figure 7. Variation trends in N:P ratios in the nSCS during 1989–2004. (a), (b), and (c) show the annual means of sea surface DIN:P (SSDIN:SSP), water column average DIN:P in the upper 200 m (DINav:Pav) and DIN:P at the 200 m layer (DIN200:P200), respectively. The lines are linear regressions.

The decrease in P concentration is probably due to uptake by phytoplankton and less P supply from deep water, due to the presence of the permanent thermocline (Shi et al., 2001; Yuan and Deng, 1997a, b). Furthermore, since 1998, the reduction in N limitation has increased the phytoplankton biomass and production (Table 3) and decreased the P concentration. Hong et al. (1983) reported that diatoms can take up 30 times more P than they need and store it for use when P is deficient. In the most years except for 1994 and 2000, the concentrations of P in the upper layer (<75 m) were even lower than the P concentration required for diatom growth (<0.1 μmol l^{-1}, Zou et al., 1983). The decreasing trends in Si concentration have probably been influenced by the decrease in Si concentration in the Pearl River runoff since the 1970s (Lei et al., 2003), but still adequate for diatom growth. At 200 m, the interannual fluctuations of both P and Si concentrations were high (Figures 8 and 9) and probably attributed to the interannual changes in upwelling of deep water. In 1996 and 1999 the depth of the 20° isopleth was shallow (average value from multi-stations was 93 m), high concentrations of P and Si occurred (Pav = 0.61 and 0.62 μmol l^{-1}, P_{200} = 1.154 and 1.31 μmol l^{-1}, Si_{av} = 25.18, and 19.55 μmol l^{-1}, and Si_{200} = 42.12 and 44.28 μmol l^{-1}, respectively, in the 2 years). In contrast, during 2004 when the depth of the 20° isopleths was deep ((the average value from multi-stations was 105 m), concentrations of P (P_{av} = 0.37 μmol l^{-1}, and P_{200} = 1.13 μmol l^{-1}), and Si (Si_{av} = 8.9 μmol l^{-1}, and Si_{200} = 20.5 μmol l^{-1}) were obviously low.

Table 3. Comparison of key nutrient concentration (μmol l^{-1}) and the ratio between the two phases (before 1997, the first phase and after 1997, the second phase).

Nutrient and ratio	mean in the mean first phase	mean in the second phase
SSDIN	1.83	5.75
DIN_{av}	4.22	7.78
DIN_{200}	12.77	20.33
SSP	0.33	0.25
P_{av}	0.56	0.48
P_{200}	0.49	0.48
SSSi	12.38	9.19
Si_{av}	14.82	12.26
Si_{200}	27.54	25.85
SSDIN:SSP	5.5	23.0
DIN_{av}:P_{av}	7.5	16.2
DIN_{200}:P_{200}	26.1	42.4
SSSi:SSDIN	6.3	1.6
Si_{av}:DIN_{av}	2.5	1.6
Si_{200}:DIN_{200}	2.2	1.3
SSSi:SSP	37.5	36.8
Si_{av}:P_{av}	26.5	25.5
Si_{200}:P_{200}	56.2	53.8

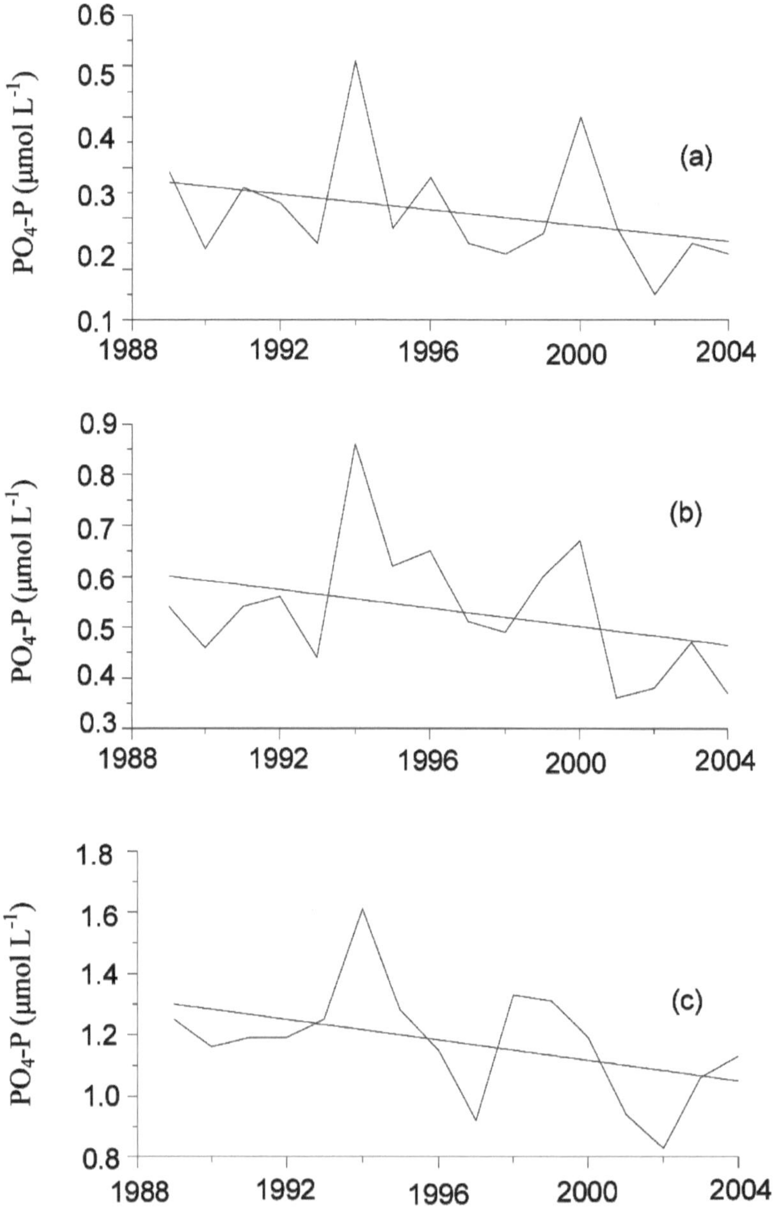

Figure 8. Variation trends in seawater salinity in the nSCS during 1976–2004. (a), (b), and (c) show the annual means of sea surface salinity (SSS), water column average salinity in the upper 200 m (Sav) and salinity at 200 m (S200), respectively. The lines are linear regressions.

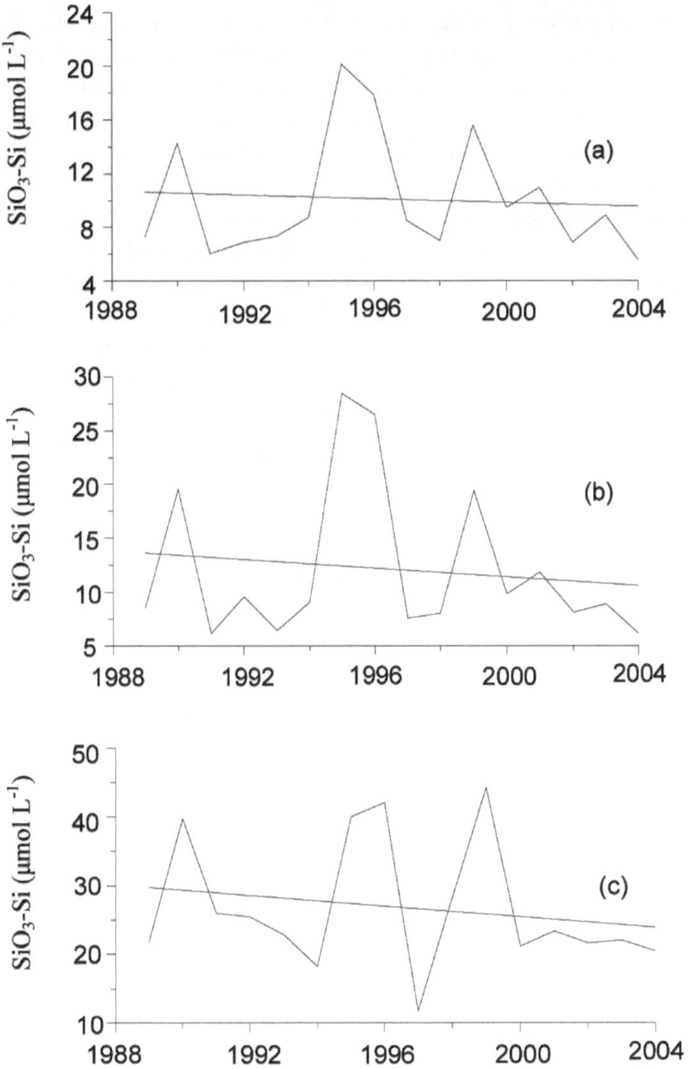

Figure 9. Variation trends of seawater temperature in the nSCS during 1976–2004. (a), (b), and (c) show the annual mean of the sea surface temperature (SST), the water column average temperature in the upper 200 m (Tav), and the temperature at the depth of 200 m (T200). The lines are the linear regressions.

Nutrient Limitation

Since 1998, DIN concentrations have exhibited a pronounced increasing trend, and therefore the key nutrient concentration and ratio can be divided into two phases, that is. before and after 1997 (Table 3). In the first phase, the average DIN concentration was very low, SSDIN and DIN_{av} were 1.83 and 4.22 μmol l^{-1}, respectively, which were lower than the low limit of suitable N concentration for diatom growth (5.71 μmol l^{-1},

Chu, 1949). The average P concentration was also low. The SSP and P_{av} were 0.33 and 0.56 μmol l^{-1}, respectively, which is closed to the low limit of suitable P concentration for diatom growth (0.48 μmol l^{-1}, Zhao et al., 2000). However, Si concentrations (SSSi and Si_{av} were 12.38 and 14.82 μmol l^{-1}, respectively) outclassed the low limit of suitable for diatom growth (4.40 μmol l^{-1}, Harvey, 1957). Therefore, Si pool was sufficient. N:P ratios were less than half the Redfield ratio of 16 (SSN:SSP and N_{av}:P_{av} were 5.5 and 7.5, respectively).

In contrast, for the second phase after 1997, the average DIN concentration has been clearly increasing (SSDIN and DIN_{av} were 5.75 and 7.78 μmol l^{-1} respectively, Table 3), and exceeded the low limit of suitable N concentration for diatom growth (5.71 μmol l^{-1}, Chu, 1949). However, the average P concentration has decreased (SSP and P_{av}, were 0.25 and 0.48 μmol l^{-1}, respectively, Table 3). Consequently, there was a rapid increase in DIN:P ratios (SSDIN:SSP and DIN_{av}:P_{av} were 23.0 and 16.2, respectively, Table 3), which were close to the Redfield ratio (16, Richardson, 1997; Hutchins et al., 1998). Therefore, it is favorable to phytoplankton growth in the second phase. The Si concentration also decreased (SSSi and Si_{av} were 9.19 and 12.26) in this phase, resulting in rapid decrease in the ratio of SSSi:SSDIN (Si_{av}:DIN_{av}), from 6.3 (2.5) to 1.6 (1.6) (Table 3). The Si concentration decrease may be due to the dam of the river courses which significantly reduced the silica delivery to the SCS.

Based on the studies on the kinetics of nutrient uptake, the thresholds of SiO_3-Si, DIN and PO_4-P for phytoplankton growth have been estimated to be 2.0, 1.0, and 0.1 μmol l^{-1}, respectively (Justic et al., 1995). In the study area, the values of all the nutrient parameters were over these threshold concentrations, except for those in the first phase, when the mean of SSDIN was close to the threshold of N (Table 3). According to chemical stoichiometry, in the first phase both SSN:SSP and DIN_{av}:P_{av} was lower than 10, and SSSi:SSDIN, Si_{av}:DIN_{av}, and Si_{200}:DIN_{200} were all over 1, indicating potential N limitation. In the second phase SSDIN:SSP and DIN_{200}:P_{200} were higher than 22, and DIN_{av}:P_{av} was equal to the Redfield ratio. All Si:DIN ratios ranged from 1.3–1.6, and Si:P ranged 25.5–53.8 (Table 3), which indicated that the potential of P limitation increased and N limitation decreased. Furthermore, Si was always sufficient during the observation period, even in the second phase when its concentration decreased (Table 3).

Response of Ecological Environment to ENSO Events

When El Nino occurs, the warm pool of the western Pacific Ocean moves eastward, whereas it moves westward during La Nina (Takeuchi, 1987; White et al., 1985; Zhang and Huang, 1993). The nSCS is located to the west of the warm pool however the response of the nSCS has not been well documented. In the present study, pronounced responses to ENSO were found. During the observation period, nine El Nino events (1976, 1982–1983, 1986–1987, 1991, 1993, 1994, 1997, 2002, and 2004) and four La Nina events (1981, 1988, 1995, and 1998–1999) occurred (Levimson, 2005; Mcphaden, 2004; Qin, 2003; Wang and Gong, 1999). In general, whenever El Nino/La Nina occurred, SST and Tav was low/high in the nSCS (Figure 6, Table 4).

In the area southwest of Dongsha Islands, near Station 4 (19–20°N, 116–117°E, Figure 1), pronounced responses to ENSO were observed. In general, whenever an El

Nino event occurred, T_{av} and DO were lower, and that the S_{av}, nutrients (PO_4–P_{av}, SiO_3–Si_{av}, and DIN), SS Chl-a were higher (Table 5, Figure 10). During a La Nina event, contrary, T_{av} and DO were higher, and that the S_{av}, nutrients (PO_4–P_{av}, SiO_3–Si_{av}, and DIN), SS Chl-a were lower (Table 5, Figure 10). The fluctuations of environmental parameters evident corresponded to El Nino/La Nina events. Furthermore, in comparison of the average values of the environmental parameters in summer of El Nino years with those of La Nina years, T_{av} and DO_{av} were lower by 1.89°C and 20.2 μmol l^{-1}, respectively; while S_{av} was higher by 0.31 psu; PO_4–P_{av}, SiO_3–Si_{av}, and DIN_{av} were higher by 0.15, 6.41, and 3.42 μmol l^{-1}, respectively. Surface Chl-a concentration was higher by 0.14 mg m^{-3}, that is higher by 1.8 times (Table 5), even higher by 0.83 times than the average value of normal years (1980, 1990, 2000, 2001, and 2003, for which mean Chl-a = 0.12 ± 0.05 mg m−3).

Table 4. Annual mean of SST and T_{av} along transect N in the nSCS in El Nino and La Nina years during 1976–2004 (unit: C).

El Nino Year	1976	1982 (1983)	1986 (1987)	1991	1993	1994	1997	2002	2004	Mean±SD[a]
SST	23.87	25.82 (25.09)	25.60 (25 .73)	25.08	25.89	26.24	25.77	26.66	26.25	25.64±0.75
T_{av}	21.13	21.40 (23.40)	23.09 (23.32)	21.46	21.65	21.79	22.25	22.98	22.62	22.27±0.83
La Nina Year		1981	1988		1995	1998	1999			
SST		25.09	25.68		25.74	26.61	26.21			25.87±0.58
T_{av}		23.28	23.72		24.00	24.07	24.51			23.92±0.46

Table 5. The mean of the ecological parameter for the water column (0–200 m) and sea surface Chl-a concentration at Station 4 in summer of El Nino/La Nina years during the study period (units: °C for temperature, psu for salinity, μmol l^{-1} for DO and nutrients and mg m^{-3} for Chl-a).

El Nino Year	1976	1982 (1983)	1986 (1987)	1991	1993	1994	1997	2002	2004	Mean±SD[a]
T_{av}	21.13	21.40 (23.40)	23.09 (23.32)	21.46	21.65	21.79	22.25	22.98	22.62	22.26±0.80
S_{av}	34.18	34.14 (34.16)	34.13 (34.2)	34.45	34.82	34.41	34.19	33.85	34.17	34.25±0.25
DO_{av}	365.6	380.1 (387.3)	355.1 (364.9)	379.2	337.3	368.5	339.0	354.2	357.8	362.6±16.0
$P0_4$-P_{av}				0.23	0.71	0.36	0.56	0.44	0.56	0.48±0.17
$Si0_3$-Si_{av}				25.36	16.22	13.38	8.96	11 .65	10.56	14.36±5.94
DIN_{av}				3.20	3.85	6.19	4.29	12.51	8.12	6.36±3.50
Chl-a[b]		0.31						0.17	0.18	0.22±0.08
La Nina year	1981	1984	1988		1995	1998	1999			
T_{av}	23.28	25.30	23.72		24.00	24.07	24.51			24.15±0.70
S_{av}	33.89	34.06	34.01		33.60	34.18	33.90			33.94±0.20
DO_{av}	391.7	431 .0	388.2		380.8	308.7	396.2			382.8±40.3
$P0_4$-P_{av}					0.20	0.23	0.56			0.33±0.20
$Si0_3$-Si_{av}					10.00	5.20	22.32			7.95±12.45
DIN_{av}					3.50	3.00	2.31			2.94±5.98
Chl-a[b]		0.08,0.06				0.08	0.09			0.08±0.01

[a] Mean±SD is the multi-year mean value ± standard deviation for the El Nino/La Nina years.
[b] The Chi-a data were derived from SeaWiFS during 1998-2004, from Nimbus-7 CZCS in 1980 and 1984, provided by X. Chen, from The Third Institute of Oceanography, SOA, China for 1982, Fan (1985) and Huang (1992) for 1990, respectively.

According to Takano et al. (1998) and Liao et al. (2006), there is a cyclonic eddy in the sea area around Station 4 (near Dongshan Islands) in summer. Whenever a medium and strong El Nino occurs, the summer monsoon is weak (Zhang et al., 2003; Zhu et al., 2000), which induces strengthened cyclonic eddy, leading to strong upwelling, resulting in low T_{av} and DO_{av}, and high S_{av}, nutrients and Chl-a induced by phytoplankton growth. During a La Nina event, the opposite occurs. This is due to ENSO/La Nina events affecting the strength of the summer monsoon related, that is both El Nino and La Nina make the anomaly of Walker circulation. During El Nino event, the heat convection of the western Pacific warm pool moves to the central Pacific. In contrast, during La Nina event, the heat convection of the western Pacific warm pool moves back to the western Pacific. That makes the anomaly of Walk circumfluence. Namely, when phase of El Nino (La Nina) fall under the influence of the anomal Walker circulation, subsidence (ascending) air current occur over low latitude and middle latitude of the east Asia. If this subsidence (ascending) air current flow over low latitude occurred in summer, and joined to west-south monsoon, it will result in weaken (strengthen) of the summer monsoon (Wang et al., 2001; Zhang et al., 2003; Zhu et al., 2000). The changes of ecological environment in the sea area around Station 4 in summer respond to ENSO events, namely, respond to the abnormity of summer monsoon.

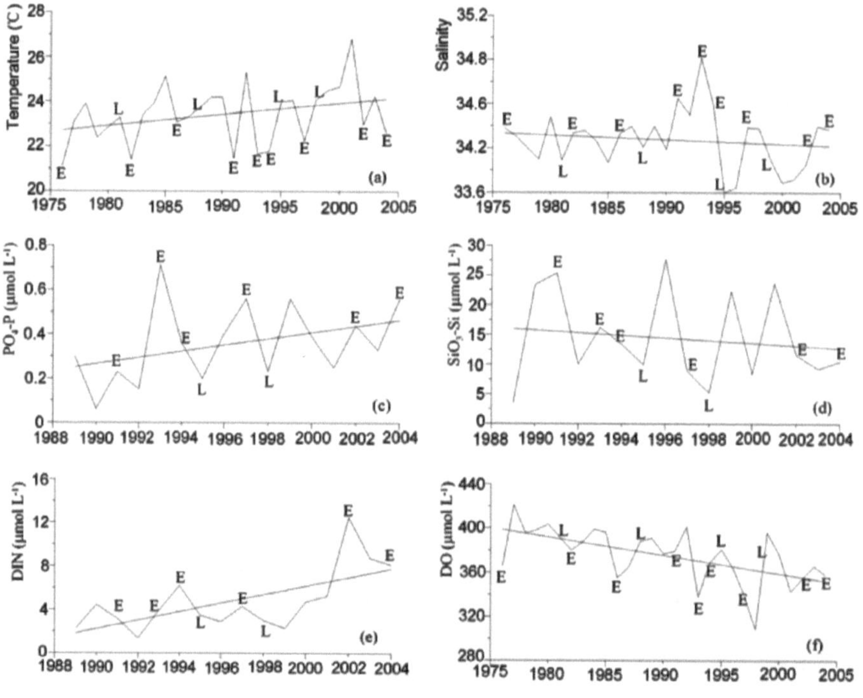

Figure 10. The interannual changes of the water column average values of the seawater temperature (a), (b) salinity, (c) PO_4-P, (d) SiO_3-Si, (e) DIN, and (f) DO at the Station 4 of the nSCS in summer during the observation period.

Response of the Ecosystem and Living Resources

Although the frequency of observation on various biology and fisheries parameters was less than that for physical and chemical parameters which reflected the long term changes in the environmental processes, some responses of the ecosystems to these environmental changes in the nSCS was still evident.

Comparing the biology and fisheries data between the two phases discussed above, the average values of Chla, PP, PA, BB, CC and DTC increased by 6.4, 1.4, 2.4, 0.7, 7.2, and 2.8 times, respectively during the second phase, except for zooplankton abundance which decreased by about 50% (Table 6). The decrease in zooplankton abundance may be related to the increase in its predators, such as fish, cephalopods, and so forth (Figure 11). The R_{xt} values of the cephalopod and DTCs were 0.89 and 0.88, respectively, and highly significant ($p \leq 0.01$). Besides, the increase in both the cephalopod and DTCs partially could be attributed to the improvement of demersal trawl fishing techniques, and also to the increase in stock and production of low trophic levels, induced by the alleviation in N limitation in the nSCS.

Table 6. Comparison of the multi–year mean values of Chl-a, primary production (PP), phytoplankton abundance (PA), zooplankton biomass (ZB), benthos biomass (BB), cephalopod catch (CC), and demersal trawl catch (DTC) in the nSCS between two periods, i.e., before and after 1997 (units: mg m^{-3}, mgC $m^{-2}d^{-1}$, x 10^3 cell m^{-3}, mgm^{-3}, gm^{-2}, x10^4t, and x10^4t, respectively).

Parameter	First phase (data source)	Second phase (data source)
Chi-a	0.29 (Fan 1985)[b]	2.16 (Cai et al., 2002a; SOAC 2003)[c]
pp	185.7 (Fan, 1985)[b]	442.5 (Ning et al., 2004; Jia et al., 2005; Hao et al., 2007)
PA	161.5 (Lin, 1985)	542.7 (Jia et al., 2005)[b]
ZB	44.5 (Zhang, 1984; Chen, 1985; Qian et al., 1990a, b; Huang et al., 1990; Chen, 1992)	22.05 (Jia et al., 2005)[d]
BB	6.6 (Shen, 1985)	11 .3 (Jia et al., 2005)
CC	1.2 (Guo and Chen, 2000)	9.8 (Guo and Chen, 2000)
DTC	89.0 (Guo and Chen, 2000)	334.8 (Guo and Chen, 2000)

a The data listed in the Table are the mean of the water column of the multi-year average for each parameter from different sources, except for BB, CC and DTC which are only treated by the multi-year average;
b The data were provided by Chen, X. from The Third Institute of Oceanography, SOA, China;
c The data was from the National Science Foundation of China (NSFC) project under the contract No. 90211021;
d The data from Jia et al. (2005) are the annual mean of the multi-year mean for each season during 1997-2002.

In addition, after 1997, phosphorus depletion in surface waters during summer coincided with a shift in the dominant species in phytoplankton community from diatoms to dinoflagellates and cyanophytes (Ning et al., 2004). Peng et al. (2006) pointed out that dinoflagellates composed more than 60% of the total PA in the HK Southeast Anti-Cyclonic Eddy and Hainan Island East Anti-Cyclonic Eddy of the nSCS, where P concentration was near detection limit in the summer of 2004.

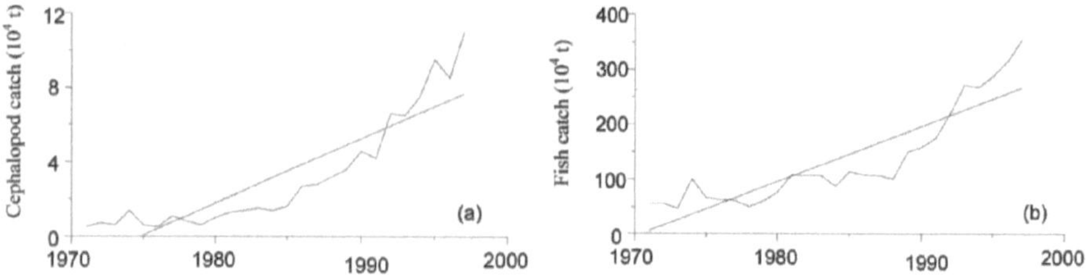

Figure 11. Variation trends in the catches of cephalopod and fishes of the nSCS. (a) Catches of cephalopod; (b) Catches of fishes (Data from Guo and Chen, 2000).

In second phase our above observations were consistent with the results obtained by the multidisciplinary investigations for assessing the environmental health status and the fisheries environment quality in the nSCS during 1997–2002 (Jia et al., 2005). They reported that both PP and PA were relatively high, averaging 409.7 mg C $m^{-2}d^{-1}$, and 837×10^3 cell m^{-3}, respectively (according to criterion of grade of primary productivity and diet organism lever of China, Jia et al., 2003). The benthic biomass (averaging 11.3 g m^{-2}) was at normal levels (the same as above row, Jia et al., 2003). The ZB (averaging 22.1 mg m^{-3}) was low (the same as above row, Jia et al., 2003). And nutrients were also low, that is the mean values of PO_4–P = 0.28 μmol l^{-1} and DIN = 3.63 μmol l^{-1} (according to criterion of Nutrition grade of sea water of China, Jia et al., 2003). The nutrients values (Jia et al., 2005, nutrients) are much lower than those we observed and the average index of ecological environment quality of fishing ground is 0.58, indicating it was in good condition (Jia et al., 2005).

The first phase (before 1997) represents the initial status of the biology and fisheries during the study period, and the second phase (after 1997) represents the response of the ecosystem to the changed environment, particularly since 1998, when the occurrence of N limitation was significantly alleviated. These ecosystem responses discussed above were clearly the result of environmental changes induced by not only climate change, but also anthropogenic activities.

RESULTS

Environmental Changes During the 29 Year Time Series

Seawater Temperature and Salinity

The annual means of SST and T_{av} (the average temperature in the water column) exhibited highly significant increasing trends ($p \leq 0.01$), their annual rates were 0.078 and 0.090°C y^{-1}, respectively. However, the temperature at 200 m (T_{200}) showed a significant decreasing trend with an annual rate of −0.108°Cy^{-1} during the 29 years of observation between 1976 and 2004 (Figure 6, Table 2). During this period, the SST and T_{av} increased by 2.26 and 2.61°C, respectively, while T_{200} decreased by 3.10°C (Figure 6, Table 2). The R_{xt} of the time series of SST, T_{av}, and T_{200} were 0.89, 0.85, and −0.71, respectively, and the variation trends for this time series were significant ($p \leq 0.01$).

During the observation period the annual rate of SSS exhibited a decreasing trend (−0.022 y^{-1}), while the annual rates of S_{av} and S_{200} exhibited increasing trend (0.007 y^{-1}

and 0.005 y^{-1}, respectively, Figure 5, Table 2). The SSS decreased by 0.653, while the S_{av} and S_{200} increased by 0.206 and 0.151. The R_{xt} values of time series of SSS, S_{av}, and S_{200} were −0.41($P \leq 0.05$), 0.38 ($P \leq 0.05$), and 0.44 ($p \leq 0.02$), respectively (Table 2).

Concentration of Dissolved Oxygen (DO)

A significant ($p \leq 0.01$) decrease in DO concentrations was observed (Figure 7, Table 2), and the annual rate was between −0.91 and −1.93 μmol l^{-1} y^{-1}. The regional mean of DO of the SS, the water column average in the upper 200 m and at 200 m significantly decreased by 26.5, 55.8 and 47.9 μmol l^{-1}. The R_{xt} values of time series for DO ranged −0.48 to −0.81 ($p \leq 0.01$, Table 2).

Concentrations of P, Si, and N

Both P and Si concentrations exhibited decreasing trends (Figures 8 and 9, Table 2). Their annual rates were −0.008 to −0.017 (for P) and −0.071 to −0.387 (for Si) μmol $l^{-1}y^{-1}$, respectively.

The concentrations of NO_3-N, NO_2-N, NH_4-N, and DIN (DIN = $NO_3 + NO_2 + NH_4$) in the nSCS exhibited obvious increasing trends during 1989–2004 (Figure 2, Table 2). The annual rates of the increases in DIN and NO3-N were 0.38–0.81 and 0.34–0.71 μmol $l^{-1}y^{-1}$, respectively. The regional means of DIN values of the SS, water column average and 200 m layer increased by 8.06, 6.10, and 13.01 μmol l^{-1}, respectively. The DIN concentrations in the 200 m layer were significantly higher than those of the average for the water column and at the SS. Its annual rate was 1.61 times greater than that of the SS. And its multi-year average value during observation period was 3.9 times higher than that of the SS layer. The R_{xt} value for the time series of NO_3-N and DIN were between 0.55–0.74. The variation trends of the NO_3-N and DIN time series were highly significant ($p \leq 0.01$) (Table 2), except for the time series of NO_3-N_{200}, which was significant ($p \leq 0.03$). It was also found that the annual rate of DIN was higher in the shallow (<200 m) than in the deep water (>200 m) areas (Table 2). The spatial-temporal distributions of the DIN_{av} (mean concentrations in the water column) indicated that since 1998 the high DIN_{av} concentrations have exhibited a pronounced large area increasing, that is after 1998 in the most study area, DIN_{av} concentration >5.71 μmol l^{-1} (Figure 3, the shadow shows the area), and exceeded the low limit of suitable N concentration for diatom growth (5.71 μmol l^{-1}, Chu, 1949). So it is favorable for phytoplankton growth in the most study area.

Nutrient Ratios

During 1989–2004, the nutrient ratios (DIN:P and Si:DIN) changed significantly. For example, the DIN: P ratios increased by 1.04–2.75 y^{-1}, and it reached 28.1 (SSDIN:SSP) and 18.0 (DIN_{av}:P_{av}) in 2004 (Figure 4, Table 2); the Si:N ratios decreased at an annual rate of −0.10~−0.44 y^{-1}, and it dropped to 1.0 (SSSi:SSDIN) and 1.4 (Siav:DIN_{av}) in 2004 (Table 2). The variation trends of all time series of N:P and Si:N were significant ($p \leq 0.01$). The annual rates of increase in DIN:P and decrease in Si:DIN were higher in the shallow coastal waters (<200 m) than in deep continental shelf waters (>200 m) (Table 2).

CONCLUSION

The fluctuations in environmental parameters in the nSCS during 1976–2004 displayed different patterns, that is temperature (SST and T_{av}), salinity (S_{av} and S_{200}), DIN and N:P annual rates increased, while DO, P, Si, Si:N, SSS, and T_{200} annual rates decreased. The climate trend coefficients, R_{xt} of these time series were all over 0.38 (n = 29) or 0.50 (n = 16), and highly significant ($p \leq 0.05$), except for the time series of P and Si.

The increasing trends in SST and T_{av} were consistent with the rise of the mean AT in the northern Hemisphere and southern China. The increase in SST and T_{av} and decrease in SSS in the nSCS led to strengthening of the thermocline and halocline, less mixing of deep water to the surface and thus a decrease in the P supply from deep waters. The increasing trend in DIN may have been influenced by the Pearl River discharge, and atmospheric dry and wet deposition, (which are related to anthropogenic activities), coastal upwelling and cyclonic eddies. The nSCS always experienced limitation of N before 1997, but the situation in the upper layer sea water has been mitigated since 1998, due to the increase in N concentration and decrease in P, which resulted in not only the positive trends in N:P ratios which are now close to the Redfield ratio, but also the decreasing trend in Si:N ratios, indicating of potential P limitation. The decrease in DO concentration may be linked to the increase in seawater temperature and the increase in the concentration of organic matter inputs mainly from the Pearl River and phytoplankton blooms, particularly since the 1990s. Chlorophyll-a, PP, PA, BB, CC, and DTC have increased, and zooplankton abundance decreased. These ecosystem responses resulted from environmental changes were induced by not only climate change, but also anthropogenic activities. After 1998, phosphorus depletion in upper layer may be associated with a shift from diatoms to dinoflagellates and cyanophytes domination.

Pronounced responses of the environmental parameters to ENSO were observed. The effects of climate change on the nSCS were mainly through changes in monsoon and its causative links, monsoon circulation nutrients PP.

It must be pointed out that although the data set we adopted in the present analysis is large and from various sources. As such the quality of data might vary throughout the long period of observation. But these data were valuable for studying the long term changes in environmental conditions and the responses of the ecosystem in this region. And using these data we can still find and understand some regulations on the response of the ecosystems to the environmental changes in the SCS. This regional response of SCS to global climate change was investigated the first which is far from well understood. The evolving nutrient environment may be related to the observed ecosystem changes in the nSCS such as increase in biological productivity. More long-time series observations on the structure and function of ecosystems and the relationships with environmental changes are needed in the SCS in the future.

KEYWORDS

- **Air temperature**
- **Dissolved oxygen**
- **Northern South China Sea**
- **Seawater temperature**
- **State Oceanic Administration**

ACKNOWLEDGMENTS

This study was supported by the National Science Foundation of China (NSFC) key projects under the contracts No. 90211021 and No. 90711006. The authors would like to thank the information center of SOA for providing the environmental data, and to Paul Harrison for his significant comments.

Chapter 12

Lower Miocene Stratigraphy Along the Panama Canal

Michael Xavier Kirby, Douglas S. Jones, and Bruce J. MacFadden

INTRODUCTION

Before the formation of the Central American Isthmus, there was a Central American Peninsula. Here we show that southern Central America existed as a peninsula as early as 19 Ma, based on new lithostratigraphic, biostratigraphic, and strontium chemostratigraphic analyses of the formations exposed along the Gaillard Cut of the Panama Canal. Land mammals found in the Miocene Cucaracha Formation have similar body sizes to conspecific taxa in North America, indicating that there existed a terrestrial connection with North America that allowed gene flow between populations during this time. How long did this peninsula last? The answer hinges on the outcome of a stratigraphic dispute: To wit, is the terrestrial Cucaracha Formation older or younger than the marine La Boca Formation? Previous stratigraphic studies of the Panama Canal Basin have suggested that the Cucaracha Formation lies stratigraphically between the shallow-marine Culebra Formation and the shallow-to-upper-bathyal La Boca Formation, the latter containing the Emperador Limestone (E.L.). If the La Boca Formation is younger than the Cucaracha Formation, as many think, then the peninsula was short-lived (1–2 million years (m.y.)), having been submerged in part by the transgression represented by the overlying La Boca Formation. On the other hand, our data support the view that the La Boca Formation is older than the Cucaracha Formation. Strontium dating shows that the La Boca Formation is older (23.07–20.62 Ma) than both the Culebra (19.83–19.12 Ma) and Cucaracha (Hemingfordian to Barstovian North American Land Mammal Ages; 19–14 Ma) formations. The E.L. is also older (21.24–20.99 Ma) than the Culebra and Cucaracha formations. What has been called the "La Boca Formation" (with the E.L.), is re-interpreted here as being the lower part of the Culebra Formation. Our new data sets demonstrate that the main axis of the volcanic arc in southern Central America more than likely existed as a peninsula connected to northern Central America and North America for much of the Miocene, which has profound implications for our understanding of the tectonic, climatic, oceanographic and biogeographic history related to the formation of the Isthmus of Panama.

The paleogeography of Central America has changed profoundly over the past 30 m.y., from a volcanic arc separated from South America by a wide seaway, to an isthmus that connected North and South America by 3 Ma (Coates and Obando, 1996; Coates et al., 1992, 2003, 2004; Duque-Caro, 1990). The formation of the Isthmus of Panama was important because it allowed the mixing of terrestrial faunas between the two continents (Webb, 1985), as well as physically separating a once continuous marine province into separate and distinct Pacific and Caribbean communities (Jackson et al., 1993, 1996; Vermeij, 1978; Vermeij and Petuch, 1986; Woodring, 1965, 1966). The formation of the Isthmus of Panama also ultimately led to profound changes in global climate (Weyl, 1968) by strengthening the Gulf Stream and thermohaline downwelling in the North Atlantic (Keigwin, 1982).

Although extensive study has constrained the timing of isthmian formation (Coates et al., 1992, 2003, 2004; Collins and Coates, 1999; Duque-Caro, 1990; Emiliani et al., 1972; Keigwin, 1978), the paleogeographic nature of southern Central America before the isthmus is still disputed. Paleobathymetric and other geologic evidence from depositional basins suggests that southern Central America arose slowly from bathyal depths during the Neogene as a result of the collision between the Panama microplate and the South American plate (Coates et al., 2003, 2004; Collins et al., 1995), suggesting that the volcanic arc during the Miocene consisted of an archipelago of volcanic islands that was slowly uplifting through the Neogene until the ultimate formation of the isthmus (Coates and Obando, 1996; Coates et al., 1992, 2003, 2004; Collins et al., 1996). For example, Coates et al. (1992) stated that (p. 816): “It is likely that during the late Neogene the Chorotega and Choco blocks formed an archipelago and there were frequent marine connections between the Caribbean and the Pacific (Duque-Caro, 1990b, his Figure 9). The topographic, tectonic, and regional geologic evidence strongly suggests that the archipelago stretched from westernmost Costa Rica to the Atrato Valley in Colombia ...” However, most of the evidence suggesting slow uplift of the volcanic arc from bathyal depths is derived from depositional basins that lie peripheral to the main axis of the volcanic arc in southern Central America (Figure 1).

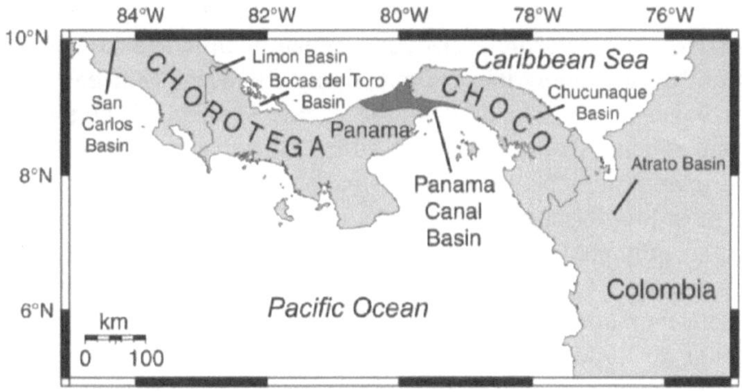

Figure 1. Location of the Panama Canal Basin and other depositional basins in southern Central America.

An alternative view is that the main axis of the volcanic arc had already arisen above sea level by the early Neogene, which would effectively make Panama a peninsula of Central America by this time (Whitmore and Stewart, 1965). Evidence supporting this latter view comes from land mammal fossils found in the Miocene Cucaracha Formation exposed in the Gaillard Cut of the Panama Canal near the center of the Panama Canal Basin (Table 1). Land mammals with only North American affinities and similar body sizes to conspecific taxa in North America suggest a terrestrial connection with North America by the early Miocene (Kirby and MacFadden, 2005; MacFadden, 2006; MacFadden and Higgins, 2004; Whitmore and Stewart, 1965). The purpose of the present study is to further resolve the Neogene paleogeography of southern Central America by placing the Cucaracha land mammals into a stratigraphic framework through lithostratigraphic, biostratigraphic, and strontium chemostratigraphic analyses that test long-standing hypotheses concerning the stratigraphy of the Gaillard Cut.

Table 1. Land mammal taxa from the Gaillard Cut Local Fauna, Cucaracha Formation, Panama.

Order	Family	Genus & species	Common name	Biogeographic affinity
Rodentia	–	Texomys steworti	Geomyoid rodent	North America
Carnivora	Canidae	Tomarctus brevirostris	Dog	North America
Carnivora	Amphicyonidae or Hemicyonidae	–	Bear dog	North America
Artiodactyla	Tayassuidae	cf. Cynorca sp.	Peccary	North America
Artiodactyla	Oreodontidae	Merycochoerus matthewi	Oreodont	North America
Artiodactyla	Protoceratidae	Paratoceras wardi	Protoceratid	North America
Perissodactyla	Equidae	Anchitherium clarencei	Horse	North America
Perissodactyla	Equidae	Archaeohippus sp.	Horse	North America
Perissodactyla	Rhinocerotidae	Menoceros barbouri	Rhinoceros	North America
Perissodactyla	Rhinocerotidae	Floridaceras whitei	Rhinoceros	North America

Excavation of the Gaillard Cut during the original construction of the Panama Canal exposed lower Neogene sediments of the Panama Canal Basin (Figures 2 and 3). As upper Neogene volcanic rocks cover much of the Panama Canal Basin, the Gaillard Cut offers a window into the underlying Oligocene-Miocene rocks beneath this volcanic cover (Figure 2). Although this excavation exposed hundreds of meters of section, the structural complexity caused by extensive faulting has obscured the stratigraphic relationships between the various formations, such that only a portion of one, or at most two, formations are present in any given fault-bounded block (Figure 3). The most recently published stratigraphy and geologic map for the Panama Canal Basin indicates that the Cucaracha Formation lies stratigraphically above the shallow-marine Culebra Formation and below the shallow-to-upper-bathyal La Boca Formation (Stewart et al., 1980). If this stratigraphic arrangement is correct, then we may conclude that the peninsula containing North American land mammals was short-lived in the early Neogene (1–2 m.y., based on the temporal duration of paleosols in the Cucaracha Formation (Retallack and Kirby, 2007)), having been submerged in part by the marine transgression represented by the overlying La Boca Formation (Whitmore and Stewart, 1965). Earlier stratigraphic arrangements, however, placed strata presently in the La Boca Formation not above the Cucaracha Formation, but below it (MacDonald, 1919; Van den Bold, 1972; Woodring and Thompson, 1949)). Given this stratigraphic arrangement, the marine transgression represented by the La Boca Formation occurred before deposition of the Cucaracha Formation, which would indicate that there is no evidence for submergence of the Central American Peninsula until

6 Ma, when there is evidence for a short-lived strait across the Panama Canal Basin (Collins et al., 1996). Was the Central American Peninsula short-lived, existing for only the 1–2 m.y. that it took to form the Cucaracha Formation? Or did the peninsula exist longer than this? Although we cannot currently determine exactly when the Central American Peninsula formed or how far east it may have extended, we can constrain the interval of time that such a peninsula may have existed by placing the land mammals of the Cucaracha Formation into a well-defined stratigraphy. Lithostratigraphic, biostratigraphic, and strontium chemostratigraphic analyses presented here allow us to test the validity of different stratigraphic models proposed for the Gaillard Cut. Correlation from lithostratigraphic analysis of 11 stratigraphic sections, biostratigraphic placement of the fossil land mammals and absolute age estimates from strontium isotopes of marine fossils are used for the first time to test the different stratigraphic models that have been proposed for the Gaillard Cut. The solution to these paleogeographic and stratigraphic problems has profound implications for our understanding of the biogeographic, paleoclimatic, tectonic, and evolutionary history of the Isthmus of Panama. Furthermore, many studies rely on a proper understanding of the stratigraphy of the Gaillard Cut (Johnson and Kirby, 2006; Kirby and MacFadden, 2005; Retallack and Kirby, 2007). The resolution of this stratigraphic problem also has important application to geotechnical studies associated with the expansion of the Panama Canal that is planned to occur between 2008 and 2014.

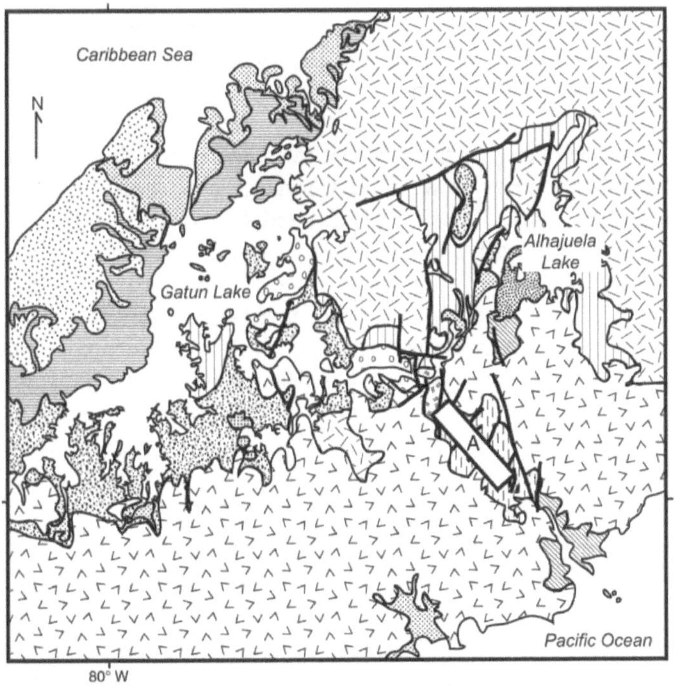

Figure 2. Geologic map of the Panama Canal Basin showing the study area (A) in relation to rock units discussed in the text.

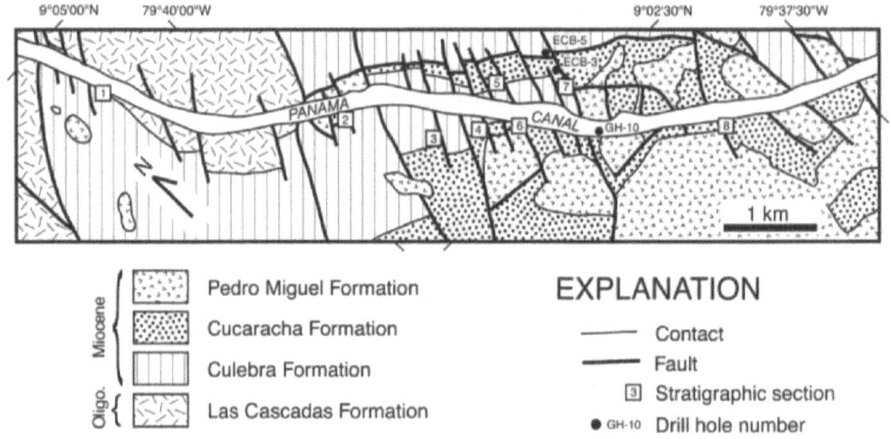

Figure 3. Geologic map of the study area along the Gaillard Cut portion of the Panama Canal, within the Panama Canal Basin (see rectangle labeled "A" in Figure 2).

Regional Geologic Setting

The Panama Canal Basin is a Tertiary structural and depositional basin that straddles the tectonic boundary between the Chorotega and Choco blocks of the Panama microplate (Figure 1) (Coates, 1999; Coates and Obando, 1996). As part of the Central American volcanic arc, the Panama microplate formed through subduction of various oceanic plates during the Cretaceous and Cenozoic (Mann, 1995). This microplate lies between the Cocos and Nazca plates to the south, the Caribbean plate to the north and the South American plate to the east (Mann, 1995). The formation of the Panama Canal Basin may be related to the hypothetical "Gatun Fault Zone," which may represent the tectonic boundary between the Chorotega and Choco blocks (Case, 1974; Coates and Obando, 1996). Case (1974) inferred the existence of a deep-shear zone trending northwest-southeast, approximately parallel to the Panama Canal, based on gravity data indicating a very steep gradient underlying the Panama Canal Basin. He speculated that this concealed fault zone may have had lateral displacement and could be of early Cenozoic or older age. Lowrie et al. (1982) also recognized a major fault zone in this area based on several lines of evidence. Later studies, however, have suggested that there is no direct evidence for the existence of the Gatun Fault Zone (Pratt et al., 2003). Nevertheless, the Panama Canal Basin exists in a structurally complex area, as indicated by thousands of northeast-southwest trending sets of faults (Pratt et al., 2003). The exact tectonic nature of this basin continues to remain unclear, but it may represent an active rift or forearc basin (Mann, 1995).

The Panama Canal Basin contains a thick sequence of sediments and volcanic rocks (>2,900 m) of Eocene to Pleistocene age (Figure 2) (Escalante, 1990; Jones, 1950; Stewart et al., 1980; Terry, 1956; Woodring, 1957–1982). The lowermost sedimentary unit is the Eocene Gatuncillo Formation, which contains marine mudstone, siltstone and limestone, and unconformably overlies pre-tertiary volcanic basement (Escalante, 1990; Woodring, 1957–1982). Overlying the Gatuncillo Formation is the Oligocene

Bohio Formation, which contains marine and non-marine conglomerate, tuffaceous sandstone and siltstone (Escalante, 1990; Woodring, 1957–1982). Stratigraphically higher are the Oligocene Bas Obispo and Las Cascadas formations (L.C.F.), both of which consist of agglomerate and tuff (Woodring, 1957–1982). Conformably overlying the L.C.F. is the lower Miocene Culebra Formation (this study), which contains marine mudstone, sandstone, limestone, conglomerate, and lignite. Portions of the La Boca, Alhajuela and Caimito formations are correlative with the Culebra Formation, as suggested by this study and previous lithostratigraphic and biostratigraphic studies (Jones, 1950; Van den Bold, 1972; Woodring, 1957–1982; Woodring and Thompson, 1949). The lower to middle Miocene Cucaracha Formation overlies the Culebra Formation and consists of subaerial claystone, sandstone, conglomerate, and lignite, all showing paleosol development (Retallack and Kirby, 2007). The middle Miocene Pedro Miguel Formation (P.M.F.) overlies conformably the Cucaracha Formation and contains basalt and agglomerate. Stratigraphically higher is the upper Miocene Gatun Formation (Coates, 1999), which contains marine siltstone, sandstone and conglomerate (Collins et al., 1996; Escalante, 1990). East of the city of Colon, the Gatun Formation overlies nonconformably unnamed Cretaceous volcanic rocks; whereas, west of Colon, the Gatun Formation overlies unconformably the Caimito Formation (Coates, 1999). The Gatun Formation is overlain disconformably by the upper Miocene Chagres Formation (Coates, 1999), which consists of conglomeratic sandstone and a basal coquina of the Toro Member (Collins et al., 1996). Unconformably above the Tertiary formations are unconsolidated Quaternary deposits, informally known as the "Pacific muck" and "Atlantic muck" (Woodring, 1957–1982).

Stratigraphic Models for the Gaillard Cut

Hill (1898) was the first to systematically name and describe formations along the Gaillard Cut (Figure 4). MacDonald (1913, 1919) later named and described several formations in the Panama Canal Basin, including the Las Cascadas and Cucaracha formations. Woodring and Thompson (1949) formally named and described the Pedro Miguel and La Boca formations. They also placed the E.L. Member within the Culebra Formation. Based on field work by R. H. Stewart of the Panama Canal Company, Woodring (1964) later restricted the Culebra Formation by placing sections containing the E.L. (i.e., the lower two-thirds of the Culebra Formation) into the La Boca Formation, which he considered younger than the Cucaracha Formation. He kept the upper one-third of the Culebra Formation stratigraphically below the Cucaracha Formation. Woodring (1964) did not state explicitly the reasons or evidence for this revision in the stratigraphy of the Gaillard Cut, only that the new evidence was derived from drill cores made by R. H. Stewart. Writing in 1964, Woodring (1964) stated that (p. 244): "After the drilling along the Empire Reach … got under way, R. H. Stewart, geologist of the Panama Canal, soon realized that the geology of the northwestern part of the Gaillard Cut area had been misinterpreted." Stewart et al. (1980) later hypothesized interfingering relationships between the Cucaracha and L.C.Fs., as well as between the Pedro Miguel and La Boca formations, in order to justify placement of sections containing the E.L. within the La Boca Formation (Figure 4) (Graham et al., 1985). However, these interfingering relationships are not apparent in outcrop exposures or in subsurface well logs. Van den Bold (1972), a noted biostratigrapher of Caribbean ostracodes, disagreed with the revised stratigraphic interpretation of Woodring (1964) by demonstrating that stratigraphic sections containing the E.L. and overlying sediments (i.e., La Boca Formation) were correlative with the Culebra Formation, based on ostracode biostratigraphy. Specifically, his zones I and IIA of the "La Boca Formation" are correlative with the Culebra Formation (Figure 4). Later studies have followed the stratigraphy as originally proposed by Woodring and Thompson (1949), based on data gathered in support of the current study (Johnson and Kirby, 2006; Retallack and Kirby, 2007).

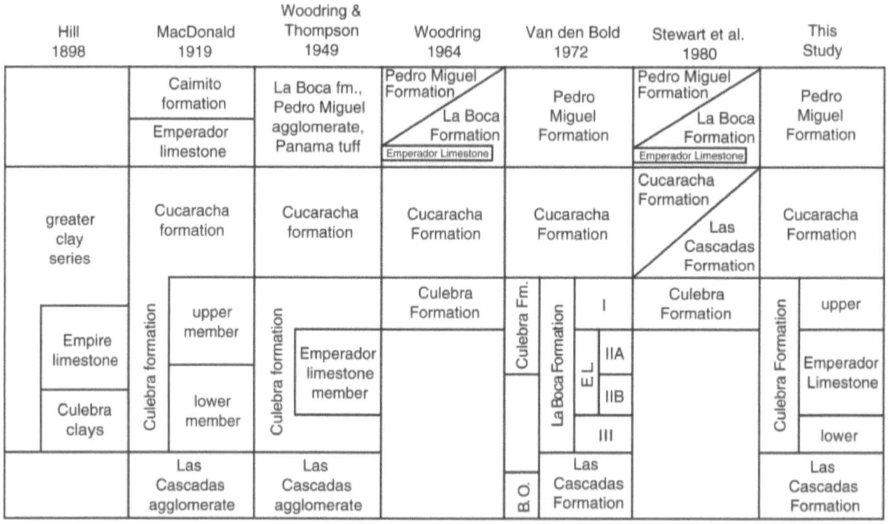

Figure 4. Summary of stratigraphic nomenclature for the formations exposed along the Gaillard Cut portion of the Panama Canal.

All of the stratigraphic models that have been proposed for the Gaillard Cut over the past 100 years can be divided into two groups, which are herein called the (1) Culebra model and the (2) La Boca model (Figure 5). The Culebra model places all of the Culebra Formation (sensu Woodring and Thompson, 1949) underneath the Cucaracha Formation (Escalante, 1990; MacDonald, 1919; Van den Bold, 1972; Woodring and Thompson, 1949). The La Boca model, on the other hand, places the lower portion of the Culebra Formation (sensu Woodring and Thompson, 1949) (marked by blue in Figure 5) into the La Boca Formation above the Cucaracha Formation (Graham et al., 1985; Stewart et al., 1980; Woodring, 1964). As the upper portion of the Culebra Formation (marked by purple in Figure 5) remains beneath the Cucaracha Formation in this model, the La Boca model considers the Culebra Formation of Woodring and Thompson (1949) to actually be two different formations (i.e., Culebra and La Boca formations), with the Cucaracha Formation lying between two (Figure 5). If the La Boca Formation (marked by blue in Figure 5) is found to be younger than the Cucaracha Formation, then the Culebra model will be rejected and the La Boca model supported. Alternatively, if the La Boca Formation is found to be older than the Cucaracha Formation, then the La Boca model will be rejected and the Culebra model supported.

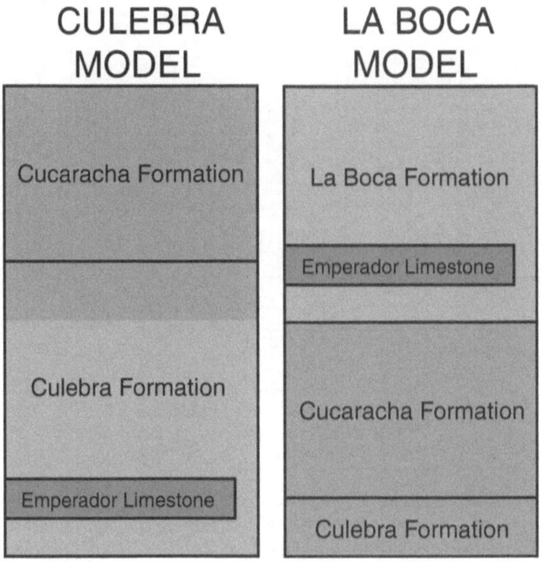

Figure 5. The two alternative stratigraphic models for the formations exposed along the Gaillard Cut portion of the Panama Canal: (1) the Culebra model (Escalante, 1990; Woodring and Thompson, 1949; Van den Bold, 1972) and (2) the La Boca model (Graham et al., 1985; Stewart et al., 1980; Woodring, 1964).

MATERIALS AND METHODS

Field work was conducted in February 2003, July 2003 to December 2004, and March 2005. This study benefited greatly from many newly exposed surface sections made by recent widening of the Panama Canal by the Panama Canal Authority (ACP) and the construction of a second bridge across the canal (Centennial Bridge). We measured eight stratigraphic sections between the towns of Pedro Miguel and Gamboa along the Gaillard Cut portion of the Panama Canal (Figure 6). These outcrop sections were measured with a Jacob Staff and Brunton compass or with a tape and Brunton compass (methods described in Compton (Compton, 1985)). We collected rock and fossil samples, recording their stratigraphic position and deposited them at the Center for Tropical Paleoecology and Archaeology, Smithsonian Tropical Research Institute, and at the Florida Museum of Natural History. Well logs derived from drill cores from the archives at the ACP were also examined in order to aid correlation of surface sections and to fill in missing intervals (ECB-3, ECB-5, and GH-10 well logs at the ACP; Figure 6). We used the biochronology of the land mammals from the Gaillard Cut local fauna, as modified by MacFaddden (2006). The land-mammal chronology and North American Land Mammal Age subdivsions follow Tedford et al. (2004).

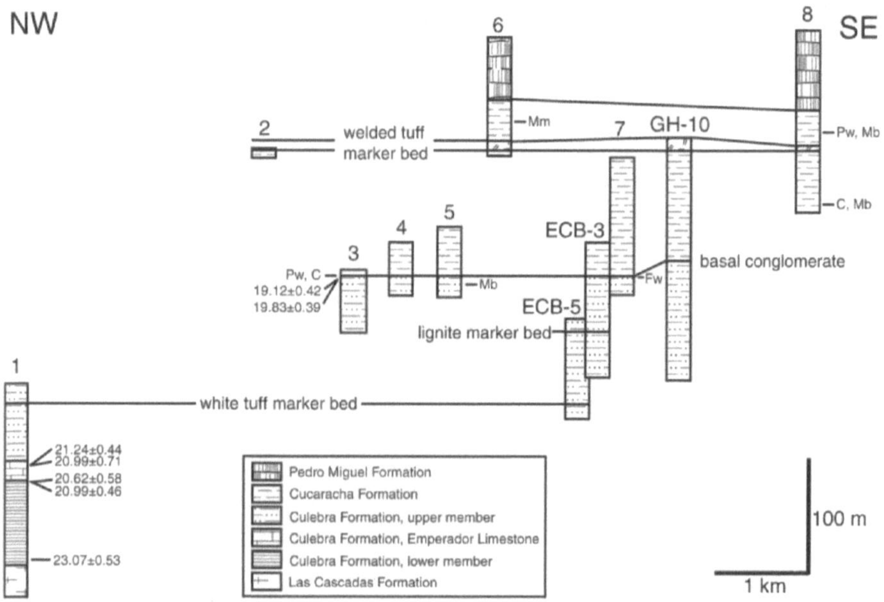

Figure 6. Lithostratigraphic correlation of stratigraphic Sections 1 to 8 and drill cores logged by R. H. Stewart of the Panama Canal Company (ECB-5, ECB-3, and GH-10) arranged northwest to southeast along the Gaillard Cut portion of the Panama Canal. Location of stratigraphic sections and drill holes in Figure 3. Absolute ages (Ma) are derived from Sr chemostratigraphic analyses (see text). C = cf. *Cynorca* sp. Mm = *Merycochoerus matthewi*. Pw = *Paratoceras wardi*. Mb = *Menoceras barbouri*. Fw = *Floridaceras whitei*.

Many studies have shown that the Neogene is generally a time of rapidly increasing 87Sr/86Sr in the global ocean and, hence, particularly amenable to dating and correlating marine sediments using strontium isotopes (Jones et al., 1993; Hodell and Woodruff, 1994; Hodell et al., 1991; Mallinson et al., 1994; Martin et al., 1999; McArthur et al., 2001; Miller et al., 1991, 1995; Oslick et al., 1994). We analyzed seven fossil specimens from the restricted, upper Culebra Formation and the La Boca Formation in the Gaillard Cut, along the Panama Canal, in order to determine the ratio of 87Sr/86Sr of the calcium carbonate composing the shell (Table 2). These data allow us to estimate the geologic age for each fossil specimen. For isotopic analyses, we first ground off a portion of the surface layer of each shell specimen to reduce possible contamination. Areas showing chalkiness or other signs of diagenetic alteration were avoided. Powdered aragonite (coral) or low-magnesium calcite (mollusc) samples were drilled from the interior of each specimen using a hand-held Dremel tool with a carbide burr. Approximately 0.01–0.03 g of powder was recovered from each fossil sample. The powdered samples were dissolved in 100 μl of 3.5 N HNO_3 and then loaded onto cation exchange columns packed with strontium-selective crown ether resin (Eichrom Technologies, Inc.) to separate Sr from other ions (Pin and Bassin, 1992). The Sr isotope analyses were performed on a Micromass Sector 54 Thermal Ionization Mass Spectrometer equipped with seven Faraday collectors and one Daly detector in the

Department of Geological Sciences at the University of Florida. The Sr was loaded onto oxidized tungsten single filaments and run in triple collector dynamic mode. Data were acquired at a beam intensity of about 1.5 V for 88Sr, with corrections for instrumental discrimination made assuming 86Sr/88Sr = 0.1194. Errors in measured 87Sr/86Sr are better than ±0.00002 (2σ), based on long-term reproducibility of NIST 987 (87Sr/86Sr = 0.71024). Age estimates were determined using the Miocene portion of Look-Up Table Version 4:08/03 associated with the strontium isotopic age model of McArthur et al. (2001).

Table 2. The Sr chemostratigraphic analyses of the Cuelbra Formation, Panama.

Sample	Taxon	Latitude	Longitude	Member	$^{87}Sr/^{86}Sr$	Std. error%	Std. error (external)	Age estimate (Ma)	Std. error (Ma)
Pan8	Coral	9' 04.661 'N	79' 40.627'W	Emperador	0.708386	0.001	0.000023	20.99	0.71
Pan9	Coral	9' 04.661 'N	79' 40.627'W	Emperador	0.708371	0.0008	0.000023	21.24	0.44
Pan6	Pectinid	9' 04.661 'N	79' 40.627'W	Lower	0.708404	0.0008	0.000023	20.62	0.58
Pan7	Pectinid	9' 04.661 'N	79' 40.627'W	Lower	0.708386	0.0008	0.000023	20.99	0.46
Pan10	Bivalve	9' 04.468'N	79' 40.522'W	Lower	0.70825	0.0008	0.000023	23.07	0.53
Pan4	Ostrea sp.	9' 03.099'N	79°39.3SO'W	Upper	0.708502	0.0008	0.000023	19.12	0.42
Pan5	Pectinid	9' 03.099'N	79°39.3SO'W	Upper	0.70845	0.0007	0.000023	19.83	0.39

DISCUSSION

We infer that the Central American Peninsula was not short-lived in the early Miocene, based on our revised stratigraphy for the Gaillard Cut (Figure 7). We instead find no evidence for the disruption of this peninsula until 6 Ma, when there is evidence for a short-lived strait across the Panama Canal Basin (Collins et al., 1996). This conclusion is different from that of Whitmore and Stewart (1965), who was the first to present evidence that Panama "was connected to North America by a land area of considerable size and stability" in the middle Miocene, based on the land mammal fossils from the Cucaracha Formation (p. 184). They also concluded, however, that the presence of the Culebra and La Boca formations indicated inundation by the sea both before and after, respectively, the time when the land mammals arrived to Panama by land from North America (Whitmore and Stewart, 1965). Returning the La Boca Formation to the lower part of the Culebra Formation removes the evidence for a transgression after deposition of the Cucaracha Formation.

The earliest evidence for a terrestrial connection to North America is 19 Ma, based on an estimated age of 19.12–19.83 Ma of the pectinid bivalves found 2 m below the land mammal fossils in the upper Culebra Formation (Section 3 in Figure 6). Given that the upper part of the Cucaracha Formation may be as young as 14 Ma, and that there are at least 355 m of undated, terrestrial, volcanic rocks of the P.M.F. overlying the Cucaracha Formation, we think it likely that the peninsula existed in this part of Central America for much of the Miocene. Low precipitation and temperature estimates derived from the paleosols of the Cucaracha Formation imply a rain shadow

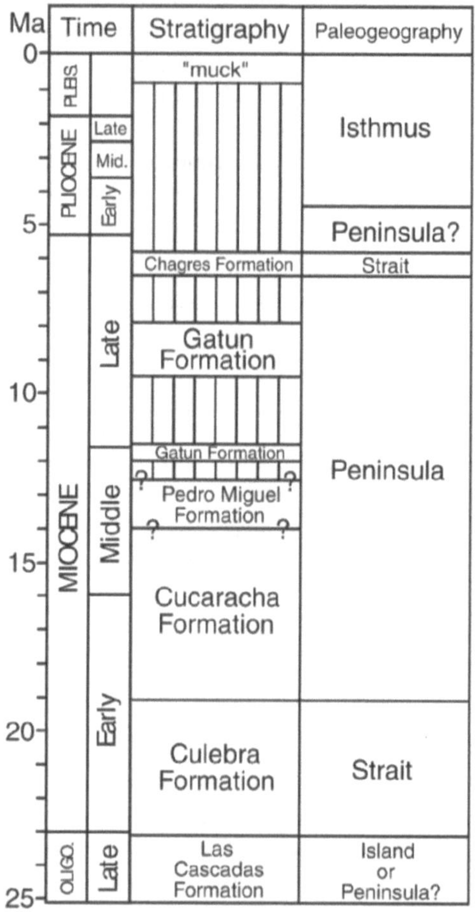

Figure 7. Biostratigraphy of the 10 taxa of land mammals found in the Gaillard Cut Local Fauna in the Culebra and Cucaracha formations, Panama, based on the North American Land Mammal Age of their conspecific taxa in North America.

from a very high volcanic mountain range (1,400–4,000 m), suggesting that this peninsula had very high relief (Retallack and Kirby, 2007). Shallowing to outer neritic depths in the northerly peripheral Limon and Bocas del Toro basins (Coates et al., 2003, 2004) (Figure 1) is also consistent with the continuing emergence of a Central American Peninsula to the south of these basins by the middle Miocene. In addition, the stratigraphically higher Gatun Formation, which is exposed in the northern part of the Panama Canal Basin, contains fossil benthic foraminifera that have a strong Caribbean affinity, indicating an effective biogeographic barrier between Caribbean and Pacific surface water in the middle to late Miocene (Collins et al., 1996), which further suggests that a peninsula existed during this time. Although we do not have evidence for a direct land connection between Panama and North America after deposition of the Cucaracha Formation, we think it likely that once a peninsula had formed, it would be more probable for its continued existence as a peninsula than for its reversion back to an archipelago. Based on all the above evidence, we think it unlikely that southern Central America existed as a complex island-arc archipelago after the early Miocene, as suggested by Coates et al. (1992), Coates and Obando (1996), and Collins et al. (1996). The evidence presented here indicates that the main axis of the volcanic arc in southern Central America had coalesced into a subaerial peninsula connected to North America by 19 Ma (Figure 8).

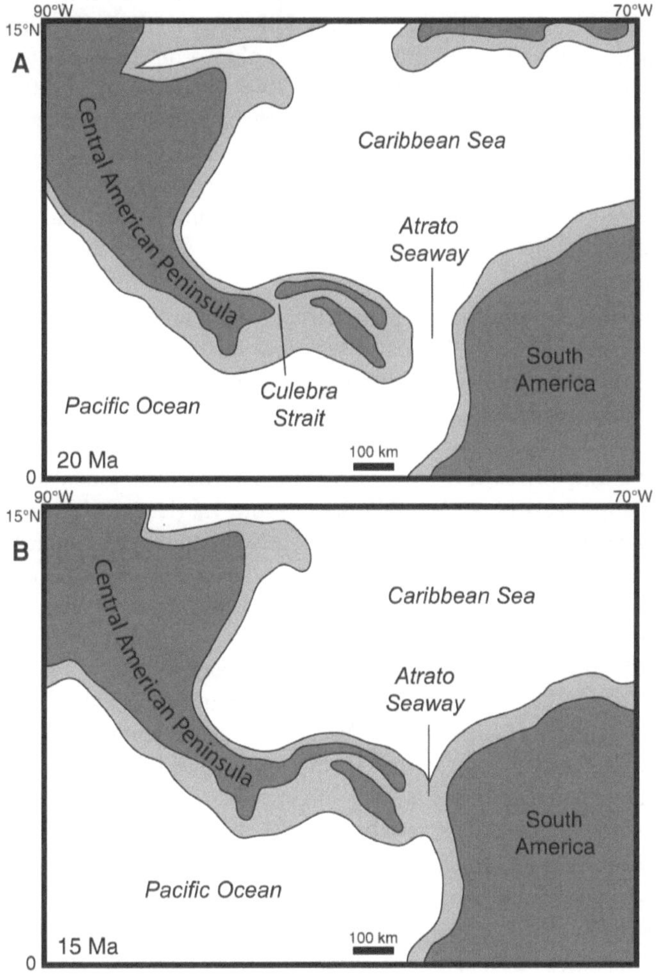

Figure 8. Composite stratigraphic section of the formations along the Gaillard Cut, Panama Canal Basin, showing lithostratigraphy and palaeoenvironmental interpretations, based on stratigraphic relationships illustrated in Figure 6."Data Source" indicates the stratigraphic sections used to compile the composite section. L.C.F. = Las Cascadas Formation. E.L. = Emperador Limestone. P.M.F. = Pedro Miguel Formation. Abbreviations at the base of the section represent grain size (C = Clay; S = Silt; Sd = S; and G = Gravel).

Nevertheless, geologically ephemeral straits across the Central American Peninsula did exist intermittently during the Neogene, as evidenced by the short-lived strait across the Panama Canal Basin 6 Ma (Collins et al., 1996). In addition, bathyal sediments in the upper member of the Culebra Formation suggest that a short-lived strait may have existed across the Panama Canal Basin between 21 and 20 Ma (Figure 8A). Other ephemeral straits may have existed intermittently across the San Carlos basin in northern Costa Rica and southern Nicaragua (Collins et al., 1996; Savin and Douglas, 1985). However, these short-lived straits probably had little impact on the long-term evolution of the marine and terrestrial biota of the Central American Peninsula (Collins et al., 1996). Of course, the Central American Seaway (also called the Atrato Seaway), located between Central and South America, remained open until the final formation of the Isthmus of Panama by 3 Ma (Coates and Obando, 1996; Coates et al., 1992, 2003, 2004; Duque-Caro, 1990). The lack of any South American land mammals in the Cucaracha Formation indicates that such a seaway must have existed in the early to middle Miocene (MacFadden, 2006; Whitmore and Stewart, 1965). The Central American Seaway was, therefore, the ultimate barrier to the migration of North American land mammals into South America, not the ephemeral straits that may have formed intermittently across the Central American Peninsula through the Neogene.

The transgressive-regressive facies pattern recorded in the Culebra Formation cannot be easily correlated with the major sea-level fluctuations of the early Miocene. For example, the transgressive-regressive pattern observed in the Culebra Formation appears to conflict with the global sea-level curve of Haq et al. (1987). The lower member of the Culebra Formation, dated between 23 and 21 Ma, indicates a local transgression during this interval, which is opposite from their global sea-level curve, which shows a lowering of sea level during this interval (Haq et al., 1987). The upper member of the Culebra Formation, dated between 21 and 19 Ma, indicates a regression during this interval, which is opposite from the sea-level curve, which shows a rising of sea level during this interval (Haq1 et al., 1987). The transgressive-regressive facies pattern in the Culebra Formation also appears to conflict with the global sea-level curve of Miller et al. (2005), which shows sea-level fluctuations of 20 m or less between 23 and 19 Ma. These fluctuations are much too small to account for the transgressive-regressive facies pattern observed in the Culebra Formation. The simplest explanation for these discrepancies is that subsidence followed by uplift resulting from regional tectonic forces had a much larger effect on relative sea level within the Panama Canal Basin than did eustasy between 23 and 19 Ma.

The existence of a Central American Peninsula containing a high volcanic mountain range for much of the Miocene has profound implications for our understanding of the tectonic, climatic, oceanographic and biogeographic history related to the formation of the Isthmus of Panama. A Central American Peninsula during the Miocene implies that: (1) uplift of the main axis of the Central American volcanic arc had already occurred near the beginning of the Neogene, (2) the changes in paleogeography thought to be responsible for intensification of the Gulf Stream and down-welling in the north Atlantic occurred much earlier than the Pliocene, (3) terrestrial communities between North and Central America were much better connected in the early to mid-Neogene than previously thought and (4) ocean circulation and biogeographic connection between the Pacific and Caribbean had to have been much more constricted in the early Neogene than previously thought.

RESULTS

Results from our lithostratigraphic, biostratigraphic and Sr chemostratigraphic analyses of the formations exposed along the Gaillard Cut allow us to reject the La Boca model. Instead, all data sets support the Culebra model as being the correct interpretation for the stratigraphy of the formations along the Gaillard Cut (Figures 6, 9, and 10). We found no evidence for interfingering relationships between the Cucaracha and L.C.Fs., or between the La Boca and P.M.Fs., as proposed by Stewart et al. (1980) and Graham et al. (1985). We did find that what has been called the La Boca Formation (i.e., lower Culebra Formation), conformably overlies the L.C.F. and that the upper Culebra Formation underlies the Cucaracha Formation, which in turn is overlain by the P.M.F. (Figure 6).

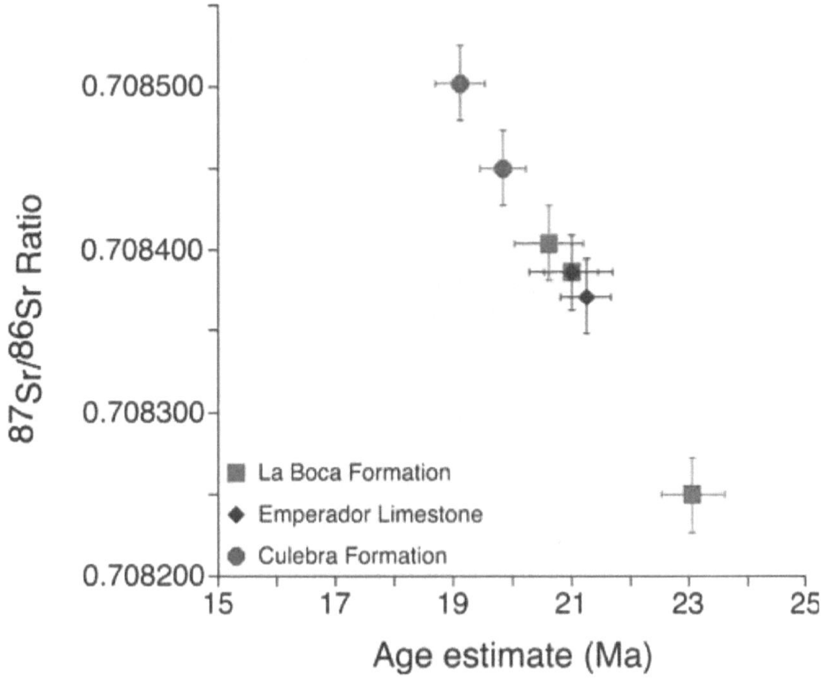

Figure 9. Geologic history of the Panama Canal Basin, showing geologic time, stratigraphy, and paleogeography. Vertical lines represent a hiatus. Question marks under stratigraphy represent uncertainty in age. Time scale from Gradstein et al. (2004). Oligo. = Oligocene. Pleis. = Pleistocene.

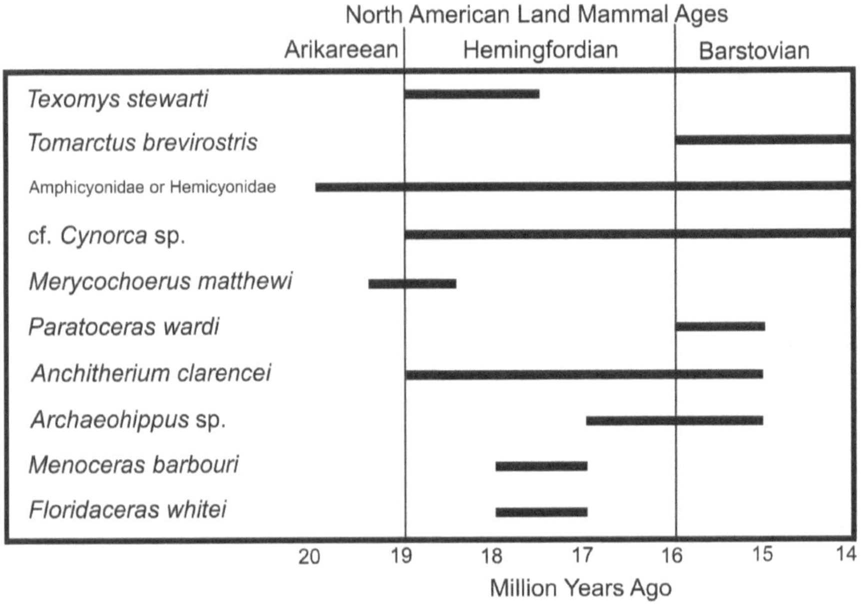

Figure 10. Paleogeographic reconstructions of Central America for (A) 20 Ma and (B) 15 Ma. Light gray represents the outline of tectonic plates containing continental or volcanic-arc crust. Dark gray represents subaerial land. Base maps for the reconstructions were derived from the ODSN Plate Tectonic Reconstruction Service (http://www.odsn.de/odsn/services/paleomap/paleomap.html). The location of subaerial land is based on this study and the distribution of Cretaceous to Tertiary continental and volcanic terranes as derived from Case and Holcombe (1980).

The Sr analyses show that the La Boca Formation measured at Section 1 is significantly older than the uppermost Culebra Formation (23.07–20.99 Ma versus 19.83–19.12 Ma) (Figure 9). If the La Boca Formation was stratigraphically higher than the uppermost Culebra Formation, then the Sr data should have indicated a younger age for the La Boca Formation. Furthermore, the upper Culebra and Cucaracha formations contain land mammal fossils that are late Hemingfordian to Barstovian in age (19.5–14 Ma), which is significantly younger than the La Boca Formation measured at Section 1 (23.07–20.99 Ma) (Figure 10). Other biostratigraphic data from benthic foraminifera, ostracodes, corals, and molluscs are consistent with an older, early Miocene age for the La Boca Formation (Blacut and Kleinpell, 1969; Johnson and Kirby, 2006; Van, 1972; Woodring, 1957–1982).

As our three data sets indicate that the Culebra model is the correct stratigraphic model for the formations exposed along the Gaillard Cut, we return the La Boca Formation containing the E.L. back to the lower Culebra Formation. This stratigraphic arrangement is consistent with that originally proposed by Woodring and Thompson (1949) (Figure 4). Our lithostratigraphic, biostratigraphic, and Sr chemostratigraphic results are discussed in greater detail below.

Culebra Formation

Lithostratigraphy

The Culebra Formation is at least 250 m thick and consists of three members that include a lower unnamed member, the E.L. and an upper unnamed member (Figure 11). The Culebra Formation contains a transgressive-regressive facies pattern, where the lower member, the E.L. and the lower part of the upper member show a transgressive pattern in which water depth deepened from intertidal at the base of the formation to upper bathyal (Blacut and Kleinpell, 1969). The upper part of the upper member shows a regressive pattern in which water depth shallowed from upper bathyal to intertidal depths. These three members are described individually below.

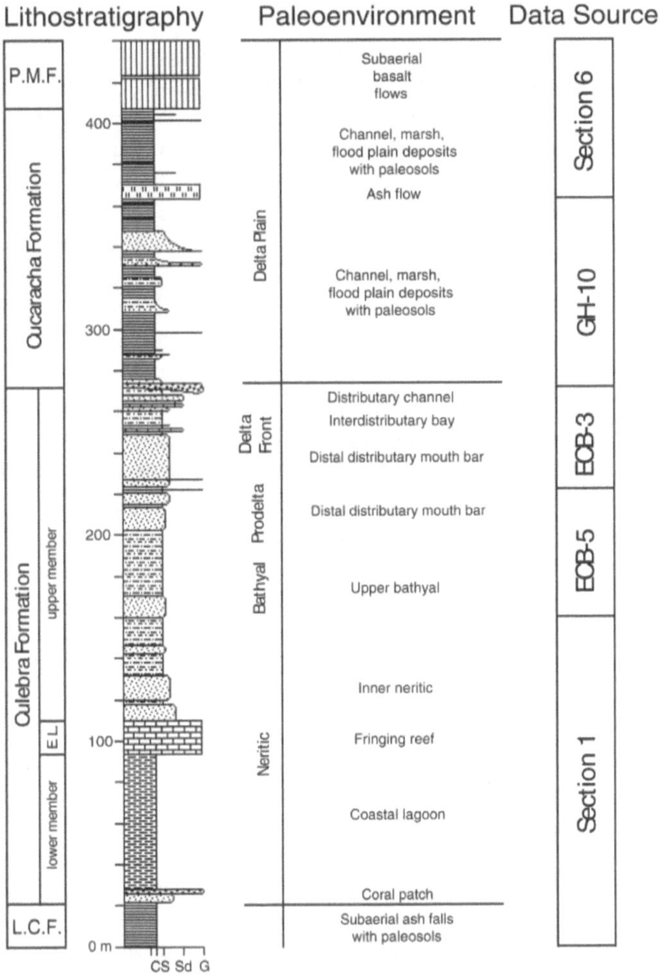

Figure 11. Scatter plot showing geologic age as a function of the 87Sr/86Sr ratio of fossil marine shells from the La Boca Formation, Emperador Limestone and upper Culebra Formation. Error bars represent standard error.

The lower member of the Culebra Formation is only exposed in Section 1 (Figure 6), where it conformably overlies the L.C.F. and conformably underlies the E.L. The L.C.F. consists of agglomerate with tuffaceous claystone interbeds representing subaerial volcanism with periods of paleosol development. The lower member of the Culebra Formation consists mostly of carbonaceous mudstone with thin tabular interbeds of fossiliferous lithic wacke. The base of the lower member is defined by a black lignitic mudstone bed that is overlain by calcarenite (calcareous sandstone) and pebble calcirudite (calcareous conglomerate). The top of the lower member is defined by a very distinctive bed of fine-grained, lithic wacke that contains pectinids (*Lepidopecten proterus*) and spondylids (*Spondylus scotti*). The lower member of the Culebra Formation represents the beginning of a transgressive sequence, with paleosols developed in volcanic sediments of the L.C.F. below the base of the Culebra Formation, and shallow-marine facies in the Culebra Formation (Figure 11). Carbonaceous mudstone and lignitic interbeds represent a shallow lagoon protected from the open ocean by a fringing reef (E.L.), which is consistent with carbonized compressions of sea grass and wood. This interpretation is also consistent with the presence of brackish Elphidium foraminifera and ostracodes in the lower member described by Blacut and Kleinpell (1969) and Van den Bold (1972), respectively. The presence of Elphidium and the absence of globigerinid foraminifera (the latter are common in siltstone in the upper member of the Culebra Formation) suggest a current-protected, nearshore environment (Blacut and Kleinpell, 1969). The lithic wacke interbeds represent storm deposits in the lagoon. Burrows in the underlying mudstone were rapidly infilled by sand during the storm events, thereby preserving the trace fossil *Thalassinoides* sp. The lowermost calcarenite and calcirudite beds within the carbonaceous mudstone represent bioclastic debris likely derived from patches of coral in the lagoon (Johnson and Kirby, 2006).

The E.L. is the middle member of the Culebra Formation and consists of five distinct facies in Section 1 (Johnson and Kirby, 2006). The base of the E.L., which overlies conformably the "pectinid-spondylid" sandstone bed in the lower member of the Culebra Formation, is defined by a branching-coral boundstone containing abundant Acropora saludensis and Montastraea canalis in a very fine-grained calcarenite matrix. The second facies consists of white, rhodolithic limestone. The third facies consists of another branching-coral boundstone dominated by *Acropora saludensis*, *Stylophora granulate*, and *Porites douvillei* as part of a diverse assemblage of corals in a mud matrix with isolated coral heads of *Montastraea imperatoris* in life position. The fourth facies consists of platy-coral boundstone. The top of the E.L. is defined by the fifth facies, which consists of calcirudite containing fragmented corals in a calcarenite matrix and displaced head corals of Montastraea species and massive Porites species. The E.L. represents a fringing reef that protected a neighboring lagoon, as represented by the carbonaceous mudstone facies in the underlying lower member (Johnson and Kirby, 2006). The E.L. is overlain conformably by alternating beds of sandstone and siltstone of the upper member of the Culebra Formation.

The upper member of the Culebra Formation consists of five distinct facies and is best exposed in Sections 1, 3–5, and 7 (Figure 6). The base is defined by alternating beds of sandstone and siltstone that conformably overlie the E.L. in Section 1. The lowermost sandstone bed contains displaced corals and abundant molluscs. Massive

Porites species and head corals of Montastraea canalis are clearly not in situ, as they show different orientations with respect to bedding. The sandstone beds become thinner and finer upsection, whereas the interbedded siltstone beds thicken. In addition, the sandstone beds grade upsection from a medium-grained calcarenite to a fine-grained, lithic wacke, such that carbonate grains decrease in abundance, whereas quartz and lithic grains increase in abundance. There is also a lateral facies change between these two facies, such that the sandstone beds become thinner and the siltstone beds thicken to the north over 1 km of exposure. The siltstone beds contain abundant foraminifera, molluscs, echinoids, shark teeth, and burrows. A distinctive white ash bed (4 cm thick) is present at the base of a tuffaceous sandstone bed (1 m thick). This ash bed and overlying tuffaceous sandstone, which serves as a useful marker bed (Figure 6), is overlain by a thick interval of siltstone (~30 m). Overlying this thick interval of siltstone is the uppermost interval of the upper member, which consists of alternating sandstone and mudstone beds with local conglomerate and lignite beds (Sections 3–5, 7, ECB-3, ECB-5, and GH-10). The sandstone beds become thicker and coarser upsection, whereas the mudstone beds become thinner and more carbonaceous upsection. A distinctive lignite bed is present in drill holes ECB-3 and ECB-5, which serves as a useful marker bed (Figure 6). Carbonaceous mudstone beds commonly contain horizons of carbonized to permineralized wood of mangrove trees. Branches and prop roots are commonly encrusted with oysters (*Crassostrea aff.* C. virginica) and are commonly bioeroded by teredinid bivalves (Kuphus "incrassatus"). Also common in the carbonaceous mudstone are poorly preserved seeds, gastropods (*Turritella venezuelana, Turritella* (Bactrospira?) *amaras, Potamides suprasulcatus*), bivalves and crustaceans. The top of the Culebra Formation is defined by the incision into either carbonaceous mudstone or sandstone (depending upon location) of a channel infilled with pebble conglomerate containing wood without teredinid borings of the Cucaracha Formation.

The lower half of the upper member represents continuing transgression through time; whereas, the upper half represents the start of a regression with shallowing water depth through time (Figure 11). The lowermost calcarenite beds above the E.L. represent open-shelf, neritic conditions. The decrease in bed thickness, in grain size and in carbonate content of these sandstone beds, as well as the thickening siltstone beds upsection, indicate increasing water depth through time. This interpretation is consistent with Blacut and Kleinpell (1969), who described benthic foraminifera representing upper bathyal depths (~200–400 m) from these siltstone beds. This interpretation is also consistent with Van den Bold (1972), who described ostracodes representing "deep-water" depths. During deposition of the middle portion of the upper member, water depth began to shallow, with sandstone beds increasing in frequency, thickening and coarsening, and mudstone beds becoming thinner and increasing in carbonaceous plant content. All of these observations suggest a small prograding, river-dominated delta (Bhattacharya and Walker, 1992; Pettijohn et al., 1987), perhaps on a similar scale to the present-day Rio Grande in Bocas del Toro, Panama. The thick siltstone interval represents the prodelta, whereas the overlying sequence of alternating mudstone and sandstone beds represent the delta front (Figure 11). Sandstone interbeds represent distal distributary mouth bars. Local lenses of pebble conglomerate represent distributary-channel deposits. Carbonaceous mudstone beds represent interdistributary bay deposits with mangal associations of mangrove and oyster.

Land Mammal Biostratigraphy

Traces of land mammals are extremely rare in the Culebra Formation. We found three molars representing three taxa in the uppermost part of the Culebra Formation in the course of our work. These discoveries are the first definite report of land mammal fossils from the marine Culebra Formation (although see Woodring (1957–1982), who reported a partial ungulate metapodial from what he described as the transition zone between the Culebra and Cucaracha formations). We found molars of the artiodactyl *Paratoceras wardi* and peccary cf. *Cynorca* sp. at the top of the Culebra Formation in Section 3 (Figure 6), as well as a molar of the rhinoceros *Menoceras barbouri* in the uppermost part of the Culebra Formation in Section 5 (Figure 6). Based on their respective ages in North America, *Paratoceras wardi* is Barstovian (16–15 Ma), cf. *Cynorca* sp. is probably early Hemingfordian to Barstovian (18.8–14 Ma) and *Menoceras barbouri* is Hemingfordian (18–17 Ma) (Figure 10) (MacFadden, 2006; Wright1 et al., 1998). Taken together, these fossils suggest an age of between 18.8 and 14 Ma for the uppermost part of the Culebra Formation.

Strontium Chemostratigraphy

In order to derive age estimates for the Culebra and La Boca formations, we collected seven samples of fossil coral and bivalves from two sections containing the La Boca and upper Culebra formations and analyzed them for their 87Sr/86Sr ratios (Figure 9; Table 2). A fossil bivalve-shell fragment from the calcirudite bed near the base of the La Boca Formation in Section 1 (i.e., the lower member of the Culebra Formation) had an estimated age of 23.07 ± 0.53 Ma, based on its 87Sr/86Sr ratio. Two pectinid bivalves from the "pectinid-spondylid" sandstone bed below the overlying E.L. had an estimated age of 20.62 ± 0.58 and 20.99 ± 0.46. Two Acropora coral specimens from the upper branching facies of the E.L. had an estimated age of 20.99 ± 0.71 and 21.24 ± 0.44. In Section 3, two pectinid bivalves collected from a fine-grained calcarenite bed (2 m below the conglomeratic sandstone containing *Paratoceras wardi* and cf. *Cynorca* sp. specimens) of the upper member of the Culebra Formation had an estimated age of 19.12 ± 0.42 and 19.83 ± 0.39. Taken together, the samples from the La Boca Formation are 1–4 m.y. older than the samples from the upper Culebra Formation (Figure 9, Table 2).

Cucaracha Formation

Lithostratigraphy

The Cucaracha Formation is about 140 m thick and consists mostly of claystone with a minor amount of conglomerate, sandstone, lignite, and welded tuff (Figure 11). Lenticular beds of conglomerate and sandstone are more common in the lower half of the formation below a distinctive welded tuff bed of volcanic origin, whereas tabular beds of claystone and lignite are more common in the upper half above the welded tuff bed (more specifically, the lower half of the formation has a sandstone/claystone ratio of 24.1%; whereas, the upper half has a sandstone/claystone ratio of only 5.7%). The base of the Cucaracha Formation is marked by a distinctive pebble conglomerate bed that lies unconformably over the Culebra Formation (Figure 6). This conglomerate bed is widely distributed and contains volcanic pebble clasts with rare fragments of carbonized wood (without teredinid borings) and oysters. This and other pebble conglomerate beds higher up in the Cucaracha Formation commonly become finer upsection, grading into lithic wacke, siltstone and claystone. Medium to coarse-grained, lithic wacke beds are commonly cross-bedded, which show an average paleocurrent direction to the east (N87°E ± 6.4° (95% confidence cone, Fisher analysis)). These interbedded channel deposits contain permineralized logs of up to 1 m in length and 30 cm in diameter oriented parallel to bedding. Pebble conglomerate and lithic wacke also contain rare fossils of land mammals. Olive-gray to blackish red claystone is the most common lithology in the Cucaracha Formation. This claystone is commonly structureless to slickensided, but may contain mottling and drab-haloed root traces. Horizons of calcite nodules and rhizoconcretions are common throughout the claystone. Two horizons contain spherical to platy barite nodules (~2 cm in diameter) in olive-gray claystone. Fossils of land mammals, turtles, fish, crocodiles, and gastropods (*Hemisinus* (Longiverena) *oeciscus*) are present locally in claystone, as noted by Whitmore and Stewart (1965), Woodring (1957–1982), and MacFadden (2006). Four lignite beds are present in the upper half of the Cucaracha Formation (Section 8; Figure 11). The Cucaracha Formation contains a distinctive bed of welded tuff 4.3–7.7 m thick (also known colloquially as the "ash flow" (Woodring, 1957–1982; Woodring and Thompson, 1949)), which is broadly distributed and serves as a useful marker bed (Figure 6).

The Cucaracha Formation represents a coastal delta plain that consists of channel, levee, flood plain and marsh deposits (Figure 11). Abundant paleosols indicate that soils commonly developed on these deposits. Retallack and Kirby (2007) recognized 12 different pedotypes that represent as many vegetation types, including mangrove, freshwater swamp, marine-influenced swamp, early successional riparian woodland, colonizing forest, dry tropical forest, and woodland. Oxygen and carbon isotopic analyses of land mammal teeth are consistent with these interpretations, as they indicate diverse, C3 plant communities, possibly ranging from dense forest to more open woodland (MacFadden and Higgins, 2004). The pebble conglomerate bed at the base of the Cucaracha Formation represents a fluvial-channel deposit that is broadly distributed (based on its geometry and sedimentology, which are typical of fluvial-channel deposits (Miall, 1978, 1992)). Incision of this channel into underlying marine mudstone and sandstone of the Culebra Formation indicates that part of the underlying section has been eroded by the channel. The pebble conglomerate contains fragments of wood that show no evidence of teredinid borings (unlike the wood found in the underlying Culebra Formation), suggesting that this basal conglomerate was deposited above sea level. The presence of oyster fragments probably represents reworking of the underlying marine Culebra Formation. Interbedded lenses of pebble conglomerate and lithic wacke further upsection represent small fluvial channels, based on their lenticular geometry and sedimentology (Miall, 1978, 1992; Pettijohn et al., 1987). The small ratio of channel deposits to claystone (the sandstone/claystone ratio for the entire formation is 18.4%) suggests that these were small meandering channels (there is generally a good correlation between channel pattern and sediment load, such that the sandstone/shale ratio provides a clear view of stream type, where meandering channels have relatively low ratios and braided channels have high ratios (Pettijohn et al., 1987)). Thick sequences of claystone represent flood-basin deposits on the coastal delta plain. Most intervals of claystone show some evidence of soil development (Retallack and Kirby, 2007). Evidence for paleosols include horizons of calcite and barite nodules, rhizoconcretions, drab-haloed root traces, mottling, relict bedding, gradational contacts between soil horizons C, B, and A, and abrupt contacts between soil horizon A and overlying sediment (criteria of (Retallack, 2001)). Paleosols indicate periods of stability in between fluvial events of 1,000–10,000 years when soils developed on flood-basin or channel deposits (Retallack, 2001). The four lignite interbeds represent histosols of tidal or poorly drained distributaries that penetrated the coastal delta plain, where thick vegetation resulted in the accumulation of much organic matter into layers of peat within marshes (Reineck and Singh, 1975). The single interbed of welded tuff represents a pyroclastic, ash-flow deposit (ignimbrite) produced by a nearby explosive eruption. Conformably overlying the Cucaracha Formation is a basalt flow of the P.M.F. (Section 8). Underlying claystone in the Cucaracha Formation shows baking and the overlying basalt shows hydrothermal alteration.

Land Mammal Biostratigraphy

We found fossils of land mammals throughout the Cucaracha Formation. Land mammal fossils of the peccary cf. *Cynora* sp., the artiodactyl *Paratoceras wardi*, the oreodont *Merycochoerus matthewi* and the rhinoceroses *Menoceras barbouri* and *Floridaceras whitei* were found in Sections 6, 7, and 8 (Figure 6). Taken together, the age of these land mammals indicates a latest Arikareean to middle Barstovian age (19.5–14 Ma), with a middle Hemingfordian age (18–17 Ma) likely (Figure 10).

CONCLUSIONS

Lithostratigraphic, biostratigraphic, and Sr chemostratigraphic analyses demonstrate for the first time that the main axis of the volcanic arc in southern Central America more than likely existed as a peninsula connected to northern Central America and North America for much of the Miocene. The Culebra Formation dates from 23–19 Ma, with the E.L. dating from 21 Ma. The overlying Cucaracha Formation dates from 19 to possibly 14 Ma. What has been called the La Boca Formation underlies, not overlies, the Cucaracha Formation. We, therefore, re-interpret the La Boca Formation (with the E.L.) as the lower part of the Culebra Formation, as originally proposed by Woodring and Thompson (1949).

Our revised stratigraphy for the Gaillard Cut shows that the Culebra Formation represents a transgressive-regressive, marine sequence with environments that include, from lowermost to uppermost: lagoon, fringing reef, neritic, upper bathyal, and prograding delta. Bathyal sediments in the upper member of the Culebra Formation suggest that a short-lived strait may have existed across the Panama Canal Basin sometime between 21 and 19 Ma. The overlying Cucaracha Formation represents a coastal delta plain with environments that include fluvial channel, overbank, flood-plain, and distributary channel marsh, all with extensive development of paleosols representing mangrove, swamp, woodland, and dry tropical forest vegetation types. Both the uppermost Culebra and Cucaracha formations contain fossil land mammals that are Hemingfordian to Barstovian in age (19.5–14 Ma).

The earliest evidence for a terrestrial connection between Panama and North America is 19 Ma, based on fossil land mammals with only North American affinities and Sr analyses of fossil corals and bivalves. Our revised stratigraphy for the Gaillard Cut demonstrates that the Central American Peninsula was not short-lived in the early Miocene. We instead find no evidence for the disruption of this peninsula until 6 Ma, when there is evidence for a short-lived strait across the Panama Canal Basin. The existence of a peninsula for much of the Miocene has profound implications for our understanding of the tectonic, climatic, oceanographic, and biogeographic history related to the formation of the Isthmus of Panama.

KEYWORDS

- **Claystone**
- **Culebra Formation**
- **Gaillard Cut**
- **Gatun Formation**
- **Panama Canal Authority**

AUTHOR' CONTRIBUTIONS

Conceived and designed the experiments: Michael Xavier Kirby, Douglas S. Jones, and Bruce J. MacFadden. Performed the experiments: Douglas S. Jones. Analyzed the data: Michael Xavier Kirby, Douglas S. Jones, and Bruce J. MacFadden. Contributed reagents/materials/analysis tools: Douglas S. Jones. Wrote the chapter: Michael Xavier Kirby.

ACKNOWLEDGMENTS

Michael Xavier Kirby is grateful to A. Coates and J. Jackson for supporting the early part of this research when Michael Xavier Kirby was a postdoctoral fellow at the Smithsonian Tropical Research Institute between 2003 and 2004. We thank the ACP for granting access to localities, F. C. de Sierra and J. Villa of the General de Recursos Minerales, Republic of Panama, for granting the necessary permits, and P. Franceschi, D. Irving, and J. Arrocha for providing subsurface data and help in the field. We also thank R. Cooke, A. Crawford, K. Johnson, E. G. Leigh, Jr., A. O'Dea, J. Ramesch, G. Retallack, F. Rodriguez, N. Smith, and S. Stanley for assistance and/or useful discussions. We thank K. Campbell, E. G. Leigh, Jr., and G. J. Vermeij for improving the presentation of our ideas.

Chapter 13

Remote Sensing Data with the Conditional Latin Hypercube Sampling

Yu-Pin Lin, Hone-Jay Chu, Cheng-Long Wang, Hsiao-Hsuan Yu, and Yung-Chieh Wang

INTRODUCTION

This study applies variogram analyses of normalized difference vegetation index (NDVI) images derived from SPOT HRV images obtained before and after the Chi-Chi earthquake in the Chenyulan watershed, Taiwan, as well as images after four large typhoons, to delineate the spatial patterns, spatial structures, and spatial variability of landscapes caused by these large disturbances. The conditional Latin hypercube sampling (LHS) approach was applied to select samples from multiple NDVI images. Kriging and sequential Gaussian simulation (SGS) with sufficient samples were then used to generate maps of NDVI images. The variography of NDVI image results demonstrate that spatial patterns of disturbed landscapes were successfully delineated by variogram analysis in study areas. The high-magnitude Chi-Chi earthquake created spatial landscape variations in the study area. After the earthquake, the cumulative impacts of typhoons on landscape patterns depended on the magnitudes and paths of typhoons, but were not always evident in the spatiotemporal variability of landscapes in the study area. The statistics and spatial structures of multiple NDVI images were captured by 3,000 samples from 62,500 grids in the NDVI images. Kriging and SGS with the 3,000 samples effectively reproduced spatial patterns of NDVI images. However, the proposed approach, which integrates the conditional LHS (cLHS) approach, variogram, kriging, and SGS in remotely sensed images, efficiently monitors, samples, and maps the effects of large chronological disturbances on spatial characteristics of landscape changes including spatial variability and heterogeneity.

The influences of large physical disturbances on ecosystem structure and function have garnered considerable attention (Foster et al., 1998; Millward and Kraft, 2004; Swanson et al., 1998; Turner and Dale, 1998). Fires, hurricanes (typhoons), tornados, ice storms, and landslides are examples of such large disturbances (Millward and Kraft, 2004). Earthquakes have long been recognized as a major cause of landslides (Keefer, 1984; Lin et al., 2008a). However, landslides are only the first in a series of processes by which materials are removed from slopes and transported out of a region by fluvial action (Keefer, 1994; Lin et al., 2008a). Additionally, typhoons are extremely important natural disturbances that characterize the structure, function, and dynamics of many tropical and temperate forest ecosystems (Lee and Lin, 2008). Taiwan, which is located in a subtropical region, sits on the Philippine plate at the Euro-Asian Plate junction (DeMets et al., 1990). Plate convergence occasionally generates earthquakes that have disastrous effects on Taiwan (Lin et al., 2006a). Moreover, typhoons that bring tremendous amounts of rainfall hit Taiwan every year from July to October (Lin et al., 2008b). During 1996–2004, large disturbances in the following sequence impacted central Taiwan: (1) typhoon Herb (August, 1996); (2) the Chi-Chi earthquake (September, 1999); (3) typhoon Xangsane (November, 2000); (4) typhoon Toraji (July, 2001); (5) typhoon Dujuan (September, 2003); and, (6) typhoon Mindulle (June, 2004) (Lin et al., 2008b). In particular, after the Chi-Chi earthquake, the expansion rate of landslide areas increased 20-fold in central Taiwan (Lin et al., 2003). Numerous extension cracks, which accelerate landslides during downpours, were generated on hill slopes during the Chi-Chi earthquake (Lin et al., 2006b). Moreover, during typhoon seasons, a massive amount of loose earth and stones accumulated on the surface of slopes, increasing the risk of debris flows, and additional landslides (Lin et al., 2001) that worsen the revegetation problem. Accordingly, monitoring, delineating and sampling landscape changes, spatial structure and spatial variation induced by large physical disturbances are essential to landscape management and restoration, and disaster management in Taiwan.

Remotely sensed data can describe surface processes, including landscape dynamics, as such data provide frequent spatial estimates of key earth surface variables (Garrigues et al., 2008b; Sellers, 1997). For example, the SPOT, LANSAT, and MODIS data sets have notable advantages that account for their use in ecological applications, including a long-running historical time-series, a special resolution appropriate to regional land-cover and land-use change investigations, and a spectral coverage appropriate to studies of vegetation properties (Cohen and Goward, 2004; Hayes and Cohen, 2007; Tarnavsky et al., 2008). The NDVI, a widely used vegetation index, is typically used to quantify landscape dynamics, including vegetation cover and landslides changes induced by large disturbances (Garrigues et al., 2008b; Lee and Lin, 2008; Lin et al., 2008a, 2008b, 2008c). Notably, NDVI images can be determined by simply geometric operations near-infrared and visible-red spectral data almost immediately after remotely sensed data is obtained. The NDVI, which is the most common vegetation index, has been extensively used to determine the vigor of plants as a surrogate measure of canopy density (Jensen, 1996a). A high NDVI indicates a high level of photosynthetic activity (Sellers, 1985). Moreover, significant differences in NDVI images before and after a natural disturbance can represent landscape changes, including vegetation and landslides induced by a disturbance that changes plant-covered land to bare lands or bare lands to plant-covered land (Lin et al., 2006c).

Spatial patterns in ecological systems are the result of an interaction among dynamic processes operating across abroad range of spatial and temporal scales (Lobo et al., 1998; Urban et al., 1987; Wiens, 1989). Ecological manifestations of large disturbances are rarely homogeneous in their spatial coverage (Millward and Kraft, 2004). Variograms are crucial to geostatistics. A variogram is a function related to the variance to spatial separation and provides a concise description of the scale and pattern of spatial variability (Curran and Atkinson, 1998). Samples of remotely sensed data (e.g., satellite or air-borne sensor imagery) can be employed to construct variograms for remotely sensed research (Curran and Atkinson, 1998). Moreover, variograms have been used widely to understand the nature and causes of spatial variation within an image (Garrigues et al., 2008a). Modeling the variogram of NDVI images with high spatial resolution is an efficient approach for characterizing and quantifying heterogeneous spatial components (spatial variability and spatial structure) of a landscape and the spatial heterogeneity of vegetation cover at the landscape level (Garrigues et al., 2006, 2008a).

Reliable data analysis of spatially distributed data requires the use of appropriate statistical tools and a sound data sampling strategy (Fortin and Edwards, 2001). Spatial sampling schemes have been developed to determine the sampling locations that cover the variation in environmental properties in a given area (Minasny and McBratney, 2007). Moreover, data samples are transformed via a series of interpretation steps to obtain complete descriptions of phenomena of interest (Edwards and Fortin, 2001). Different sampling schemes are, say, random, systematic, stratified, or nested schemes (Edwards and Fortin, 2001; Thompson, 1992). The LHS is a stratified random procedure that is an efficient way of sampling variables from their multivariate distributions (Minasny and McBratney, 2006). Initially developed for Monte-Carlo simulation, LHS efficiently selects input variables for computer models (Iman and Conover, 1980; McKay et al., 1979). Kriging, a geostatistical method, is a linear interpolation approach that provides a best linear unbiased estimator (BLUE) for quantities that vary spatially (Lin et al., 2008d). However, kriging interpolate algorithms generate maps of best local estimate and generally smooth out the local details of the spatial variation of an attribute (Goovaters, 1997). For sampled data, a geostatistical conditional simulation technique, such as SGS, can be applied to generate multiple realizations, including an error component, which is absent from classical interpolation approaches (Lin et al., 2008d). In such conditional simulations, all generated realizations reproduce available data at measurement locations, and, on average, reproduce a data histogram and a model of spatial correlations (i.e., variogram) between observations (Kyriakidis, 2001). In SGS, Gaussian transformation of available measurements is simulated, such that each simulated value is conditional on original data and all previously simulated values (Deutsch and Journel, 1992; Goovaters, 1997; Kyriakidis, 2001; Lin et al., 2008d). Geostatistical conditional simulations have been widely applied to simulate the spatial variability and spatial distribution of interest in many fields. Moreover, geostatistical simulation techniques with LHS have been applied to simulate Gaussian random fields (Kyriakidis, 2001; Pebesma and Heuvelink, 1999; Xu et al., 2005; Zhang and Pinder, 2004).

This study applied variogram analysis to delineate spatial variations of NDVI images before and after large physical disturbances in central Taiwan. The NDVI data derived from SPOT images before and after the Chi-Chi earthquake (ML = 7.3 on the Richter scale) in the Chenyulan basin, Taiwan, as well as images before and after four large typhoons (Xangsane, Toraji, Dujuan, and Mindulle) were analyzed to identify the spatial patterns of landscapes caused by these major disturbances. Landscape spatial patterns of different disturbance regimes were discussed. Moreover, cLHS schemes with NDVI images were used to select spatial samples from actual NDVI images to detect landscape changes induced by a series of large disturbances. The best cLHS samples selected with the NDVI values were used to estimate and simulate NDVI distributions using kriging and SGS. The simulated NDVI images were compared with actual NDVI images induced by the disturbances.

MATERIALS AND METHODS

Study Area and Remote Sensing Data

The Chenyulan watershed, located in central Taiwan, is a classical intermountain watershed, and has an average altitude of 1,540 m and an area of 449 km^2 (Figure 1). The Chenyulan stream, which coincides with the Chenyulan fault, flows from south to north and elongates the watershed in the same direction. Differences in uplifting along the fault generated abundant fractures over the watershed and resulted in an average slope of 62.5% and relief of 585 m/km^2. Moreover, the main course of the Chenyulan stream had a gradient of 6.1%, and more than 60% of its tributaries had gradients exceeding 20%. The special geological and geographical characteristics of the watershed result in frequent landsides and debris flows (Lin et al., 2003). The September 21, 1999, Chi-Chi earthquake occurred at 1:47 am local time (17:47:18 GMT the previous day) at an epicentral location of 23.85_N and 120.78_E and at a depth 6.99 km (Figure 1). It was caused by a rupture in the Chelungpu Fault. The magnitude of the earthquake was estimated to be ML = 7.3 (ML: Local Magnitude or Richter Magnitude), and the rupture zone, defined by the aftershocks, measured about 80 km north-south by 25–30 km downdip (Lin et al., 2006a; Roger and Yu, 2000). Iso-contour maps of the earthquake's magnitude were reproduced from the Central Weather Bureau (Figure 1) (Central Weather Bureau, 1999). After the earthquake, from October 31, 2000 to November 1,2000, the center of typhoon Xiangsane moved from south to north through eastern Taiwan (Central Weather Bureau, 2000), with a maximum wind speed of 138.9 km/hr and a radius of 250 km (Figure 1). The maximum daily rainfall was 550 mm/day. On July 30, 2001, the Toraji typhoon swept across central Taiwan from east to west (Central Weather Bureau, 2001), with a maximum wind speed of 138.9 km/hr and a radius of 180 km (Figure 1). The typhoon brought extremely heavy rainfall, from 230 to 650 mm/day, and triggered more than 6,000 landslides in Taiwan. After crossing Taiwan, typhoon Toraji became a tropical storm; however it brought 339–757 mm of total accumulated rainfall in the watershed (Central Weather Bureau, 2001) (Figure 1). After typhoon Toraji, typhoons Dujuan with a maximum wind speed of 165.0 km/hr, a radius of 200 km and maximum rainfall 200 mm/hr (August 31, 2003–September 2, 2003) and Mindulle with maximum wind speed of 200.0 km/hr, a radius of 200 km and maximum rainfall 166 mm/hr (June 29, 2004–July 2, 2004) chronologically produced heavy rainfall that fell across the eastern and central parts of Taiwan on September 2003 and June 2004 (Central Weather Bureau, 2008) (Figure 1). The two study area with dimensions of 50×50 km^2 (250×250 pixels) was selected from the upstream of the large debris flood announced in the watershed, as shown in Figure 1.

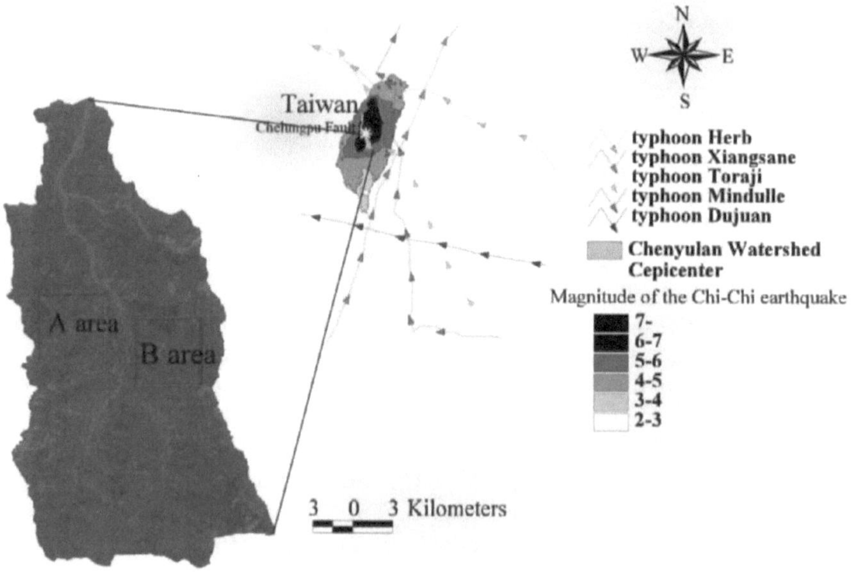

Figure 1. Location of the study areas.

NORMALIZED DIFFERENCE VEGETATION INDEX (NDVI)

Seven cloud-free SPOT images (1996/11/08, 1999/03/06, 1999/10/31, 2000/11/27, 2001/11/20, 2003/12/17, and 2004/11/19) of the Chenyulan watershed were purchased from the Space and Remote-sensing Research Center, Taiwan. The NDVI images of the study area were generated from SPOT HRV images with a resolution of 20 m according to the following equation:

$$NDVI = \frac{NIR - R}{NIR + R} \tag{1}$$

where NIR and R are near-infrared and visible-red spectral data, respectively. The NDVI values range from 1 to +1; a high NDVI value represents a large amount of high photosynthesizing vegetation (Jensen, 1996b).

Variogram and Kriging Estimation

In geostatistical methods, variograms can be used to quantify the observed relationship between the values of samples and the proximity of samples (Lin et al., 2008d). Following the work of Garrigues et al. (2006), Garrigues et al. (2008b), and Lin et al. (2008), NDVI data are considered values of punctual regionalized variable. An experimental variogram for interval lag distance class h, γ(h), is represented by

$$\gamma(h) = \frac{1}{2n(h)} \sum_{i=1}^{n(h)} \left[Z(x_i + h) - Z(x_i) \right]^2 \tag{2}$$

where h is the lag distance that separates pairs of points; $Z(x)$ is bird diversity at location x, and $Z(x + h)$ is bird diversity at location $x + h$; $n(h)$ is the number of pairs separated by lag distance h.

Kriging is estimated using weighted sums of adjacent sampled concentrations. The weights depend on the correlation structure exhibited. The weights are determined by minimizing estimated variance. In this context, kriging estimates (BLUE) are the most accurate of all linear estimators. Accordingly, kriging estimates the value of the random variable at unsampled location X_0 based on measured values in a linear form:

$$Z^*(x_0)=\sum_{i=1}^{N}\lambda_{i0}Z(x_i) \tag{3}$$

where $Z^*(x_0)$ is the estimated value at location x_0, λ_{i0} is the estimation weight of $Z(x_i)$, x_i is the location of sampling point for variable Z, and N is the number of the variable Z involved in the estimation. Based on non-biased constraints and minimizing estimation variance, estimated kriging variance can be presented as:

$$\sigma^2_{kriging}=\sum_{i=1}^{N}\lambda_{i0}\gamma_{zz}(x_i-x_0)+\mu \tag{4}$$

where μ is the Lagrange multiplier.

Conditional Latin Hypercube

The cLHS, which is based on the empirical distribution of original data, provides a full coverage of range each variable by maximally stratifying the marginal distribution and ensuring a good spread of sampling points (Minasny and McBratney, 2006). This sampling procedure represents an optimization problem: given N sites with ancillary data (Z), select n sample sites ($n << N$) such that the sampled sites z form a Latin hypercube. For continuous variables, each component of X (size, $N \times k$) is divided into n (sample size) equally probable strata based on their distributions, and x (size $n \times k$) is a sub-sample of X. The procedures of the cLHS algorithm (Minasny and McBratney, 2006) are follows.

1. Divide the quantile distribution of X into n strata, and calculate the quantile distribution for each variable, $q^i_j, \ldots, q^{n+1}_j$. Calculate the correlation matrix for Z (C).
2. Pick n random samples from N; z (i=1,..., n) are the sampled sites. Calculate the correlation matrix of x (T).
3. Calculate the objective function. The overall objective function is $O = w_1O_1 + w_2O_2 + w_3O_3$, where w is the weight given to each component of the objective function. For general applications, w is set to 1 for all components of the objective function.

A. For continuous variables,

$$O_1 = \sum_{i=1}^{n}\sum_{j=1}^{k}\left|\eta\left(q_j \le x_j \le q_j^{i+1}\right) - 1\right| \tag{5}$$

where $\eta\left(q_j \le x_j \le q_j^{i+1}\right)$ is the number of x_i that falls between quantiles q^i_j and q^{i+1}_j

B. For categorical data, the objective function is to match the probability distribution for each class of:

$$O_2 = \sum_{j=1}^{c}\left|\frac{\eta'(z_j)}{n} - k_j\right| \tag{6}$$

where $\eta'(x_i)$ is the number of x that belongs to class j in sampled data, and k_i is the proportion of class j in X.

C. To ensure that the correlation of the sampled variables will replicate the original data, another objective function is added:

$$O_3 = \sum_{i=1}^{k}\sum_{j=1}^{k}\left|c_{ij} - t_{ij}\right| \tag{7}$$

where c is the element of C, the correlation matrix of X, and t is the equivalent element of T, the correlation matrix of x.

4. Perform an annealing schedule (Press et al., 1992): M = exp[-ΔO/T], where ΔO is the change in the objective function, and T is a cooling temperature (between 0 and 1), which is decreased by a factor d during each iteration.
5. Generate a uniform random number between 0 and 1. If rand <M, accept the new values; otherwise, discard changes.
6. Try to perform changes: Generate a uniform random number rand. If rand < P, pick a sample randomly from x and swap it with a random site from unsampled sites r. Otherwise, remove the sample(s) from x that has the largest $\eta\left(q_j^i \le x_j \le q^{i+1}\right)$ and replace it with a random site(s) from unsampled sites r. End when the value of P is between 0 and 1, indicating that the probability of the search is a random search or systematically replacing the samples that have the worst fit with the strata.
7. Go to step 3 Repeat steps 3–7 until the objective function value falls beyond a given stop criterion or a specified number of iterations.

Sequential Gaussian Simulation

In sequential simulation algorithm, modeling of the N-point cumulative density function (ccdf) is a sequence of N univariate ccdfs at each node (grid cell) along a random path (Kyriakidis, 2001). The sequential simulation algorithm has the following steps (Kyriakidis, 2001):

1. Establish a random path that is visited once and only once, all nodes {u_i, i = 1, Λ, N} discretizing the domain of interest Doman. A random visiting sequence ensures that no spatial continuity artifact is introduced into the simulation by a specific path visiting N nodes.

2. At the first visited N nodes u_1:
   A. Model, using either a parametric or nonparametric approach, the local ccdf of $Z(u_1)$ conditional on n original data $\{Z(u_\alpha), \alpha = 1,\Lambda, n\}$ $F_Z(u_1; z_1|(n)) = \text{prob}\{Z(u_1) \leq z_1|(n)\}$
   B. Generate, via the Monte Carlo drawing relation, a simulated value $z^{(l)}(u_1)$ from this ccdf $F_Z(u_1: z_1|(n))$, and add it to the conditioning data set, now of dimension n + 1, to be used for all subsequent local ccdf determinations.
3. At the i_{th} node u_i along the random path:
   A. Model the local ccdf of $Z(u_i)$ conditional on n original data and the i -1 near previously simulated values $\{ z^{(l)}(u_j), j = 1,\Lambda, i -1\}$:

$$F_Z(u_1; z_{i|}\ (n+i\ \ 1)) = \text{prob}\{Z(u_i) \leq z_i\ (n+i\ 1)\} \quad (8)$$

   B. Generate a simulated value $z^{(l)}(u_i)$ from this ccdf and add it to the conditioning data set, now of dimension n + i.
4. Repeat step 3 until all N nodes along the random path are visited.

The SGS assumes a Gaussian random field, such that the mean value and covariance completely characterize the ccdf (Press et al., 1998). During the SGS process, Gaussian transformation of available measurements is simulated, such that each simulated value is conditional on original data and all previously simulated values (Deutsch and Journel, 1992; Lin et al., 2001). A value simulated at a one location is randomly selected from the normal distribution function defined by the kriging mean and variance based on neighborhood values. Finally, simulated normal values are back-transformed into simulated values to yield the original variable. The simulated value at the new randomly visited point value depends on both original data and previously simulated values. This process is repeated until all points have been simulated.

RESULTS AND DISCUSSION

Statistics and Spatial Structures of NDVI Images

Statistics of remotely sensed images can be used as a basic tool to characterize landscape changes (Akiwumi and Butler, 2008; Fox et al., 2008; Giriraj et al., 2008; Ward et al., 2000; Zomeni et al., 2008). Table 1 summarizes the statistics for seven actual NDVI images of areas A and B before and after disturbances. The lowest mean and minimum NDVI values in 1996–2004 occurred on March 6, 1999, after the Chi-Chi earthquake in both areas A and B areas. Moreover, the largest range between minimum and maximum NDVI values also occurred on March 6, 1999, after the Chi-Chi earthquake in both areas A and B. The most negative minimum NDVI values occurred on November 27, 2000, and December 17, 2003, in both areas A and B. On these dates, the standard deviations of NDVI values were slightly larger than those on other dates. These statistical results illustrate that the Chi-Chi earthquake had the largest impact on all landscapes represented by NDVI images for areas A and B. The second and third greatest impacts on all landscapes are from typhoons Xangsane (November, 2000) and Dujuan (September, 2003) in areas A and B, respectively (Figures 2 and 3 and Table 1). Particularly, typhoon Xangsane right after the Chi-Chi earthquake was the second disturbance to impact landscape changes in the study areas. Numerous extension cracks, which increase the number of landslides during downpours, were generated on hill slopes during the Chi-Chi earthquake (Lin et al., 2006b). Statistical results illustrate that the effects of disturbances on the watershed landscape in the study areas were cumulative, but were not always evident in space and time over the entire landscape (Lin et al., 2006b). The effects of the Chi-Chi earthquake on the landscapes of the study areas gradually declined; this finding was also obtained by Chang et al. (2007). However, in the Chenyulan watershed, as the landslide ratio increased with successive rainstorms and strong earthquakes, the NDVI values decreased (Lin et al., 2008b). Hence, subsequent rainstorms cause divergent destruction of vegetation; this destruction may be influenced by the precipitation distribution and typhoon path (Lin et al., 2008b) (Table 1).

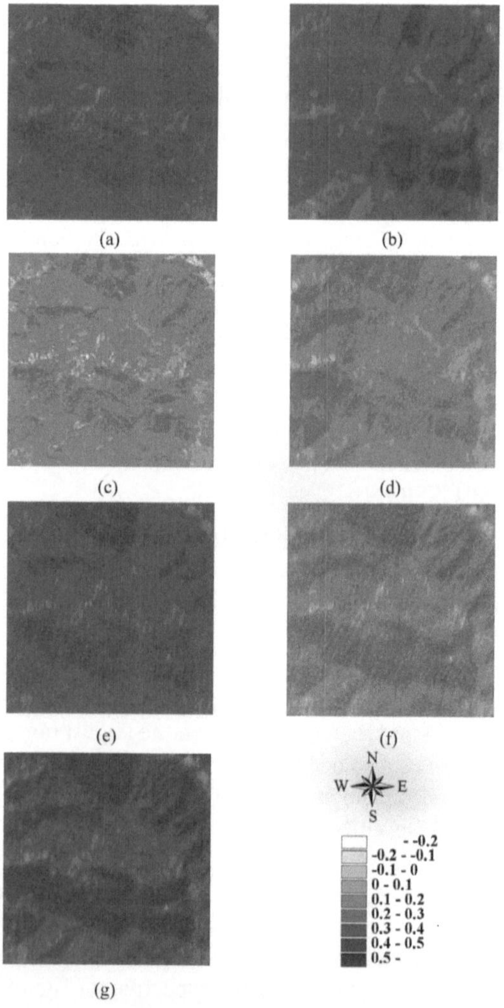

Figure 2. NDVI images of area A on (a) 1996/11/08, (b) 1999/03/06, (c) 1999/10/31, (d) 2000/11/27, (e) 2001/11/20, (f) 2003/12/17, and (g) 2004/11/19.

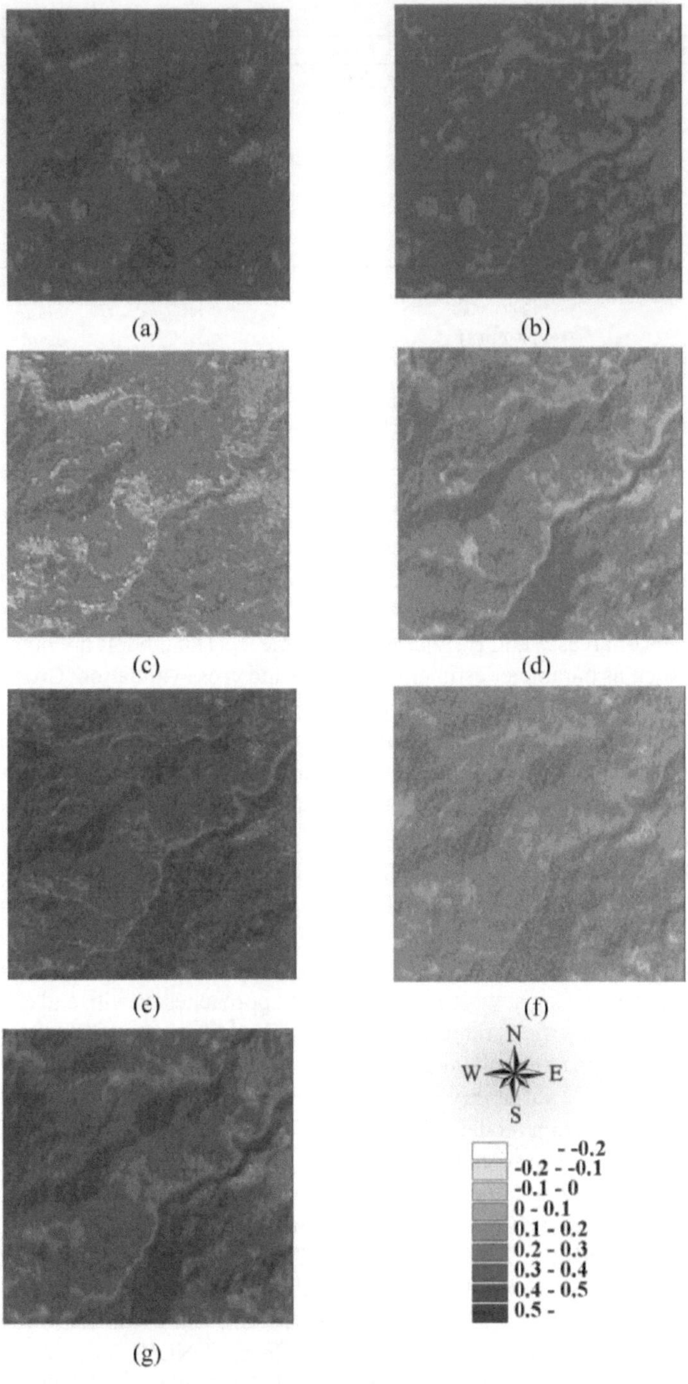

Figure 3. NDVI images of area B on (a) 1996/11/08, (b) 1999/03/06, (c) 1999/10/31, (d) 2000/11/27, (e) 2001/11/20, (f) 2003/12/17, and (g) 2004/11/19.

Table 1. Statistics of NDVI images.

Area	Date	Mean	Std.	Min.	Max.
	1996/11/08	0.36	0.04	0.11	0.48
	1999/03/06	0.32	0.04	0.13	0.43
	1999/10/31	0.14	0.07	-0.22	0.33
A	2000/11/27	0.15	0.07	-0.14	0.35
	2001/11/20	0.37	0.05	0.03	0.50
	2003/12/17	0.15	0.06	-0.12	0.33
	2004/11/19	0.35	0.06	0.05	0.54
	1996/ 11/08	0.36	0.03	0.13	0.47
	1999/03/06	0.36	0.04	0.14	0.48
	1999/10/31	0.16	0.05	-0.20	0.38
B	2000/11/127	0.17	0.05	-0.09	0.33
	2001/11/20	0.37	0.04	0.14	0.48
	2003/12/17	0.20	0.06	-0.08	0.44
	2004/11/19	0.39	0.05	0.10	0.57

Previous studies that quantified the impact of large disturbances did not evaluate the spatial structures of NDVI images in the study areas. To demonstrate the ability of the variogram to depict landscape heterogeneity, spatial variability and patterns, experimental variograms, and their variogram models were first analyzed and fit to seven images of areas A and B (Figure 4 and Table 2). The models are obtained in two processes such as parameter estimation (fitting) and cross validation. Cross-validation in Table 2 is a means for evaluating effective parameters for kriging interpolations. In cross-validation analysis each measured point in a spatial domain is individually removed from the domain and its value estimated via kriging as though it were never there. In this way a graph can be constructed of the estimated versus actual values for each sample location in the domain.

The three main features of a typical variogram model are (1) the range, (2) the sill, and (3) the nugget effect. The sill is the upper limit that a variogram approaches at a large distance, and is a measure of the variability of the investigated variable: a higher sill corresponds to greater variability in the variable. The range of a variogram model is the distance lag at which the variogram approaches the sill, and can reveal the distance above which the variables become spatially independent. The nugget effect is exhibited by the apparent non-zero value of the variogram at the origin, which may be due to the small-scale variability of the investigated process and/or measured errors. In this study, the variogram models of the seven NDVI images for areas A and B areas are exponential models. The spatial variations (Sill; $C_0 + C$) of NDVI images from high to low are in 2003/12/17, 2004/11/19, 1999/10, 2000/11/27, 1999/03/06, 2001/11/20, and 1996/11/08 in area A. The spatial variations (Sill; $C_0 + C$) of NDVI images from high to low in area B are in 1999/10/31, 2000/11/27, 2004/11/19, 2003/12/17, 2001/11/20, 1999/03/06, and 1996/11/08. The spatial variations of NDVI images increase considerably from 1996/11/08 to 1999/10/31 (after the Chi-Chi earthquake) in both areas A and B. Similarly, small variations (Nugget effect) of NDVI images in 2003/12/17 (after typhoon Dujuan), 1999/10 (after the Chi-Chi earthquake) and 2000/11/27 in area A are larger than those in 1999/03/06, 2001/11/20, 1996/11/08 and 2004/11/19. In area

B, small variations (Nugget effects) of NDVI images in 1999/10/31 are larger than those in other images. As the range of a variogram model increases, the continuity of an NDVI image increases. The ranges of NDVI variogram models in area A from long range to short range are in 2000/11/27, 2001/11/20, 1999/03/06, 1996/11/08, 1999/10, 2003/12/17, and 2004/11/19. In area B, the ranges of NDVI variogram models from long range to short range are in 1999/03/06, 2000/11/27, 2003/12/17, 2004/11/19, 2001/11/20, 1996/11/08, and 1999/10. However, exponential models with large sills, large nugget effects and short-range NDVI images are indicative of significant spatial heterogeneous landscapes induced by the Chi-Chi earthquake in areas A and B. Moreover, typhoons Xangsane and Dujuan generated heterogeneous landscapes in area A.

Table 2. Variogram models of NDVI images.

Area	Date	Model	Parameters	The fit	Cross-validate
A	1996/11/08	Exponential model	Co=0.000453, Co+C=0.001212, R=1204.000	(SS=7.774E-08; r^2 =0.832, Co/Co+C=0.374)	r^2=0.722
	1999/03/06	Exponential model	Co=0.000147, C0+C=0.001744; R=1278.000	(SS=3 .490E-08; r^2=0.873, Co/Co+C=0.352)	r^2=0.893
	1999/10/31	Exponential model	Co=0.000878, Co+C=0.002496; R=1020.000	(SS=18.597E-08; r =0.961, Co/Co+C=0.310)	r^2=0.839
	2000/11/27	Exponential model	Co=0.000761, Co+C=0.002452; R=1881.000	(SS=5.124E-08; r =0.878, Co/Co+C=0.400)	r^2=0.894
	2001/11/20	Exponential model	Co=0.000518, Co+C=0.001294; R=1497.000	(SS=5.124E-08; r =0.878, Co/Co+C=0.400)	r^2=0.723
	2003/12/17	Exponential model	Co=0.000700, Co+C=0.003370; R=981.000	(SS=3.420E-07; r^2 = 0.893, Co/Co+C=0.208)	r^2=0.737
	2004/11/19	Exponential model	Co=0.000229, Co+C=0.002878; R=918.000	(SS=1.918E-07; r^2 = 0.930, Co/Co+C=0.080)	r^2=0.862
B	1996/ 11/08	Exponential model	Co=0.000138, Co+C=0.001326; R=654.000	(SS=1.610E-08; r^2=0.953, Co/Co+C=0.104)	r^2=0.781
	1999/03/06	Exponential model	Co=0.000712, Co+C=0.001814; R=4620.000	(SS=6.070E-08; r^2=0.945, Co/Co+C=0.393)	r^2=0.901
	1999/10/31	Exponential model	Co=0.000590, Co+C=0.004440; R=564.000	(SS= 1.678E-07; r^2=0.939, Co/Co+C=0.133)	r^2=0.849
	2000/11/127	Exponential model	Co=0.0001863 Co+C=0.004676; R=2646.000,	(SS=2.474E-07; r^2=0.952, Co/Co+C=0.398)	r^2=0.908
	2001/11/20	Exponential model	Co=0.0001205, Co+C=0.002429; R=1281.000	(SS=5 .621E-08; r^2=0.933, Co/Co+C=0.498)	r^2=0.728
	2003/12/17	Exponential model	Co=0.0001258, Co+C=0.003126; R=2298.000	(SS= 1.567E-07; r^2=0.949, Co/Co+C=0.402)	r^2=0.820
	2004/11/19	Exponential model	Co=0.0001161, Co+C=0.003832; R=1680.000	(SS= 1.186E-07; r^2 =0.977, Co/Co+C=0.303)	r^2=0.902

Co=Nugget; Co+C=Sill; R= Range

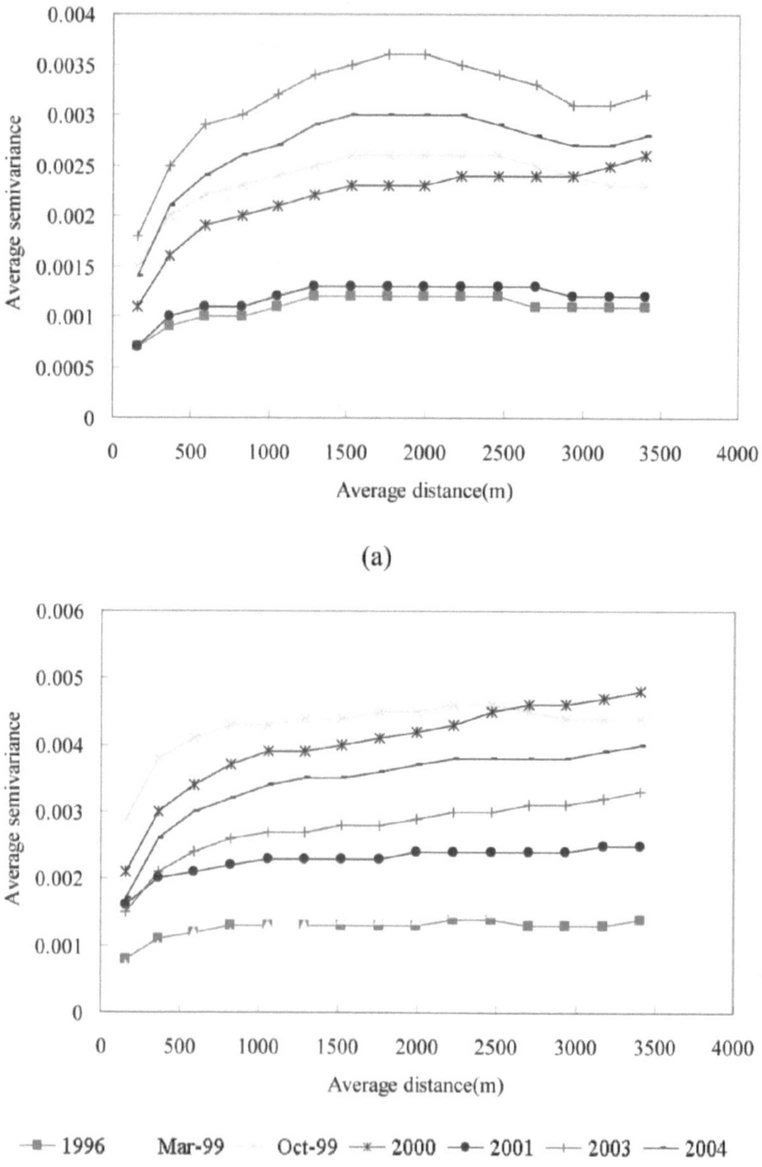

(b)

Figure 4. Experimental variograms of NDVI images before and after disturbances in areas (a) A and (b) B.

High-spatial-resolution observations (e.g., SPOT-HRV, pixel size of 20 m) capture most landscape spatial heterogeneity and are thus can be used to quantify the spatial heterogeneity within moderate spatial resolution pixels (Garrigues et al., 2006, 2008b).

The shape of variograms can be used to understand the NDVI spatial structures within an image domain (Garrigues et al., 2006). Millward and Kraft (2004) applied variograms to evaluate the impacts of disturbances on landscapes. In this study, experimental variogram and modeling results indicate that large disturbances, such as the Chi-Chi earthquake, created extremely complex heterogeneous patterns across the landscape. Notably, a disturbance may affect some areas but not others, and disturbance severity often varies considerably within an affected area on the landscape level (Lin et al., 2006b; Turner and Dale, 1998). Variography results illustrate that NDVI discontinuities between fields create a mosaic spatial structure resulting primarily from large disturbances, such as the Chi-Chi earthquake, in the study areas. Moreover, the high-magnitude Chi-Chi earthquake created these landscape variations in space in the Chenyulan watershed (Lin et al., 2006b). Previous studies (Chang et al., 2007; Lin et al., 2006b, 2008b) indicated that landslides in the Chenyulan watershed were impacted by the Chi-Chi earthquake; however, the effect of the earthquake decreased as the time between a typhoon and the Chi-Chi earthquake increased (Chang et al., 2007). Moreover, variography results confirm that the impacts of disturbances on the watershed landscape pattern were cumulative, but were not always evident in space and time in the entire landscape (Chang et al., 2007; Lin et al., 2006c). Moreover, landslides induced by earthquakes and typhoons have distinct spatial patterns (Lin et al., 2008b). Typhoons significantly influence NDVI variations via the flow of accumulated rainfall and wind gradients (Lin et al., 2008d). The statistical and variogram results also indicate that basic statistics without variograms of NDVI images may not sufficient to present landscape changes induced by disturbances, particularly via spatial structure, variability, and heterogeneity analysis. Moreover, variogram modeling results also support the above statistical results, indicating that subsequent rainstorms may cause divergent destruction of vegetation, and then this destruction may be influenced by the precipitation distribution and typhoon path (Lin et al., 2003, 2006b).

Latin Hypercube Sampling for Multiple Images

Sampling strategies are typically based on an assumed theoretical framework (Edwards and Fortin, 2001). Sampling under such a framework involves attempting to locate samples to capture the possible variations and fluctuations in a gradient field (Edwards and Fortin, 2001). An efficient sampling method is therefore needed to cover the entire range of ancillary variables (Minasny and McBratney, 2006). In this study, experimental variograms of cLHS samples with their NDVI values were constructed using the same lag interval to compare the spatial structures of the actual NDVI images. Figures 5 and 6 show experimental variograms for 100, 500, 1,000, 2,000, 2,500, and 3,000 cLHS samples in 1996/11/08, 1999/03/06, 1999/10/31, 2000/11/27, 2001/11/20, 2003/12/17, and 2004/11/19, respectively. These experimental variograms show that as the number of samples increased from 100 to 3,000, the ability of experimental variograms to capture the spatial structure of actual NDVI images increased. These variography results show that the cLHS approach can simultaneously select samples from multiple NDVI images to capture spatial structures of all NDVI spatial structures.

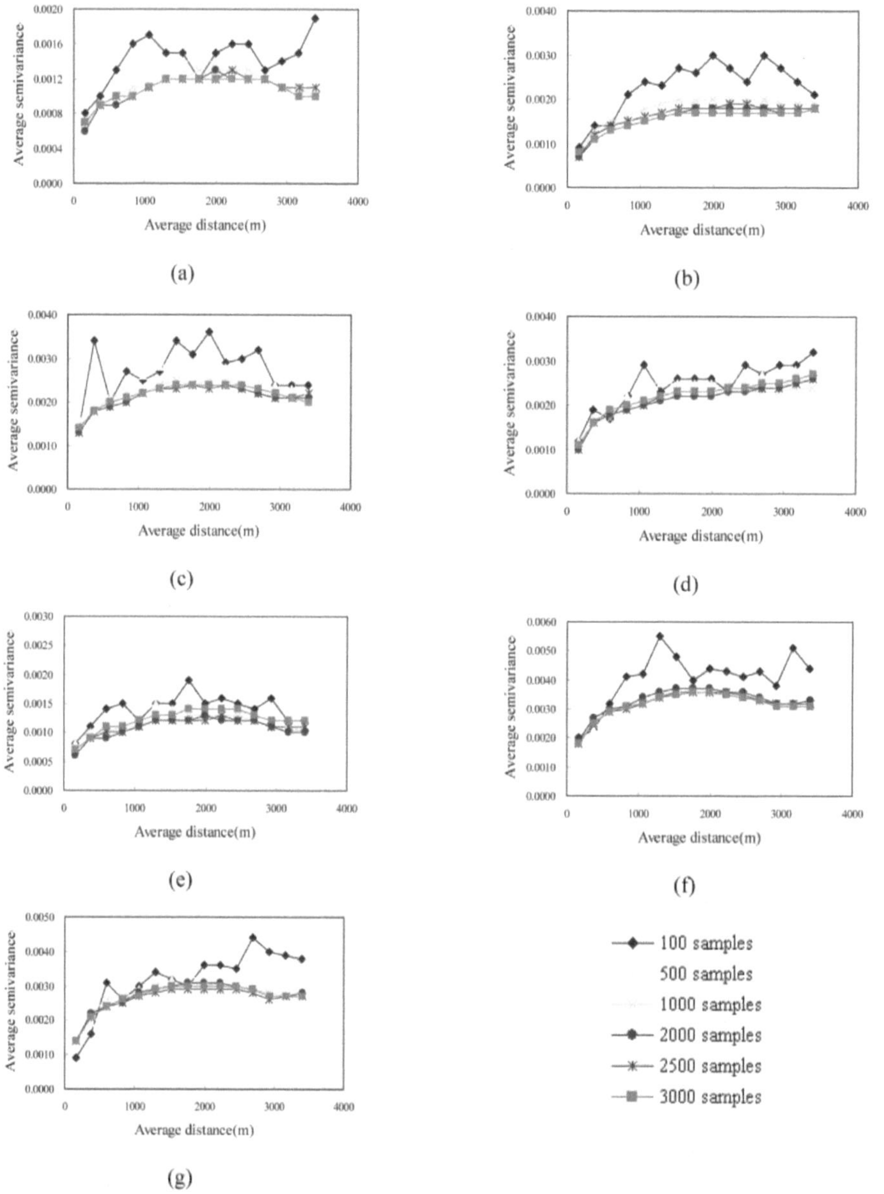

Figure 5. Experimental variograms of NDVI samples for area A on (a) 1996/11/08, (b) 1999/03/06, (c) 1999/10/31, (d) 2000/11/27, (e) 2001/11/20 (f) 2003/12/17, and (g) 2004/11/19.

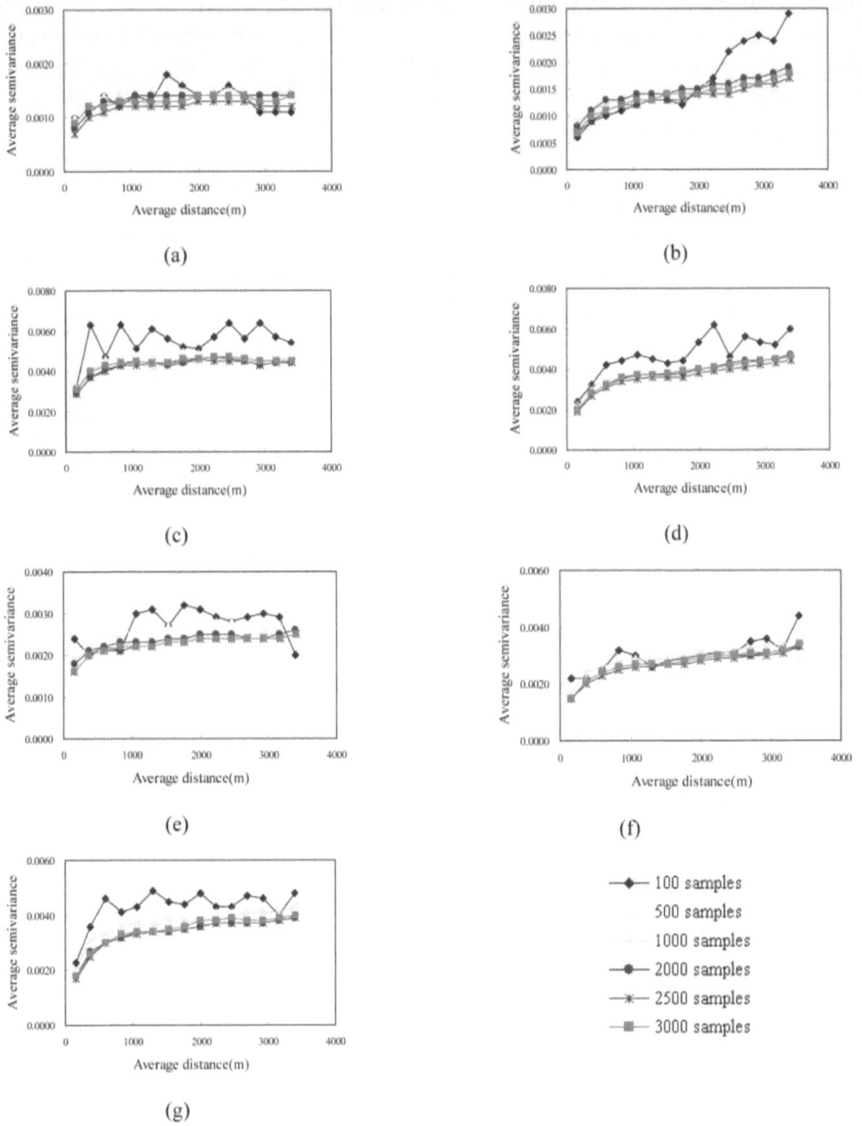

Figure 6. Experimental variograms of NDVI samples for area B on (a) 1996/11/08, (b) 1999/03/06, (c) 1999/10/31, (d) 2000/11/27, (e) 2001/11/20, (f) 2003/12/17, and (g) 2004/11/19.

Table 3 lists statistics for 100, 500, 1,000, and 3,000 samples from multiple NDVI images with 62,500 grids using the cLHS approach. Figure 7 shows the 3,000 samples selected using cLHS in each area. The distributions of selected samples confirm that samples selected using cLHS provide a good coverage of the study area and are well spread and partially clustered in the study areas (Minasny and McBratney, 2006). The statistics for these 3,000 samples indicate that the statistics obtained by cLHS can capture statistics of all actual NDVI images. The statistical and variogram analyses of cLHS samples also illustrate that the cLHS approach can be applied to select samples and capture the spatial structures of multiple historically accurate NDVI images. These samples can be used in further monitoring and to determine the impacts of disturbances on study landscapes in the future.

Table 3. Statistics of 100, 500, 1,000 and 3,000 samples from NDVI images.

Area		Date	Mean	Std.	Min.	Max.	Area		Date	Mean	Std.	Min.	Max.
100	A	1996/11/08	0.36	0.04	0.22	0.44	1,000	A	1996/11/08	0.36	0.03	0.17	0.45
		1999/03/06	0.36	0.05	0.22	0.45			1999/03/06	0.36	0.04	0.17	0.47
		1999/10/31	0.16	0.05	0.00	0.24			1999/10/31	0.16	0.05	-0.10	0.33
		2000/11/27	0.17	0.05	0.00	0.28			2000/11/27	0.17	0.05	0.00	0.33
		2001/11/20	0.37	0.04	0.19	0.44			2001/11/20	0.37	0.04	0.20	0.46
		2003/12/17	0.19	0.06	0.01	0.33			2003/12/17	0.20	0.06	0.00	0.38
		2004/11/19	0.39	0.05	0.19	0.05			2004/11/19	0.40	0.05	0.17	0.54
	B	1996/ 11/08	0.36	0.04	0.24	0.44		B	1996/ 11/08	0.16	0.07	0.00	0.30
		1999/03/06	0.31	0.05	0.20	0.38			1999/03/06	0.36	0.05	0.15	0.47
		1999/10/31	0.13	0.08	-0.08	0.28			1999/10/31	0.15	0.06	-0.04	0.29
		2000/11/27	0.15	0.07	0.00	0.28			2000/11/27	0.36	0.06	0.12	0.50
		2001/11/20	0.35	0.06	0.20	0.46			2001/11/20	0.36	0.04	0.17	0.44
		2003/12/17	0.15	0.07	-0.05	0.29			2003/12/17	0.32	0.04	0.16	0.41
		2004/11/19	0.35	0.08	0.16	0.49			2004/11/19	0.14	0.07	-0.12	0.29
500	A	1996/11/08	0.37	0.04	0.17	0.44	3,000	A	1996/11/08	0.36	0.04	0.15	0.46
		1999/03/06	0.36	0.04	0.19	0.46			1999/03/06	0.36	0.04	0.16	0.48
		1999/10/31	0.16	0.05	-0.20	0.26			1999/10/31	0.16	0.05	-0.10	0.30
		2000/11/27	0.17	0.05	0.00	0.31			2000/11/27	0.17	0.05	0.00	0.33
		2001/11/20	0.37	0.04	0.19	0.45			2001/11/20	0.37	0.04	0.15	0.48
		2003/12/17	0.20	0.06	0.00	0.36			2003/12/17	0.20	0.06	0.00	0.44
		2004/11/19	0.40	0.06	0.17	0.53			2004/11/19	0.39	0.06	0.13	0.57
	B	1996/ 11/08	0.35	0.04	0.17	0.44		B	1996/ 11/08	0.36	0.04	0.20	0.46
		1999/03/06	0.32	0.04	0.18	0.40			1999/03/06	0.32	0.04	0.16	0.41
		1999/10/31	0.13	0.07	-0.15	0.25			1999/10/31	0.14	0.07	-0.19	0.33
		2000/11/27	0.15	0.06	0.00	0.30			2000/11/27	0.15	0.07	0.00	0.32
		2001/11/20	0.36	0.05	0.17	0.46			2001/11/20	0.36	0.05	0.07	0.47
		2003/12/17	0.14	0.06	-0.05	0.31			2003/12/17	0.15	0.06	-0.11	0.31
		2004/11/19	0.35	0.06	0.15	0.49			2004/11/19	0.35	0.06	0.12	0.52

Figure 7. Locations of the 3,000 samples in areas (a) A and (b) B.

Estimations and Conditional Simulations with Selected Samples

The LHS approach can also be used in SGS (Pebesma and Heuvelink, 1999; Xu et al., 2005) and kriging estimation. However, because the LHS is conducted by shifting simple random sampling, meaningful deviations exist when sample size is small (Pebesma and Heuvelink, 1999; Xu et al., 2005). In this study, ordinary kriging estimates and SGS simulations were performed based on the above variogram models of 3,000 samples for 7 NDVI images in areas A and B. Figures 8–11 show the maps of kriging and averages of 1,000 realizations of SGS of NDVI images in 62,500 grids in areas A and B. A comparison of actual NDVI images and NDVI images estimated by kriging indicates that kriging estimation with sufficient samples provides the best local estimates to capture actual NDVI images, but generally smoothed extreme values of the actual NDVI images in areas A and B (Figures 8–11). Figures 12 and 13 show NDVI maps produced by SGS simulations with 100, 500, and 1000 cLHS samples in 1999/10/31 for areas A and B. The kriging estimation results illustrate that interpolation techniques such as kriging typically ignore phase information, which can result in an over-smoothed view of the distribution of spatial variables and remove important information about spatial discontinuities in a pattern (Deutsch and Journel, 1992; Lin et al., 2001, 2008a, 2008c; Lobo et al., 1998), particular with an insufficient number of samples (Figures 813). However, kriging interpolation algorithms produce maps of the best local estimate and generally smooth local details of spatial variation of an attribute (Goovaters, 1997).

The SGS results verify that the limits of spatial analysis and interpolations of landscape variables are based on semivariograms (or autocorrelation functions) solely, stressing the need to account for spatial discontinuities (Lobo et al., 1998), particularly in highly heterogeneous landscapes induced by large physical disturbances such as the Chi-Chi earthquake. Therefore, procedures for interpolation of ecological variables must include information on spatial discontinuities, either directly from remotely sensed images (assuming the phase pattern in the image and ecological variables are equivalent) or indirectly by sampling with sufficient intensity spatial variables in the field that have a known functional relationship with the variable of interest (assuming the variable of interest is too difficult or expensive to sample in the field) (Lobo et al., 1998). The simulated NDVI images show that kriging and SGS and the cLHS approach provide effective tools for monitoring, sampling, and mapping landscape changes induced by large disturbances.

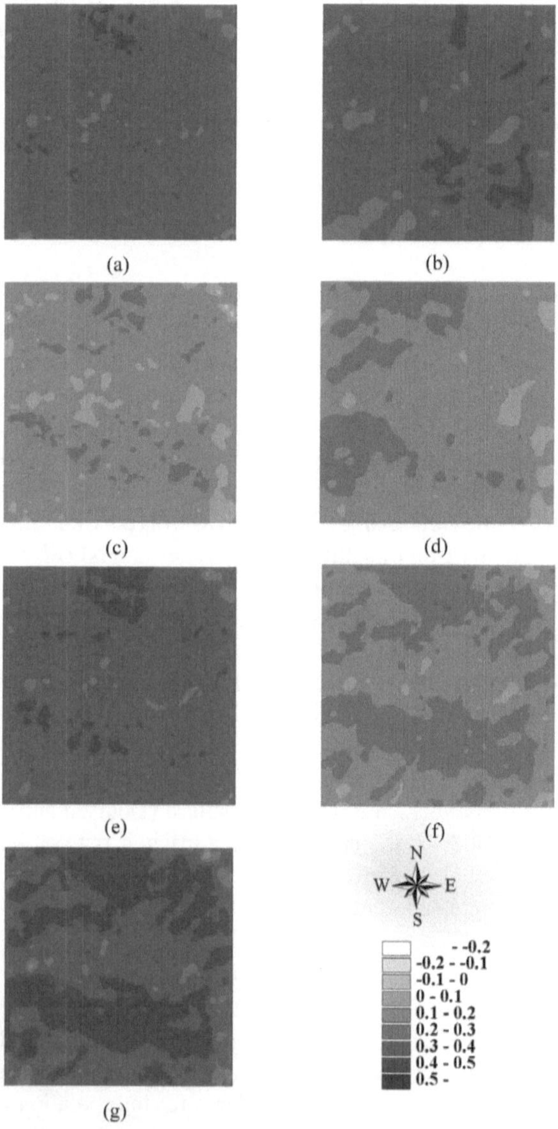

Figure 8. Kriging estimated NDVI images based on 3,000 samples in area A on (a) 1996/11/08, (b) 1999/03/06, (c) 1999/10/31, (d) 2000/11/27, (e) 2001/11/20, (f) 2003/12/17, and (g) 2004/11/19.

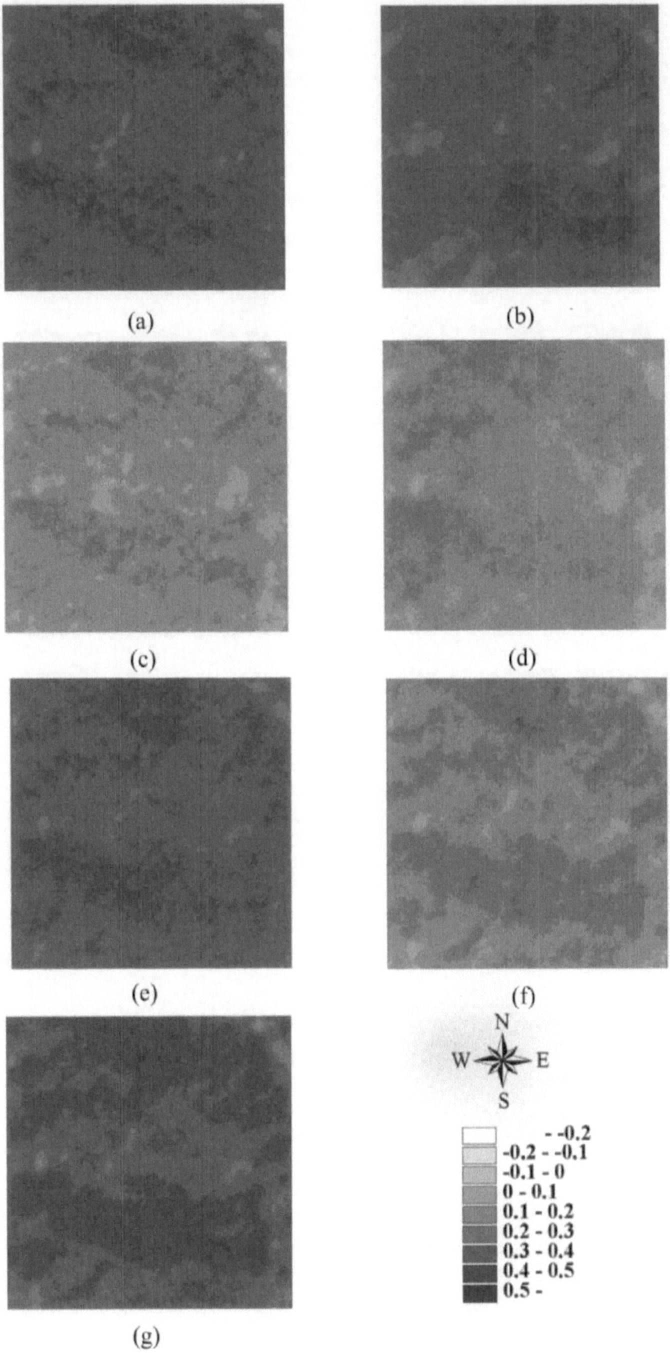

Figure 9. Conditional simulated NDVI images based on 3,000 samples in area A on (a) 1996/11/08, (b) 1999/03/06, (c) 1999/10/31 (d) 2000/11/27, (e) 2001/11/20, (f) 2003/12/17, and (g) 2004/11/19.

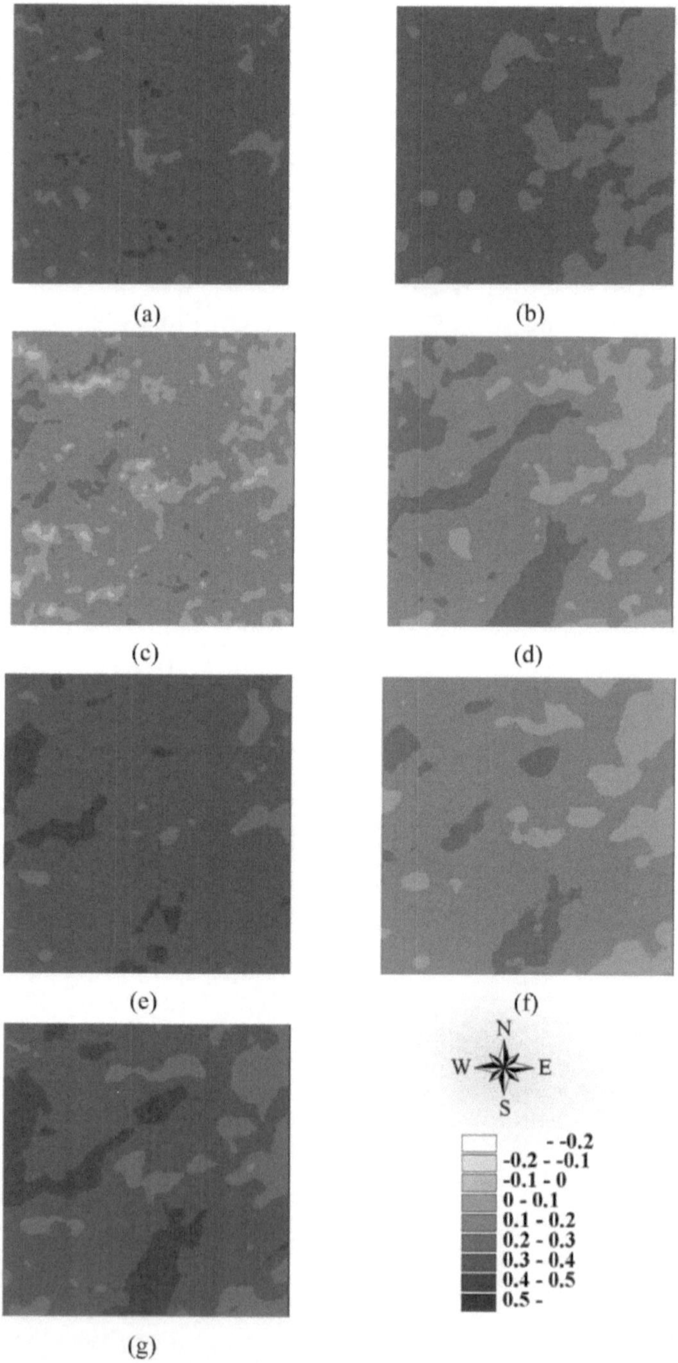

Figure 10. The NDVI images estimated by kriging based on 3,000 samples in area B on (a) 1996/11/08, (b) 1999/03/06, (c) 1999/10/31, (d) 2000/11/27, (e) 2001/11/20, (f) 2003/12/17, and (g) 2004/11/19.

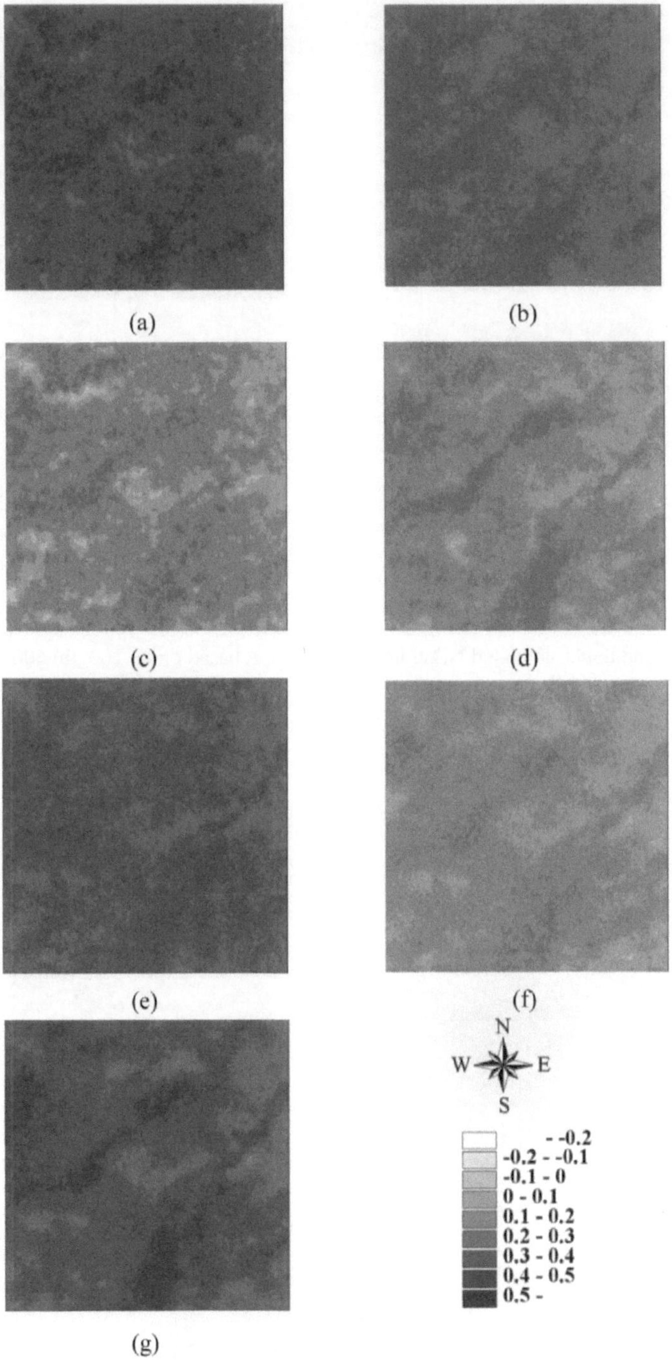

Figure 11. Conditional simulated NDVI images based on 3,000 samples for area B on (a) 1996/11/08, (b) 1999/03/06, (c) 1999/10/31, (d) 2000/11/27, (e) 2001/11/20, (f) 2003/12/17, and (g) 2004/11/19.

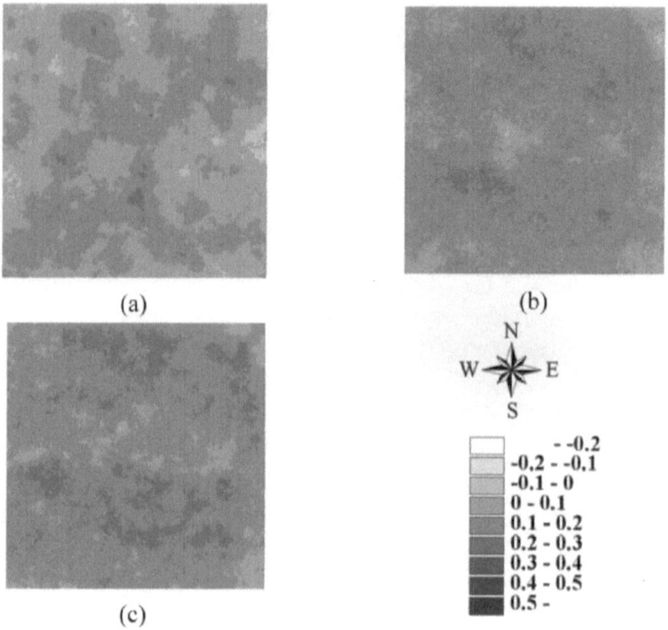

Figure 12. Conditional simulated NDVI images for area A based on (a) 100, (b) 500, and (c) 1,000 cLHS samples on 1999/10/31.

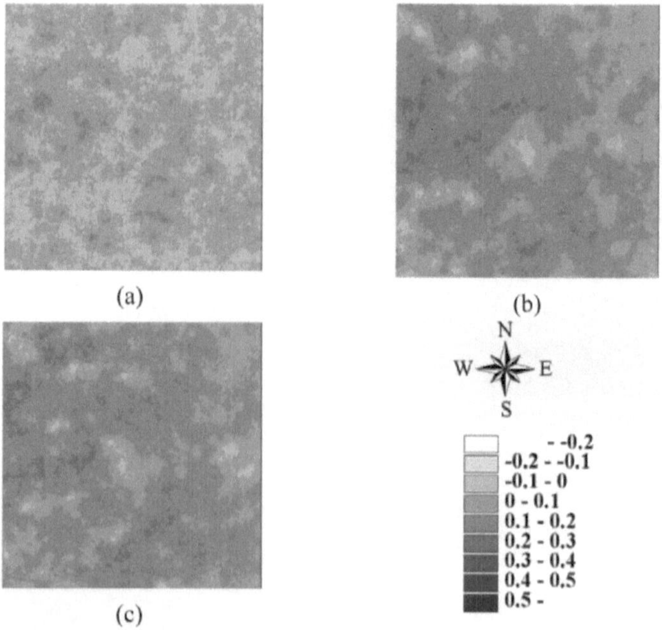

Figure 13. Conditional simulated NDVI images for area B based on (a) 100, (b) 500, and (c) 1,000 cLHS samples on 1999/10/31.

CONCLUSIONS

This study presents a novel and effective approach that integrates cLHS, variograms, kriging, and SGS in remotely sensed images for efficient monitoring, sampling and mapping of the impacts of chronologically ordered large disturbances on spatial characteristics of landscape changes to spatial structure, variability and heterogeneity. The NDVI images, which can be generated almost immediately after the remotely sensed data are acquired, were used as the inferential index because landscape changes induced by a large disturbance are easily recognized by changes in NDVI images. Variography of multiple NDVI images before and after a large disturbance is an essential method for characterizing and quantifying the spatial variability, structure and heterogeneity of landscapes induced by a disturbance. The variography results illustrated that cumulative impacts of disturbances on spatial variability existed and depended on disturbance magnitudes and paths, but were not always evident in spatiotemporal variability of landscapes in the study areas. Moreover, the cLHS approach is an effective sampling approach for multiple true NDVI images from their multivariate distributions to replicate the statistical distribution and spatial structures of the NDVI images. The sufficient number of NDVI samples using cLHS can be used to monitor and sample changes in landscapes induced by large physical disturbances. Kriging and SGS combined with the sufficient number of cLHS samples can be used to estimate and simulate NDVI images to generate maps that capture the spatial pattern and variability of actual NDVI images of disturbed landscapes. Kriging with sufficient number of NDVI cLHS samples produces NDVI maps with the best local estimates to identify patterns of NDVI images of disturbed landscapes. SGS with sufficient cLHS samples generate multiple realizations and an average of the realizations of NDVI and captures the spatial variability and heterogeneity of disturbed landscapes.

KEYWORDS

- **Chi-Chi earthquake**
- **Latin hypercube sampling**
- **Normalized difference vegetation index**
- **Sequential Gaussian simulation**
- **Variogram**

ACKNOWLEDGMENTS

The authors thank the Soil and Water Conservation Bureau of Taiwan for providing field data and financially supporting this research under Contract No. SWCB-92-026-08. The authors also would like to thanks Mr. Deng for treatments of remote sensing data.

Chapter 14

The Global Sweep of Pollution

Bob Weinhold

INTRODUCTION

Towering smokestacks were a popular mid-20th century "remedy" for industrial emissions. Pump the stuff high enough into the air, went the thinking, and the problem would go away. But evidence collected since then has strongly suggested that tall smokestacks are not sufficient to mitigate the effects of pollution—those pollutants eventually came down somewhere, dozens or thousands of miles away. In the November 2006 issue of EHP, for example, Morton Lippmann of the New York University School of Medicine and colleagues reported a strong link between nickel emitted from a very tall smokestack at a smelter in Sudbury, Canada, and acute heart rate changes in mice some 500 miles away. At the same time, we also now know that tall stacks are not necessary for pollutant emissions to waft great distances, as verified by scores of individual studies showing that one pollutant or another—such as ozone, particulate matter (PM), and sulfur dioxide (SO_2)—can blow from country to country, and continent to continent.

Beginning in the 1980s and accelerating since about 2000, satellite surveillance with increasingly sophisticated instruments has enabled us to better visualize the complex fluctuations of several important pollutants as they ebb and flow around the planet. This new capability is partly serendipitous. "Most of these satellites weren't designed to have an air quality focus," says Terry Keating, an environmental scientist with the US Environmental Protection Agency (EPA) and co-chair of the Task Force on Hemispheric Transport of Air Pollution, which was created in 2004 by the United Nations Economic Commission for Europe's Convention on Long-range Transboundary Air Pollution. "But we find ourselves with this stream of data, and we are figuring out how to use it," he says.

One use has been a handful of pilot projects directly linking satellite-observed column-wise concentrations of atmospheric pollutants—that is, the concentration from the Earth's surface to the top of the atmosphere—with concentrations at the ground level. "Only in the past ten years have we been able to advance epidemiological science with satellites," says John Haynes, program manager for public health and aviation applications at the National Aeronautics and Space Administration (NASA). "This is truly a leap forward into the twenty-first-century science of epidemiology."

Figure 1. An artist's rendering of the Aura satellite, which is dedicated to collecting data on air pollutants (image courtesy of NASA).

LAUNCH OF A NEW FIELD

Steady progress in the capabilities of satellites during the 20th century provided glimpses of the planet from ever higher elevations. With the 1957 launch of the first satellite, Sputnik, by the Soviet Union, we gained our first view of the Earth from above the troposphere, the layer of the atmosphere that shimmies around the planet from the ground to elevations of about 4–12 miles.

The following year the US created NASA. The study of atmospheric pollutants was far from the first priority in NASA's early years; instead, the program focused on keeping the US ahead of the Soviets in the Space Race. But the technological capabilities developed as part of the US space exploration program have provided critical tools for beginning to observe the complex gyrations of various pollutants in the atmosphere.

Among the many other countries now involved in satellite programs targeting air pollutants are Brazil, Canada, China, India, Japan, and the 20 European countries that are the primary supporters of the European Space Agency (ESA). "More and more countries are getting aware of this problem all over the world," says Claus Zehner, an ESA Earth observation applications engineer. "It is a growing theme in the satellite world."

Tracking individual air pollutants can be quite a challenge. Satellite instruments use the unique spectral signature of a single chemical or class of chemicals to distinguish it from all the other substances floating around it. But scientists must also account for variables such as clouds, humidity, wind, and landforms; reflectance from land and ocean surfaces (which form the backdrop of what the instrument sees); time of day and season (which can affect formation of various compounds); and chemical reactions (which can alter the tracked compound before it reaches the ground, making it difficult to predict the exact chemical culprits of environmental health concern). In addition, there's horizontal and vertical mixing of air masses at a wide range of temporal and spatial scales, including that caused when the upper troposphere mingles with the "planetary boundary layer," where friction with the Earth's surface alters airflow. And there's one last complication—the mixing within the planetary boundary layer that determines the level of pollutants near the ground.

So far, scientists have begun to get the best handle on these variables for PM, nitrogen dioxide (NO_2), carbon monoxide (CO), and formaldehyde. They are also making some progress with ozone, but SO_2 and most volatile organic compounds (VOCs) still pose a significant challenge. Other important pollutants that are known to be widely distributed—among them polycyclic aromatic hydrocarbons, ammonia, pesticides, and metals such as lead and mercury—are hard to distinguish from the PM to which they are attached or have spectral signatures that are difficult to detect amidst all the complicating atmospheric conditions, says John Burrows, co-director of the Institute of Remote Sensing and Environmental Physics at the University of Bremen. Tracking such pollutants will require better instruments, Burrows says.

Indeed, we still have a long way to go before we can fully document and accurately forecast pollution streams and their effects on the ground. But data from satellites, aircraft, balloons, ground-level monitors, chemical transport models, and other tools are already improving our understanding of the global transport of pollutants as individual nations and international collaborations struggle to better address emissions and the effects they have on downwind countries and continents.

A FINE VIEW OF PARTICULATE MATTER

The PM is created by combustion of fuels in vehicles, power plants, domestic cooking, and industrial processes, and by natural processes such as wildfires, volcanic eruptions, wind-blown dust, and bursting of sea salt-laden bubbles at the ocean's surface. The PM can cause premature death and a range of respiratory and cardiovascular problems. Because of its various sources and residence time in the atmosphere, PM is highly variable in space and time. In contrast to gaseous pollutants, PM characterization must also include composition or type, particle size, and shape—all important dimensions for assessing detrimental effects of particulates on human health.

By carefully quantifying how much visible and near-infrared sunlight a PM plume reflects back up into space, scientists can accurately estimate the average size of the individual particles within the plume. Particles of natural origin (such as windblown dust) tend to be larger in size than human-produced particles, which mostly originate from the process of combustion (from fires or factories) and are therefore broken down into smaller particles.

The PM was not considered a major long-range transport issue until about the mid-1990s. Even now, the 51 countries that participate in the United Nations Convention on Long-Range Transboundary Air Pollution are just beginning to address PM, discussing whether and how to add policies for this pollutant to the convention, says Andre Zuber, co-chair with Keating and a policy officer with the European Commission.

Satellite images indicate how far PM can travel. Almost all continents and regions can be a PM hotspot at one time or another. Some areas are a major source almost year-round, including central and southern Africa, eastern Asia, Indonesia, Europe, the eastern US, and northern and central South America.

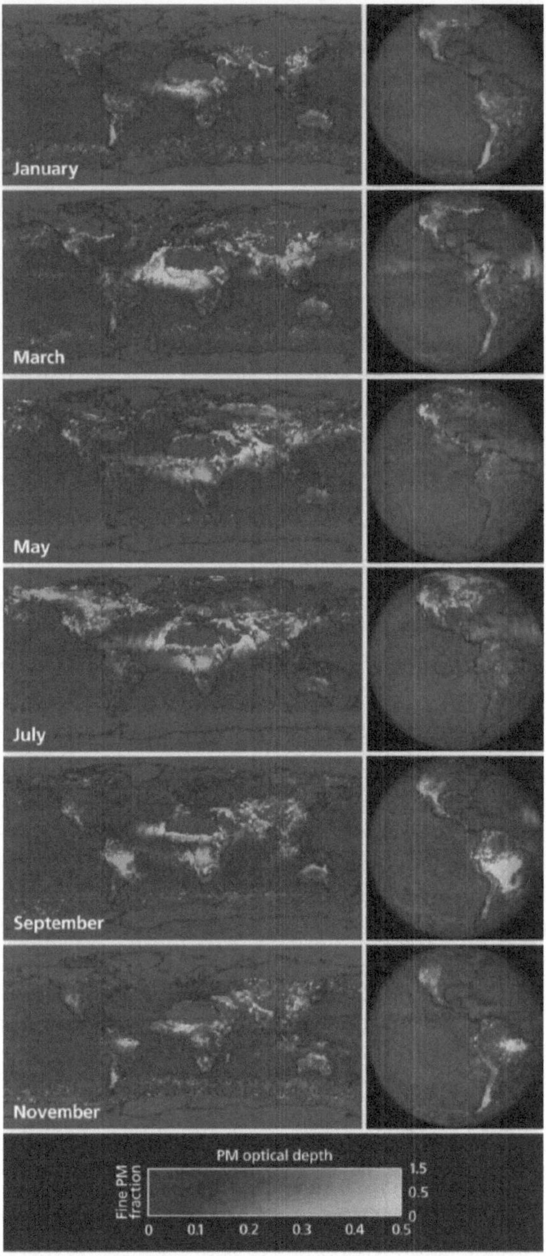

Figure 2. The PM travels hundreds and thousands of miles, as seen in these satellite images for 2004 that illustrate PM density from the Earth's surface to the top of the troposphere (which ranges from about 4 to 12 miles). Reds generally indicate human-generated sources, greens indicate primarily natural sources, and browns indicate a blend. Darker colors indicate greater concentrations. Atmospheric concentrations tend to be strongly to moderately correlated with ground-level concentrations, depending on local conditions.

Source: NASA.

Satellites are also being used to quantify PM movement between continents. Hongbin Yu, an associate research scientist at the Goddard Earth Sciences and Technology Center, a joint research center of NASA and the University of Maryland Baltimore County, and his colleagues found that East Asian pollution sources added an average of about 15% to the non-dust PM burden already being generated in the US and Canada from 2002 to 2005. These findings, published in the April 22, 2008 *Journal of Geophysical Research*, provide the first satellite-based estimate of transport of PM from East Asia to North America.

The study did not calculate the effects at ground level, which requires enhanced satellite capability to assess on a daily basis vertical structures of pollution plumes with high accuracy. "[That] process is still a bit fuzzy," says Solar Smith, a research associate at the Center for Space Research at the University of Texas at Austin, who did not participate in the Yu study. "On a long-term scale, you can take the average over time and get a fairly reliable understanding of how [ground-level] air quality is changing. The challenge is individual scene analysis on individual days. We're always working to improve algorithms."

Using existing algorithms, an international team studied 26 locations in Sydney, Delhi, Hong Kong, New York City, and Switzerland, and found an overall 96% correlation between satellite measurements of atmospheric column PM loading and ground measurements of PM concentration, although there was significant variation caused by factors such as cloud cover, relative humidity, and circulation within the boundary layer. These results were published in the September 2006 issue of *Atmospheric Environment*. In the June 2008 issue of the same journal, Klaus Schäfer and colleagues reported a 90% correlation between satellite and ground measurements for fine particulates 2.5 µm in diameter or smaller ($PM_{2.5}$) in the winter, but in the summer only a 48% correlation for PM smaller than 1 µm in diameter.

These variable results and those of other studies confirm that there is indeed still a bit of fuzziness in the processes of both measurement and calculation. The emerging data from lidar (the optical analog of radar) measurements taken aboard the Cloud-Aerosol Lidar and Infrared Pathfinder Satellite Observation (CALIPSO) satellite will help to get a clearer picture, says Yu. Satellite measurements with more extensive coverage than CALIPSO are needed in the future. But the results are becoming reliable enough—at least in some regions and seasons—that a handful of local projects are using or considering using PM data derived from satellites.

The HELIX-Atlanta (Health and Environment Linked for Information Exchange in Atlanta, Georgia) is a 5-county demonstration project whose principal investigator for its first half was Amanda Niskar, then with the Centers for Disease Control and Prevention (CDC). The project's partners include NASA, the CDC, the EPA, the Georgia Environmental Protection Division, Kaiser Permanente of Georgia, the Georgia Institute of Technology, and Emory University. The goal is to improve knowledge of the link between $PM_{2.5}$ and respiratory diseases among residents of the 5-county area and improve forecasting of dangerous $PM_{2.5}$ events as well as government, medical, and individual responses to them. Eventually this information could be extrapolated to other settings.

Satellites allow interpolation of $PM_{2.5}$ data between EPA monitors on the ground. "That's the great revolution, really," Haynes says. "You get much better representation of $PM_{2.5}$ concentrations over an area." Another key benefit of the study is that the detailed individual-level health data provided by Kaiser Permanente (a health care provider) enables researchers to more accurately evaluate the role $PM_{2.5}$ may be playing in complex medical conditions. The HELIX researchers expect to soon publish their findings on the links between satellite and ground monitor data.

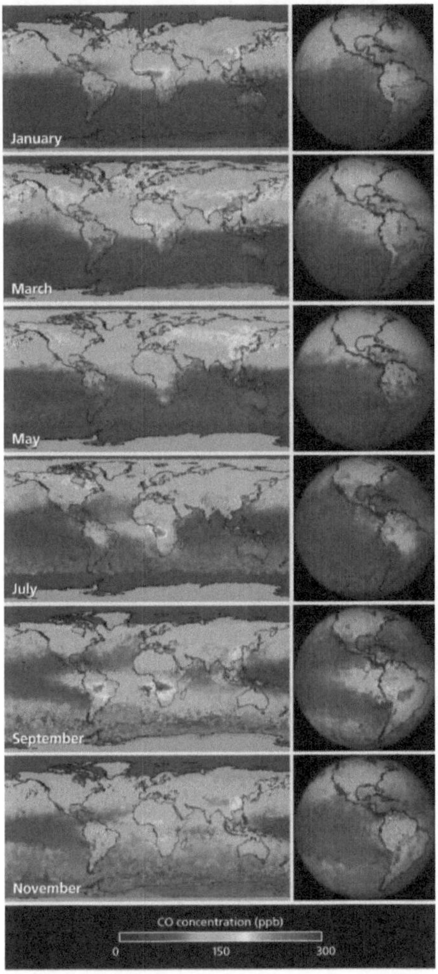

Figure 3. The CO spreads hundreds and thousands of miles, as seen in these images of concentrations at about 12,000 feet. The data were collected in 2004 by the MOPITT (Measurements of Pollution in the Troposphere) sensor on NASA's Terra satellite. Red indicates higher concentrations, and purple indicates lower concentrations. These concentrations tend not to pose a direct health threat, but serve as a tracer for a variety of other pollutants that can. The CO, produced when carbon-based fuels burn incompletely or inefficiently, is one of the six major air pollutants regulated in the US and in many other nations around the world.

Source: NASA

The amounts and sources of atmospheric CO change with locale and season. In Africa, for example, the seasonal shifts in CO are tied to the widespread agricultural burning that shifts north and south of the Equator with the seasons.

The related HELIX-Israel project is still in its organizational and fund-raising stages. Niskar, now a faculty member at the Tel Aviv University School of Public Health and director of HELIX-Israel, says the portion of the project that would use satellite data on air pollutants may begin as early as 2009. She says the Israeli project could be even more revealing than its Atlanta counterpart because the public health data available in Israel are of higher quality than those in the US, and the network of ground monitors is denser, allowing for better cross-checking of satellite and ground data. If the project is successful, Niskar says she is eager to help other countries set up similar systems. She would also like to include other health outcomes beyond the initial respiratory and cardiovascular targets and other pollutants (such as ozone) once NASA and others develop accurate methods for determining ground-level concentrations.

Another project will look at links between $PM_{2.5}$ and health outcomes such as stroke, blood pressure changes, cholesterol levels, deep vein thrombosis, and cognitive function. The REGARDS project (Reasons for Geographic and Racial Differences in Stroke), a multicenter study sponsored by the National Institute of Neurological Disorders and Stroke, already has 30,228 volunteers signed up in the lower 48 states and may be able to begin the satellite-related portion of its work as soon as 2009 if NASA funding comes through, says Leslie McClure, an assistant professor of biostatistics at The University of Alabama at Birmingham. "This is all very cutting-edge," she says. "To be able to estimate exposure for someone in the plains of Kansas where monitors are scarce is an extraordinary power."

CLOSING IN ON GASES

Satellite tracking of several gaseous pollutants also is improving. One class of such pollutants is the nitrogen oxides (NO_x), which are created primarily through vehicle, power plant, and industrial combustion processes, as well as by natural sources such as fires, soils, and lightning. The NO_x, despite their short lifespan in the lower atmosphere, play a key role in the formation of ground-level ozone, acid rain, and greenhouse gases. The NO_2, along with other toxic by-products of NO_x reactions, can cause health problems such as premature death, cancer, and respiratory and cardiovascular effects. The NO_2 is typically used by researchers and regulatory agencies as a surrogate for NO_x reaction by-products.

The NO_2 is one of the easier pollutants to track via satellite because of its spectral signature. "You can see it all the way down to the surface," says Mark Schoeberl, project scientist for NASA's Aura satellite, which carries several pollutant monitoring instruments. This has made it relatively simple to capture extensive satellite data about the global movement of this compound. There are a number of hotspots for the origin of NO_2, and the pollutant can travel long distances. Portions of eastern China are major generators all year long, and moderately high emissions come from areas in Europe, the eastern US, and central and southern Africa during much of the year.

In a study published in the February 22, 2008 *Journal of Geophysical Research*, Ronald van der A, a senior project scientist at the Royal Netherlands Meteorological Institute, and his colleagues reviewed satellite data from 1996 to 2006 for NO_2 and found that some areas in China had seen an average annual increase of up to 29%. Portions of India, Iran, Russia, South Africa, and the central US also had annual increases. But pollution reduction efforts may have contributed to small annual decreases in much of Europe and portions of the eastern US and the Philippines.

In a presentation at the July 2006 *Third Annual Dragon Programme Symposium* in Lijiang, China, van der A and colleagues including Jeroen Kuenen reported that NO_x emissions from China alone between 1997 and 2005 contributed to a small but important increase in ground-level ozone around the entire northern hemisphere, averaging roughly 0.3–0.5 ppb by volume when the air mass reached western North America, and about 0.2 ppb by volume when it drifted to Greenland, Europe, and northern Africa.

Several studies have found a strong correlation between satellite measurements and ground-level concentrations of NO_2, says Randall Martin, an associate professor of physics and atmospheric science at Dalhousie University in Halifax, Nova Scotia. However, he says more study is needed to cross-check satellite data with ground monitors, develop better algorithms, and better address variables such as geographic location, seasonal effects, and atmospheric conditions. But Martin and colleague Lok Nath Lamsal say they're getting close to more accurately determining ground-level concentrations via satellite, having developed algorithms that resulted in an 86% correlation between satellite and ground measurements of NO_2 in favorable circumstances. Their report of these data has been accepted for publication in the *Journal of Geophysical Research.*

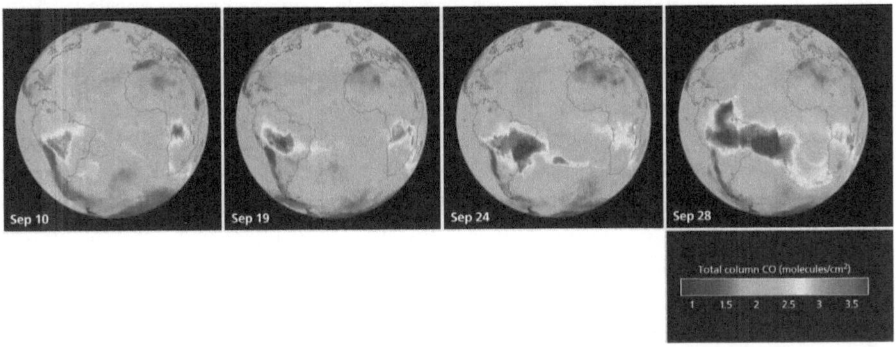

Figure 4. This image sequence taken over an 18-day period in September 2005 shows CO from agricultural fires blooming repeatedly over the Amazon basin, then traveling across the Atlantic Ocean, where it meets CO from fires in sub-Saharan Africa. These images came from the Atmospheric Infrared Sounder Experiment, whose visible, infrared, and microwave detectors provide a three-dimensional map of temperature, humidity, cloud cover, greenhouse gases, and other atmospheric phenomena.

Source: NASA/JPL

Because of its direct and indirect impacts on health and the environment, and spectral properties that make it relatively easy to track, CO has been monitored globally for a number of years. All or parts of all continents contain major emission generators at one time of year or another, and the transported CO adds to the load at certain times on all continents.

The CO is created primarily by incomplete fuel combustion. Primary sources, especially in developed areas, include vehicles and various industrial processes. Natural and human-set vegetation fires also are a significant source. The concentrations of CO that are added to local settings via long-distance transport usually are not considered an added health burden on their own, Schoeberl says, but he notes that CO is "a great tracer for human activities and biomass burning."

The CO also plays a role in ground-level ozone formation. Several studies have found that transport of CO from one location is associated with increased ground-level ozone thousands of miles away. Examples include transport from North America to Europe, from Alaska and western Canada to Houston, Texas, and from Asia to western North America. Atmospheric CO also affects a number of other chemical reactions in the atmosphere.

Ground-level ozone is created primarily through reactions among NO_x, VOCs, and other chemicals in the presence of sunlight. Ozone can cause health problems such as premature death and a range of respiratory and cardiovascular disorders.

Studies conducted over the past 15 years or so that are based on airplane data, ground monitors, and other instruments have provided substantial evidence that long-distance transport of ozone can affect other countries and continents. For instance, a report by Arlene M. Fiore and colleagues in the August 15, 2002 *Journal of Geophysical Research* noted that transport from outside North America can boost ground-level ozone by 15–35 ppb on summer afternoons in the US. These imports of ozone can, in some places on some days, spike levels in some counties above the EPA standard of 75 ppb.

The long-distance spread of ozone has played a role in the large increase in ozone concentrations in many areas of the planet since about 1950, according to Roxanne Vingarzan, a senior scientist with Environment Canada. In a July 2004 *Atmospheric Environment* article she reported that ozone concentrations around the globe have roughly doubled since then. As scientists become more adept at using satellite imagery to track long-distance ozone movement within the troposphere and down to ground level, more detail should become available.

But the spectral properties of ozone make it difficult to take advantage of satellite instruments to track ground-level concentrations. "We have had satellites measuring total column ozone for some time," Keating says. The problem, he explains, is that 90% of the ozone is in the stratosphere, so "measuring ozone in the troposphere requires looking through the stratosphere for the proverbial needle in a haystack."

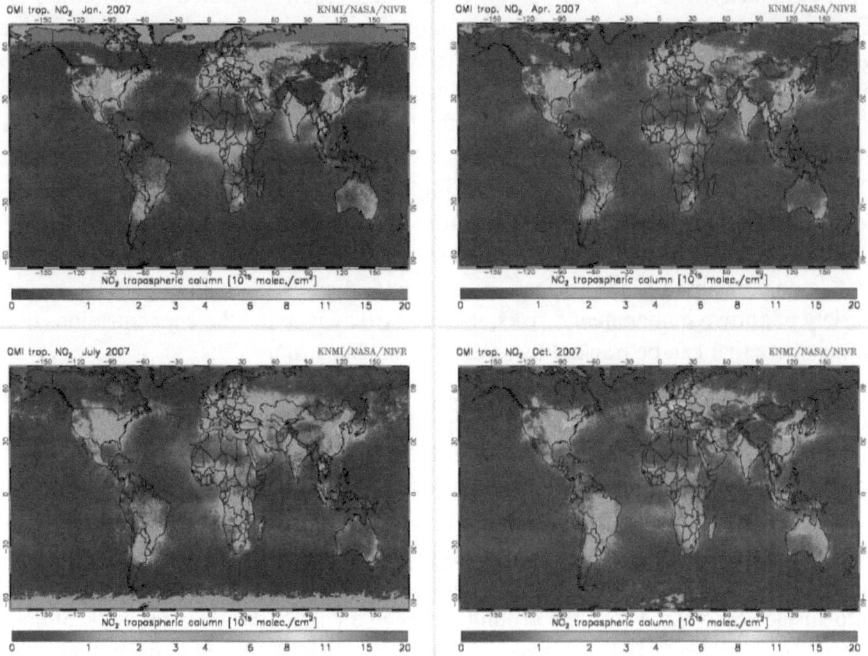

Figure 5. The NO_2 readily travels hundreds of miles at higher concentrations, and can travel thousands of miles at moderate or lower concentrations, as shown in this series of images from 2007. Red indicates the highest density, and purple the lowest, for the full column from the surface to the top of the troposphere. There can be a fairly strong correlation between atmospheric and ground-level concentrations, depending on local conditions.

Source: ESA

Schoeberl says the solutions to this problem likely lie in better algorithms to analyze the satellite data, as well as in new instruments aboard satellites. But he expects it may be a decade or so before there is a new ozone instrument in orbit.

For now, satellite images tracking ozone at an altitude of about 3–8 miles suggest that ozone is coming largely from developed countries in the northern hemisphere and from biomass burning (sometimes linked with human activities such as agriculture) in the southern hemisphere. The highest concentrations tend to occur in the summer months, although some areas—such as eastern China, California, and southwestern Africa—can have elevated concentrations in other months.

A GLIMPSE OF THE ELUSIVE VOCS AND SO_2

Vehicles, power plants, industrial processes, and consumer products are major human sources of VOCs, and vegetation is an important natural source. Most VOCs considered important from a health perspective—such as benzene, trichloroethylene, and chloroform—tend to have spectral properties that are hard to distinguish by satellite.

Formaldehyde is the only major VOC that's readily detected via satellite, says Gunnar Schade, an assistant professor in atmospheric sciences at Texas A&M University in College Station. Its presence generally signals hydrocarbon sources such as forests, biomass burning, vehicles, and industrial processes that emit formaldehyde precursors such as ethene, isoprene, and methane. Deciduous forests tend to be the largest source of formaldehyde detected by satellites, though urban hotspots also show up.

Formaldehyde is short-lived in the atmosphere, but global images show modest long-distance transport at lower concentrations. The concentrations tracked by satellite "are not a real health concern," Schoeberl says. However, formaldehyde and other VOCs do contribute to the formation of ground-level ozone. Although VOCs pose a variety of known human health risks, it is hard to say whether the concentrations in the atmosphere that can be tracked by satellite pose those same risks.

The SO_2 is created primarily through combustion processes in power plants and various industrial processes such as pulp and paper mills. Important natural sources include volcanoes and biomass burning. Health effects include premature mortality, multiple respiratory problems, headache, nausea, and thyroid system disruptions.

Like VOCs, SO_2 is difficult to track via satellite because of its spectral properties. Only major events such as volcanic eruptions and concentrated sources such as major urban areas tend to show up. It is even harder translating atmospheric contamination into surface levels. "SO_2 on the ground has been a challenge," Haynes says, although recent advancements have allowed near-ground retrievals of SO_2 from such major events.

Significant breakthroughs in satellite instruments or algorithms do not appear to be on the near horizon. However, SO_2 remains a concern for long-range transport. In an article by Chulkyu Lee and colleagues in the March 2008 issue of Atmospheric Environment, a team of Korean and German researchers using satellite data as one tool concluded that Chinese sources of SO_2 boosted the pollutant at ground level at a Korean location in May 2005 by up to 7.8 ppb. That's a significant portion of the EPA annual standard of 30 ppb or the 24-hour standard of 140 ppb.

THE GOLDEN AGE?

Satellite instruments may become increasingly important for tracking long-distance air pollutant transport, filling in data gaps between monitors, cross-checking ground-based emissions estimates and measurements, forecasting air quality, providing advance alerts for health care professionals, and supplying additional information for regulators as well as people and organizations trying to meet regulations. Along with information on air pollutants, satellite information is also being used to help with other important environmental health concerns, such as the spread of infectious diseases (as reflected by changes in vegetation and temperature, which can influence disease vector populations), dust, heat, land use changes, and climate change.

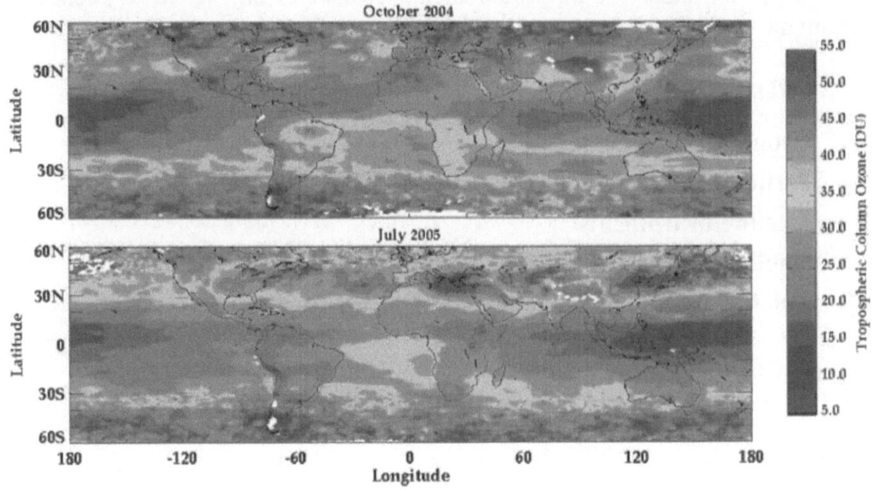

Figure 6. These images reflect relative ozone concentration in the middle troposphere, about 3–8 miles above the Earth's surface. Red indicates higher concentrations, and purple indicates lower concentrations. It remains challenging to extend these data to layers closer than about 1 mile above the surface, but these images illustrate the major sources and general movement patterns of ozone, which blankets areas for up to thousands of miles. Short-term experimental global forecasts of surface-level ozone are available at http://gems.ecmwf.int/d/products/grg/realtime/daily_fields/.
Source: NASA

On the basis of what he's seen so far, Keating says satellites just might pan out for the EPA in terms of monitoring air pollution. "I think there are some potential national applications. But it's still in the developmental process."

Part of that process includes plans by NASA, ESA, and others for about a dozen new satellites in various stages of conceptualization, planning, or design that would include air pollutant—tracking instruments. These could be launched anytime from 2009 to 2020 and beyond, and could replace and possibly improve upon the current fleet of at least nine satellites that track pollutants. Keating says the EPA is currently talking with NASA about specific pollutant tracking needs.

Nonetheless, competing interests and funding limitations may hinder the growth of the Earth-observation field, including pollutant-tracking efforts. On January 14, 2004 President Bush announced "a new vision for the Nation's space exploration program," emphasizing human and unmanned exploration of our solar system. Much of the initial focus is on returning to the moon and staying for extended periods, then going to Mars and eventually elsewhere. Little additional money has been budgeted for these projects so far, so the funds will come primarily out of existing NASA programs. "Earth science is a little on the back burner," says Schoeberl. "NASA isn't quite sure what it wants to do [given] the White House's new focus on Mars."

With launch dates for satellites routinely being pushed back years at a time, and with changing government priorities, Keating says the current fleet may be as good as it gets. "We're already in the golden age of atmospheric chemistry information," he says. "[But] we wonder how long these satellites will be in play. We're very concerned there might be a dry period."

KEYWORDS

- **Ground-level ozone**
- **Particulate matter**
- **Satellite instruments**
- **Satellite programs**
- **Volatile organic compounds**

Chapter 15

Geodetic Network Design for Disaster Management

Kerem Halicioglu and Haluk Ozener

INTRODUCTION

Both seismological and geodynamic research emphasize that the Aegean Region, which comprises the Hellenic Arc, the Greek mainland, and Western Turkey is the most seismically active region in Western Eurasia. The convergence of the Eurasian and African lithospheric plates forces a westward motion on the Anatolian plate relative to the Eurasian one. Western Anatolia is a valuable laboratory for Earth Science research because of its complex geological structure. Izmir is a large city in Turkey with a population of about 2.5 million that is at great risk from big earthquakes. Unfortunately, previous geodynamics studies performed in this region are insufficient or cover large areas instead of specific faults. The Tuzla fault (TF), which is aligned trending NE–SW between the town of Menderes and Cape Doganbey, is an important fault in terms of seismic activity and its proximity to the city of Izmir. This study aims to perform a large scale investigation focusing on the TF and its vicinity for better understanding of the region's tectonics. In order to investigate the crustal deformation along the TF and Izmir Bay, a geodetic network has been designed and optimizations were performed. This chapter suggests a schedule for a crustal deformation monitoring study which includes research on the tectonics of the region, network design and optimization strategies, theory, and practice of processing. The study is also open for extension in terms of monitoring different types of fault characteristics. A one-dimensional fault model with two parametersstandard strike-slip model of dislocation theory in an elastic half-spaceis formulated in order to determine which sites are suitable for the campaign based geodetic GPS measurements. Geodetic results can be used as a background data for disaster management systems.

This study suggests a plan for a large scale crustal deformation monitoring project including the relations between the global tectonics, the interpretation of seismicity and tectonics of the study area, appropriate geodetic techniques for deformation monitoring, combination of different techniques, geodetic network design, and optimization. Deformation measurements performed using geodetic techniques include some critical steps in the processing and design stages. Moreover, other parameters such as the location of deformed area or the deformation type should also be taken into consideration. An appropriate technique should be chosen considering the deformation type, the proximity of the deforming area or object to urban areas and suitable processing techniques.

The Aegean Region and Western Anatolia are one of the most seismically active and deforming parts of the Alpine-Himalayan orogenic belt. Consequently high seismic activity has been observed in this region. An extensional deformation regime has led to subsidence of the continental crust over all regions behind the south Aegean. The region is mainly under pure shear stress from an internally deforming counter-clockwise rotation of the Anatolian Plate relative to the Eurasian one. There is a multi disciplinary research report in the literature concerning the plate interactions through the whole Arabia–Africa and Eurasian plates performed for several periods (Reilinger et al., 2006). Figure 1 shows the result of this study, performed by Reilinger et al. The Aegean Region has been suffering active N-S extensional tectonics, under the control of two main motions. One of the motions is the westward escape at a rate of 2025 mm/year of the Anatolian plate, bound by the North Anatolian Fault and East Anatolian Fault, and intersecting at the Karliova depression of the East Anatolia. The westward motions change direction in West Anatolia with a rather abrupt counter-clockwise rotation towards the southwest over the Hellenic Trench. The other motion is the N-S extension of the Western Anatolian and the Aegean plates with a rate of about 36 cm/year. As a result of these motions a group of E-W trending grabens have been developing. These grabens are bound by E-W trending normal fault zones which extend about 10,0150 km. These fault zones are generally segmented and each segment is no longer than 810 km (Yilmaz, 2000).

The complicated geology of the region has given rise to disagreements on the source or beginning of the extension of the region. McKenzie (McKenzie, 1978) suggests the beginning time of the extension as 5 Ma, while by other researchers have suggested 1311 Ma (Mercier, 1989). This variety in the suggestions concerning the beginning of the N-S extension for the Aegean Region may be due to on the insufficient accuracy of the methods used to determine the beginning time or lack of information about the previous geological researches that preclude accurate estimations.

Consequently, the result has been a focus of the geological investigations on the Aegean Region in order to understand the tectonics of the area. Geodesy and geodynamics can also contribute additional information (Segall and Davis, 1997). Geodesy builds its investigations on the information gathered from the seismological studies. Therefore, interpretation of earthquake distributions, determination of focal mechanisms, and field studies that aim to define fault traces provide valuable data for geodetic crustal deformation studies. Thus, the project area that is to be monitored with geodetic techniques has to be evaluated in terms of the project area's seismicity. A complete picture for deformation monitoring studies using geodetic techniques has to be formed including definitions on tectonics of the study area, network design regarding to the geological and geophysical parameters of the region, and approaches to the combination of different geodetic techniques. The chapter discusses the possible extensions in the size of the network depending on the fault characteristics. Monitoring two or more faults together, for instance, can be a better solution to understand the characteristics of the region in some cases so the study area needs to be extended during the geodetic observations. This chapter uses the formulations in one-dimensional fault model with two parameters standard strike-slip model of dislocation theory in an elastic half-space for selecting suitable site locations of the network.

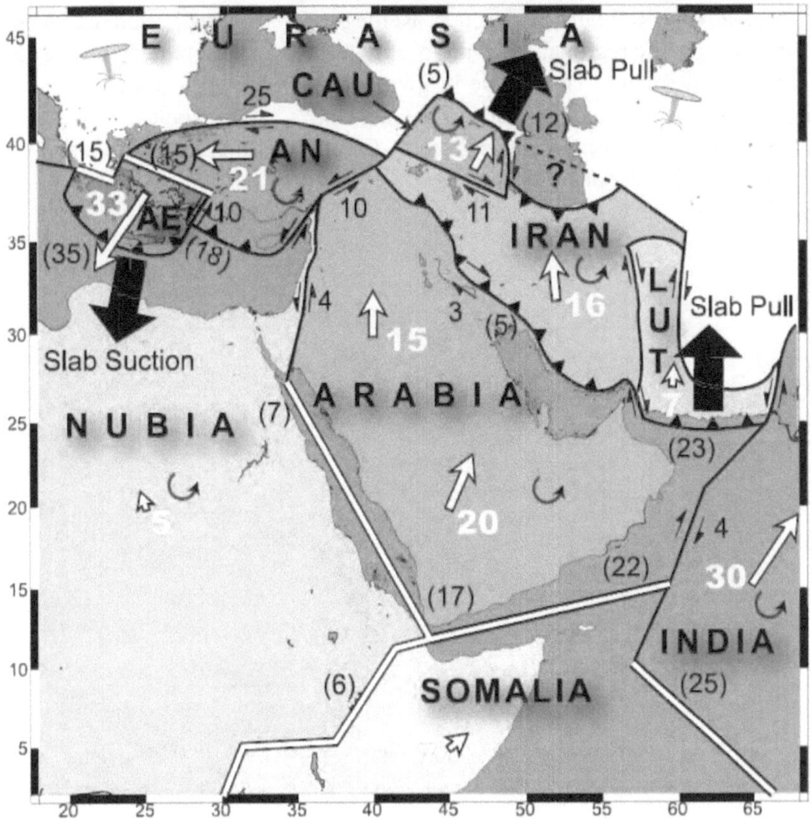

Figure 1. Plate interactions of Arabia-Africa-Eurasia zone (Reilinger et al., 2006).

MICRO-TECTONICS FEATURES OF THE REGION OF INTEREST

The studied region has a high seismic activity due to the extensional regime of the Aegean Region. Thus Western Anatolia contributes greatly to Turkey's earthquake activity and neotectonics. Ozmen et al. (1997) produced a seismicity map considering the data beginning from the instrumental time to present that indicates the different perspective of western Anatolia than the Turkey's total activity. Figure 2 shows the seismic risk zones and the study area which is in the high risk-zone I.

There are two main seismic belts within the boundaries of the region. One of them lies in the CreteRhodes-Fethiye and Burdur direction and the other one is in a direction along the Simav-Emet-Gediz and Afyon locations. These two belts have the highest seismicity in the whole Aegean Region (Bagci, 2000; KOERI Earthquake Catalog, 2008). Geodynamic studies show that the Aegean Region needs to be investigated continuously with different scientific techniques. This study is going to subject the geodetic contribution to regional tectonics with some geodetic optimization techniques using gathered information from different sources. Although the tectonics of

Eastern Mediterranean have been explained by long term episodic and continuous GPS observations (Reilinger et al., 2006; Reilinger, 2000) some special cases need to be defined in specific regional deformations. Izmir as a high populated city settled on seismically active faults. Thus, there is always high seismic risk underlined in many studies (Aktug and Kilicoglu, 2006; Kreemer and Chamot-Rooke, 2004; Nur and Cline, 2000; Ocakoglu et al., 2005; Ozcep, 2000; Zhu et al., 2006) in Izmir, like the North Anatolian Fault Zone.

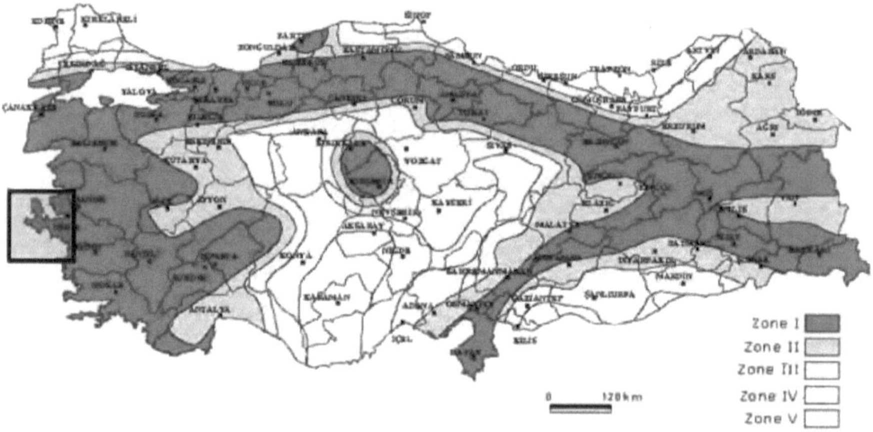

Figure 2. Turkey earthquake hazard map and study area (Ozmen et al., 1997).

Several GPS network optimization studies have also been published during the last decade (Blewitt 2000; Gerasimenko et al., 2000; Wu et al., 2003). Therefore, there is a need to perform a large scale crustal deformation monitoring study using the results of previous studies mentioned above, in order to evaluate regional tectonics. However, the tectonics of Izmir and its vicinity is very complex in the geological sense and should be investigated in detail to understand long and short term geodynamic activities.

The deformation pattern in the Mediterranean region which forms a low elevated part of the Alpine Himalayan belt is rather complex, and usually occurs in the continental collision zones. The Aegean region is bounded to the north by the stable continental Eurasian plate, to the west by the Adriatic region, to the east by the central Anatolian plate, and to the south by the oceanic material beneath the Mediterranean Sea, which is northern edge of the African plate. The Black and Mediterranean Sea floors have mean depths of 1,500 and 1,300 meters, respectively, while the Aegean Sea floor has a mean depth of 350 m. In other words, the Aegean Sea floor may be seen as a high plateau between the deeper Black Sea and Mediterranean Sea floors. The Aegean is characterized by a relatively thicker crust (2,530 km) than a typical oceanic crust, which might conversely be interpreted as a thinned continental crust. The Aegean is also situated in the convergent boundary between the African plate and Eurasian plate. The African plate has rotated counter-clockwise with respect to Eurasian plate during the last 92 Ma (Müller et al., 1997). The spatial distribution of earthquakes and detailed topographic studies indicate the existence of a northward-dipping subducted slab beneath this region (African plate beneath Eurasian plate). However, according to Müller et al. (1997), a roughly N-S directed lithosphere shortening rate is increasing from west to east in the Aegean region. The region is also characterized by high heat flow, which is related to thin and deformed (stretched) continental crust. This thinning is continuing until now and for this reason, it is the worldwide most seismically active and internally deforming area of the entire Alpine-Himalayan belt and at of all continents (Jackson et al., 1994; Mercier et al., 1989).

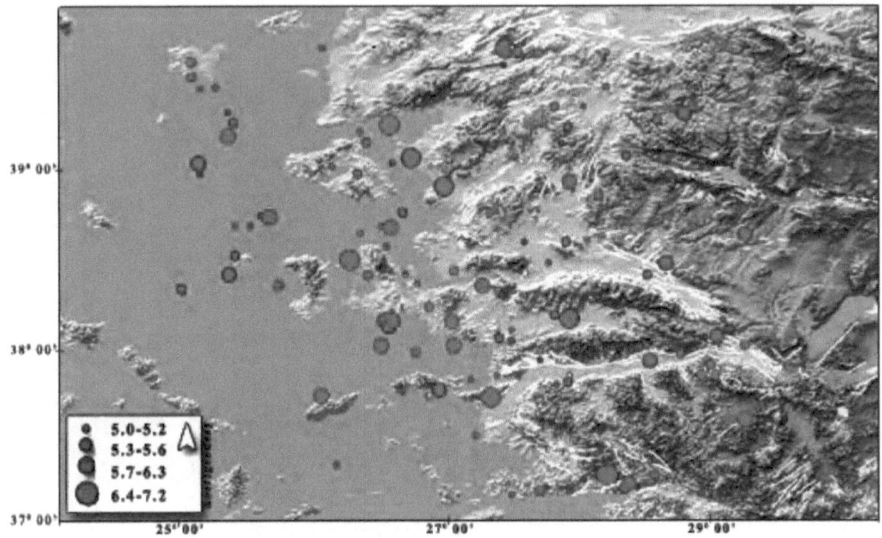

Figure 3. Faults and earthquakes, M>5 in 19002006 (KOERI Earthquake Catalog, 2008).

Papazachos (1999) defines the northern and eastern boundaries of the Aegean plates which comprises the Hellenic Arc, Greek mainland, and western Turkey. The Anatolian plate has a relative motion of 2225 mm/year with respect to the Eurasian Plate according to McClusky et al. (2000). The focal mechanism solutions of earthquakes indicate that the faulting in the western part of the Aegean region is mostly extensional in nature on normal faults, with a NW to WNW strike and slip vectors directed NW to N (Taymaz, 2001). The evidences from paleomagnetism show that this region rotates clockwise relative to a stable Eurasia. According to Piper et al. (2001), paleomagnetic data in the eastern Aegean Region is consistent with very small or no rotations in the northern part and possibly counter-clockwise rotations in the south relative to the Europe, including some ambiguities. The strike-slip faulting that lying through the central Aegean from the east appears to end abruptly in the SW against the NW trending normal faults of Greece.

The extension tectonic regime affected Western Anatolia in the neotectonic age. Izmir lies on the west side of the Gediz Graben and bound by the Gulf of Izmir. There are several active faults that have triggered the dense earthquake activity recorded beginning from the 20th century as shown in Figure 3. In addition some major faults have the capacity to produce big earthquakes. According to the report on Active Faults and Seismicity in Izmir and its vicinity (Emre et al., 2005), there is not enough investigation on the earthquake activity potential except for the Gediz Graben. The report defines active faults within a 50 km semi-diameter area which has an origin at central Izmir. Emre et al. (2005) defined the 14 active faults shown in Figure 4 through the region. These faults are Guzelhisar, Menemen, Yenifoca, Izmir, Bornova, Tuzla, Seferihisar, Gulbahce, Gumuldur, Gediz Graben detachment faults, Daglikizca, Kemalpasa, and Manisa Faults. The following paragraphs give brief explanations about these active faults and focus on the TF in detail.

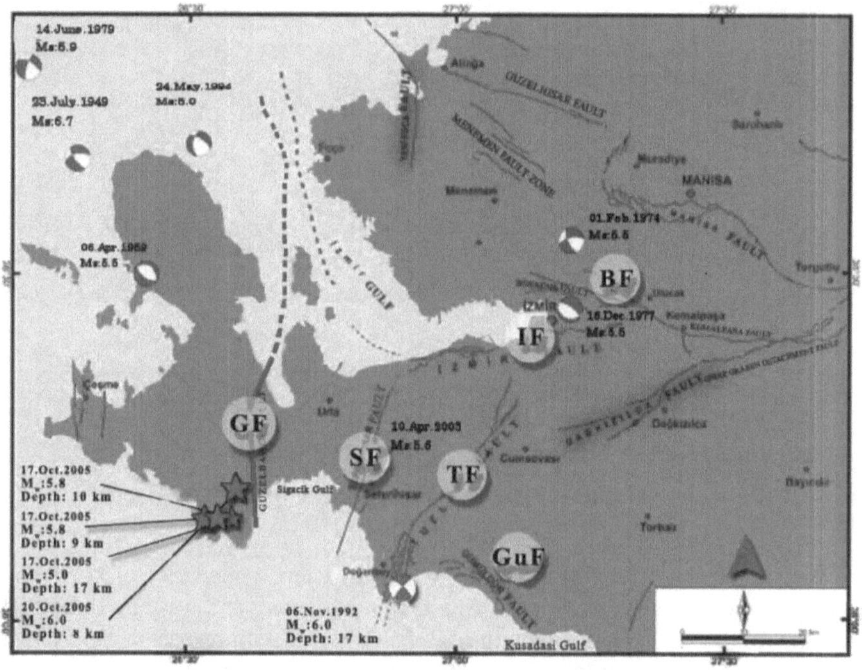

Figure 4. Important faults of Izmir and its vicinity modified from (Emre et al., 2005) (GF: Guzelhisar Fault, IF: Izmir Fault, BF: Bornova Fault, TF: Tuzla Fault, SF: Seferihisar Fault, GuF: Gumuldur Fault).

The TF is in the southwest of Izmir, between Cape Doganbey, and Gaziemir counties with an alignment trending NE-SW. It has been known by different names in the literature, such as Cumaovasi and Orhanli faults (Emre and Barka, 2000; Saroglu et al., 1992). The fault is 42 km long through the land side. However, in 2004 and 2005, after the investigations performed by GDMRE Sismik-1 research ship in Cape Doganbey the total length published was more than 50 km. The TF has three main segments that have different directions. Emre et al. (2005), named these three parts the Catalca, Orhanli, and Cumali sections, arranged from north to south, respectively. Therefore, the right-lateral strike slip TF, with its 50 km length (including the undersea segments) is considered as an active and important tectonic phenomenon of Izmir and its vicinity. On the other hand, TF is the main element that defines the paleo-geography of the region during the Miocene period (Genc et al., 2001). Genc et al. also claimed that the fault has left-lateral strike slip behavior. To the contrary, some other studies (Emre and Barka, 2000; Emre et al., 2005; McCluskyet al., 2000) propose that the fault had a right-lateral strike slip behavior during the Quaternary. The fault plane solutions determined by Turkelli et al. (1992, 1995) also confirm this theory. Important earthquakes within last 2 decades which confirms the seismic risk of the Aegean Region are shown in Table 1.

Table 1. Important earthquakes in the region within the last 2 decades.

Day	Year	Lat.	Lon.	Depth	M
06.Nov	1992	38.16	26.99	17	6.0
28.Jan	1994	38.69	27.49	5	5.2
24.May	1994	38.66	26.54	17	5.0
10.Apr	2003	38.26	26.83	16	5.6
17.Apr	2003	38.24	26.86	6	4.8
17.0ct	2005	38.15	26.54	10	5.8
17.0ct	2005	38.15	26.53	9	5.8
17.0ct	2005	38.15	26.58	17	5.0
20.0ct	2005	38.18	26.59	8	6.0

DESIGN AND IMPLEMENTATION

Observation techniques, selected equipment, and surveying interval of any project have to be optimized in terms of several parameters. These optimizations, in general, are realized to achieve a desired precision. Besides, reliability is also as important as precision. One should trust not only the results but also the reliability of a network which can be expressed as mathematical relations. The precision, reliability and economical parameters in a geodetic network can be arranged in order to achieve the optimum solution which is defined as the optimization of geodetic networks (Ayan, 1981). In order to determine the deformation, generally local networks are preferred. A deforming area is generally covered by a number of control points. These points constitute a geodetic network and their location or structure is defined by the topographic and geological parameters. The number of points is directly related with the deforming object and the deformation accepted in the area. The ideal approach is an interdisciplinary study to define the number of points and locations for these "control networks". Not only geodesy but also geology, geophysics, and disaster management should contribute deformation monitoring studies. Ayan (1981) has suggested three sets of control points for deformation monitoring which are deformation points, reference points, and orientation points.

In order to contribute geodynamic studies, after the 2005 M: 5.9 Sigacik earthquake, this region's seismic risk was considered. Izmir has a very complex geological structure of faults with different characteristics. This variety in fault characteristics also made it difficult to select the project area. For this reason, at the beginning the information collected about the region covered the whole Western Anatolia region, but then the research had focused on the most important section of the region. The TF and its vicinity coincide with the aims defined at the beginning of the study because of the active behavior of the fault, its closeness to Izmir, and big earthquakes recorded in the area.

Geodetic deformation analysis requires a stable, continuously or periodically observed network. Moreover, in order to estimate the small amount of deformations, some additional techniques such as precise leveling and gravimetric, or astrogeodetic techniques are generally considered. Leveling routes are generally designed in perpendicular lines with respect to the fault trace. This chapter, which focused on the TF, designed a micro-geodetic network considering the valuable information gathered from different resources such as municipalities, and then state the theoretical background with some scientists' approaches in terms of geodetic optimization.

The general plan for the network design performed on several parameters which are the available data collected from local resources, the topographic, and economic situations, equipment which is going to be used and the fault geometry. The outputs of these parameters are the approximate locations of the geodetic control points, the number of the stations, and the observation and processing strategies.

Especially over the last two decades studies in the area of crustal deformation along plate boundaries and individual fault traces have grown, so the interest in an optimal design of monitoring schemes has increased. Because of the effectiveness of GPS for crustal deformation monitoring processes, the optimal design of monitoring network becomes a great practical interest. Designing a geodetic network can be generally divided into four main stages. The zero-order design (ZOD) which generally deals with the definition of the optimum reference system of the network. The first-order design (FOD) involves the geometric shape of the network including the optimum number and locations of the geodetic stations. The second-order design (SOD) deals with the determination of the weights of network measurements. The SOD interested in which observations and with what precision should be achieved in the network. Finally, the third-order design (TOD) considers the improvement of an existing network including the additional measurements that has to be made with the desired precision and what weights are selected for the improvement of network. Schmitt (Schmitt, 1985) claimed that in cases where the period of time between consecutive observations is taken into account, the term fourth-order design maybe used.

In order to define the number of station that should be added into a deformation network or which sites should be used for that purpose is directly concerned with the phenomenon understanding fault mechanics. Gerasimenko et al. (2000) conceived a model for this purpose using a simple strike-slip fault model in which the deformations are parallel to the fault trace, in order to facilitate the solution. A one-dimensional fault model with two parameters standard strike-slip model of dislocation theory in an elastic half-space can be formulated as:

$$d(x) = -\frac{V}{\pi}\arctan\left(\frac{x}{H}\right)$$

where x is the distance perpendicular to the fault, and the fault plane extends from the surface of the half space to infinite depth, locked from the surface to H km, and freely slipping below this depth V millimeter per year. The method suggested by Blewitt (2000) leads to exact analytical solutions for the ideal transform fault locked down to depth D. According to this method, to resolve the depth of locking D and the location of the fault simultaneously, optimal station locations are at D/√3 from the a priori fault plane. The seismogenic zone which is obtained as 12 km, derived from earthquake depths using the information taken from KOERI earthquake catalogs (2008). In other words, geodetic sites which are chosen and established are around 7 km away from the fault trace. On the other hand, analysis of slip partitioning in two-fault system shows that the resolution is optimized by including a station between faults. If the distance between faults is greater than 2D which is approximately 30 km the resolution is limited. Design is also suitable for precise leveling on short baselines of the network in order to increase the vertical component accuracy of position by using precise leveling technique.

According to the optimization strategies, performed experiments, and collected information stated above, a geodetic network has been designed in order to monitor TF and its vicinity and interpretation strategies are discussed.

The network was designed based on the information from existing control points and the fault trace geometry. Some additional stations were established in order to define the locking depth and slip rate of the fault trace according to the conclusions defined above. Moreover, because of the possibility of the extension of the study area, other active faults were taken into consideration in the design process. The station names are identified using four character Turkish National Fundamental GPS Network (TNFGN) station names.

After discussions with the local administrations, 14 control stations were selected for the network from among hundreds of possible sites. Numerous station points have been established throughout the region, especially in last 3 years for cadastre projects.

Figure 5 shows the locations of the sites and an approximate trace of Tuzla and Seferihisar faults (SFs). Stations are distributed both on the fault trace and some 20 km away from the fault. The stations are close to each other along the south segment of TF because the fault has a very complex and sectional structure in that area. This complexity, named the Cumali segment, is a zone of several faults that are parallel to each other (Emre et al., 2005). The length of this segment is 15 km and it has 10 km long undersea part (Ocakoglu et al., 2005). Moreover, this segment is very close to a 15 km long normal fault, Gumuldur fault (GuF), so this complexity has to be considered in any design process. Adding extra control points to the network would be a solution for monitoring this dynamic region of study area. There are some short baselines in the network such as CCEK-GMDR baseline because of the adjacency of two active faults. There is another fault very near to TF and GMDR and CCEK points are very close to that GuF. The WGS84 coordinates of the control station shown in Table 2.

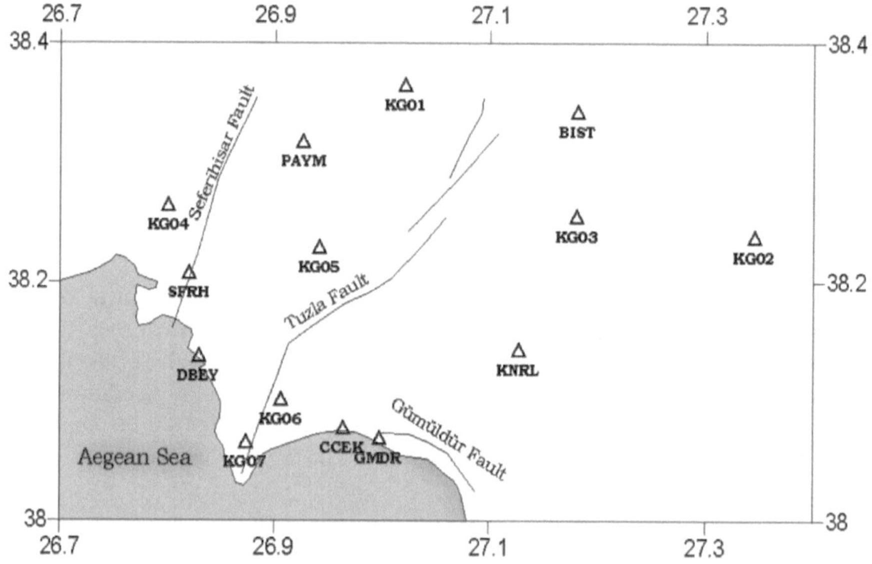

Figure 5. Locations of the sites of Izmir microgeodetic network.

Table 2. Locations of network stations.

Site name	q> Latitude (in Degrees)	A. Longitude (in Degrees)
KG01	38.36416	27.02072
KG02	38.23637	27.34451
KG03	38.25372	27.18084
KG04	38.26320	26.80143
KG05	38.22815	26.94152
KG06	38.10099	26.90646
KG07	38.06433	26.87388
PAYM	38.31700	26.92600
KNRL	38.14244	27.12700
GMDR	38.06800	26.99700

Table 2. *(Continued)*

Site name	q> Latitude (in Degrees)	A. Longitude (in Degrees)
DBEY	38.13700	26.83000
SFRH	38.20700	26.82100
BIST	38.34200	27.18100
CCEK	38.07659	26.96351

In summary, the locations of the station points of the microgeodetic network are distributed on both sides of the fault. Moreover, some stations are located very near to the fault trace and some others as far as 20 km away from the fault trace, according to the distribution of the surface deformation with respect to the distance from the fault trace.

The network is compatible with the studies performed in first order network design studies. Generally the lines connecting GPS stations are in alignment with the direction of extension or compression, the angles of triangles composed by GPS stations are generally between 30 and 130 degrees (Wu et al., 2003). On the other hand, some additional points that were added to the network like GMDR, KG07, KG06, and KG02 do not satisfy the above rules. However, those points were selected deliberately because of the very complex structure of the southern segment of the faults, composed of several pieces. The KG02 was selected because we desired to evaluate the results in terms of short and long baselines and for various perpendicular distances to the fault trace.

Moreover, a block exists in the middle of Karaburun peninsula that has a differential motion at a rate of $3 - 5 \pm 1$ mm/year to the east and $5 - 6 \pm 1$ mm/year to the south (Aktug and Kilicoglu, 2006). Therefore, 14 points were thought to be enough for determining the slip rate, which is not as small, as stated by Gerasimenko et al. (2000). The network designed to be suitable for future studies which have a possibility to enlarge the project area, so the suggestions mentioned in Blewitt (2000) are taken into consideration. The sites were also selected according to the transportation possibilities and visibility of open sky. The reconnaissance performed in the region made it easy to define those site properties.

RESULTS AND DISCUSSION

This study focused on the idea of dealing with a crustal deformation monitoring project on a particular fault which has a high-seismic risk using geodetic techniques. Moreover, this study attempted to establish interactions between geosciences and geodesy in terms of deformation monitoring projects. The chapter explains the tectonics of the eastern Mediterranean and Aegean Region in general and the tectonics of Izmir and its surrounding area in more detail. Important faults are underlined from a recent study performed by General Directorate of Mineral Research and Exploration in 2005 (Emre et al., 2005). Some projects that have geodetic components were also investigated (Aktug and Kilicoglu, 2006; McClusky et al., 2000; Reilinger et al., 2000, 2006) to focus on the movements of Anatolia and western Turkey.

According to the study of McClusky et al. (2000), the high rate of velocity vectors especially in the Aegean Region is pointed out. Moreover, Reilinger et al. (2006), mentioned the high rate of movement of western Anatolia according to the Anatolian plate. Another recent study (Aktug and Kilicoglu, 2006) that covers an area between latitudes 37°45′ and 39°00′, and longitudes 26°00′ and 28°00′ mention the high rate of velocities especially near TF. The velocities from two different studies can be seen in the Figure 6, where the black arrows indicates the residual velocities obtained by differentiating ITRF2000 and Eurasia plate velocities by using the following formula. On the other hand, red colored arrows indicate the Eurasia fixed velocity vectors. Figure 6 indicates an important deformation rate especially around TF.

$$V_r = \hat{V}_{ITRF2000} - \hat{V}_{PLATE}$$

In order to contribute these projects by performing large scale fault based deformation monitoring study, TF, and its vicinity was selected considering its high seismic risk. Therefore, a reconnaissance was planned after the literature research in order to investigate the field and collect necessary information from local resources. Thus, this reconnaissance to the region was performed, the information collected, evaluated, and analyzed within this study. Moreover, first order network design problems are quoted to create a harmony between microgeodetic networks. Network stations are selected from a large set of control points according to the suggestions mentioned in several studies. These whole processes produced a microgeodetic network that is selected from a huge set of information.

The network has an open end for future studies. In other words, there is a possibility of an extension for the network in order to monitor some additional faults. The TF exists in the center of the region and is very near to the big metropolitan city, Izmir. Thus the origin of the study is selected near this fault. Some researchers also mention the high seismic risk of the region including TF (Ocakoglu et al., 2005; Zhu et al., 2006). On the other hand, it is certain that, the area should be monitored by a larger and dense network with continuously operating GPS stations. For further studies, campaign based GPS observations are planned beginning from the current network designed in this study and will extend to the west to the Karaburun Peninsula, and to the east to the eastern Aegean region. According to the results achieved from some researches (Akyol et al., 2006; Kaymakci, 2006; Zhu et al., 2006), there is a great seismic risk through the transform faults to the east near Pamukkale-Denizli. However, in this study, because of the topography related effects such as high mountains and the small rate vertical deformation make it nearly impossible to study with GPS or precise leveling techniques. For the reasons mentioned above, the network established to the area that is covering the TF.

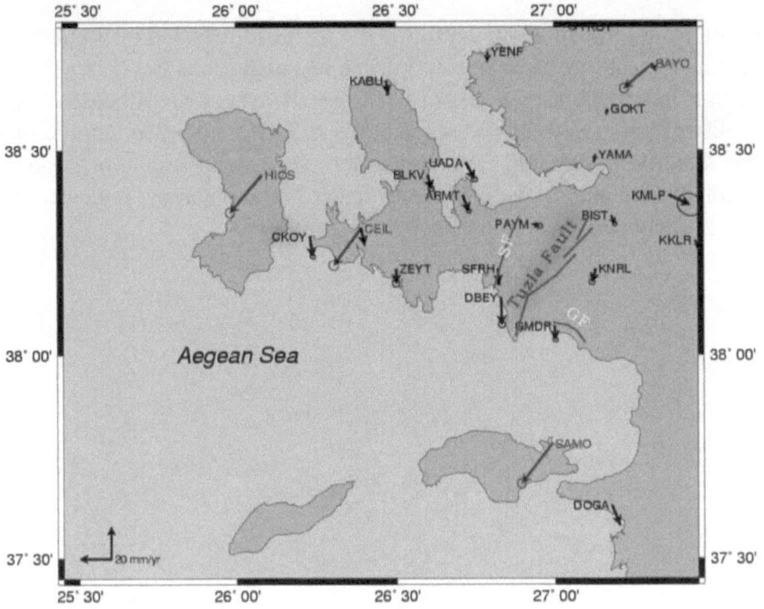

Figure 6. Red colored arrows indicate the Eurasia fixed velocity vectors, black arrows indicate the Anatolia fixed velocities (Aktug and Kilicoglu, 2006; McClusky et al., 2000), blue lines indicate faults

In conclusion, this chapter, a plan for deformation monitoring studies using geodetic techniques including network design and optimization was prepared. The next step of this study will be three GPS campaigns on the designed network in two periods. In addition to GPS technique, conventional geodetic techniques such as precise leveling technique would be a choice for normal faults where small vertical deformations need to be determined. Further studies will be built on the information and techniques introduced in this study. It is certain that geodetic techniques are capable of determining small movements which are quite valuable information for earth sciences. Moreover, geodetic results can be valuable information for management systems in terms of the decision making based on characteristics of the geological features of the study area.

KEYWORDS

- **Geodetic crustal deformation**
- **GPS measurements**
- **Network design**
- **Seismic risk**
- **Seismicity**

ACKNOWLEDGMENTS

The authors would like to thank the following organizations and institutions for their invaluable and great efforts in support of this research and *MSc Thesis* (Halicioglu, 2007); Izmir and Istanbul branches of Chamber of Survey and Cadastre Engineers, Municipality of Izmir, Seferihisar branch of General Directorate of Land Registry and Cadastre, Kandilli Observatory, and Earthquake Research Institute of Bogazici University. Finally, the authors would like to thank the anonymous reviewers for their valuable comments that help to improve the manuscript.

Permissions

Chapter 1: Cyber Infrastructure and Data Services for Earth System Science Education and Research was originally published as "A new generation of cyberinfrastructure and data services for earth system science education and research" in *Adv. Geosci., 8, 69–78, 2006 www.adv-geosci.net/8/69/2006*. Reprinted with permission under the Creative Commons Attribution License or equivalent.

Chapter 2: Health Benefits of Geologic Materials and Geologic Processes was originally published as "Health Benefits of Geologic Materials and Geologic Processes" in *Int. J. Environ. Res. Public Health* 2006, *3(4)*, 338-342. Reprinted with permission under the Creative Commons Attribution License or equivalent.

Chapter 3: Cloud Statistics from Spaceborne Lidar Systems was originally published as "Comparison of cloud statistics from spaceborne lidar systems" in *Atmos. Chem. Phys. Discuss, 8, 5269–5304, 2008 www.atmos-chem-phys-discuss.net/8/5269/2008*. Reprinted with permission under the Creative Commons Attribution License or equivalent.

Chapter 4: Pre-earthquake Signals was originally published as "Towards a Unified Theory for Pre-Earthquake Signals" in *Work supported by NASA through "Earth Surface and Interior" program and by a grant from the NASA Ames Research Centre 2:4, 2009*. Reprinted with permission under the Creative Commons Attribution License or equivalent.

Chapter 5: Using Earthquakes to Uncover the Earth's Inner Secrets was originally published as "Using earthquakes to uncover the Earth's inner secrets: interactive exhibits for geophysical education" in *Advances in Geosciences, 3, 15–18, 2005 SRef-ID: 1680-7359/adgeo/2005-3-15 European Geosciences Union*. Reprinted with permission under the Creative Commons Attribution License or equivalent.

Chapter 6: Ground Water Desalination was originally published as "Desalination of Ground Water: Earth Science Perspectives" in *USGS Fact Sheet 075-03*. Reprinted with permission under the Creative Commons Attribution License or equivalent.

Chapter 7: Deforestation Prediction for Different Carbon-prices was originally published as "Predicting the deforestation-trend under different carbon-prices" in *BioMed Central Ltd. 12:6, 2006*. Reprinted with permission under the Creative Commons Attribution License or equivalent.

Chapter 8: Volcanic Hazards and Evacuation Procedures was originally published as "Resident perception of volcanic hazards and evacuation procedures" in *Nat. Hazards Earth Syst. Sci., 9, 251-266, 2009 www.nat-hazards-earth-syst-sci.net/9/251/2009/doi: 10.5194/nhess-9-251-2009*. Reprinted with permission under the Creative Commons Attribution License or equivalent.

Chapter 9: Identification of Earthquake Induced Damage Areas was originally published as "Identification of Earthquake Induced Damage Areas Using Fourier Transform and SPOT HRVIR Pan Images" in *Sensors 2009, 9, 1471-1484; doi: 10.3390/s90301471*. Reprinted with permission under the Creative Commons Attribution License or equivalent.

Chapter 10: Ionospheric Quasi-static Electric Field Anomalies During Seismic Activity was originally published as "Ionospheric quasi-static electric field anomalies during seismic activity in August–September 1981" in *Nat. Hazards Earth Syst. Sci., 9, 3-15, 2009 www.nat-hazards-earth-syst-sci.net/9/3/2009/doi: 10.5194/nhess-9-3-2009*. Reprinted with permission under the Creative Commons Attribution License or equivalent.

Chapter 11: Ecosystem Changes in the Northern South China Sea was originally published as "Long term changes in the ecosystem in the northern South China Sea during 1976–2004" in *Biogeosciences, 6, 2227–2243, 2009 www.biogeosciences.net/6/2227/2009*. Reprinted with permission under the Creative Commons Attribution License or equivalent.

Chapter 12: Lower Miocene Stratigraphy Along the Panama Canal was originally published as "Lower Miocene Stratigraphy along the Panama Canal and Its Bearing on the Central American Peninsula" in *PLoS ONE 7:30, 2008*. Reprinted with permission under the Creative Commons Attribution License or equivalent.

Chapter 13: Remote Sensing Data with the Conditional Latin Hypercube Sampling was originally published as "Remote Sensing Data with the Conditional Latin Hypercube Sampling and Geostatistical Approach to Delineate Landscape Changes Induced by Large Chronological Physical Disturbances" in *Sensors 2009, 9, 148-174; doi: 10.3390/s90100148*. Reprinted with permission under the Creative Commons Attribution License or equivalent.

Chapter 14: The Global Sweep of Pollution was originally published as "The Global Sweep of Pollution: Satellite Snapshots Capture Long-Distance Movement" in *Environmental Health Perspectives, 2008 August; 116(8): A338–A345*. Reprinted with permission under the Creative Commons Attribution License or equivalent.

Chapter 15: Geodetic Network Design for Disaster Management was originally published as "Geodetic Network Design and Optimization on the Active Tuzla Fault (Izmir, Turkey) for Disaster Management" in *Sensors 2008, 8, 4742-4757; DOI: 10.3390/s8084742*. Reprinted with permission under the Creative Commons Attribution License or equivalent.

References

1

AGU (1997). *Shaping the Future of Undergraduate Earth Science Education: Innovation and Change Using an Earth System Approach.*

Channabasavaiah, K., Holley, K., and Tuggle, E. (2003). Migrating to a service-oriented architecture. *Parts I & II, IBM Developer-Works White Paper.* Retrieved from http://www-128.ibm.com/developerworks/webservices/library/ws-migratesoa/ and http://www-128.ibm.com/developerworks/library/ws-migratesoa2/

CIERE (2003). *Cyberinfrastructure for Environmental Research and Education, Workshop sponsored by the National Science Foundation,* Boulder, Colorado, 2002. Retrieved from url-http://www.ncar.ucar.edu/cyber/

Colwell (1998). The National Science Foundation's Role in the Arctic. Dr. R. R. Colwell, *Opportunities in Arctic Research: A Community Workshop.* Arlington, Virginia, 1998. Retrieved from http://www.nsf. gov/od/lpa/forum/colwell/rc80903.htm

Domenico, B., Caron, J., Davis, E., Kambic, R., and Nativi, S. (2002). Thematic Real-time Environmental Distributed Data Services (THREDDS): Incorporating Interactive Analysis Tools into NSDL. *J. Digital Info.* **2**, 4. Retrieved from http://jodi.ecs.soton.ac.uk/Articles/v02/i04/Domenico/

Foster, I. (2002). What is the Grid? A three point checklist. *Grid Today* **1**, 6. Retrieved from http://www.gridtoday.com/02/0722/100136.html

Foster, I., Alpert, E., Chervenak, A., Drach, B., Kasselman, C., Nefedova, V., Middleton, D., Shoshani, A., Sim, A., and Williams, D. (2002). *The Earth System Grid II: Turning Climate Datasets into Community Resources.* 19th Conference on Interactive Information Processing Systems, American Meteorological Society, Long Beach, CA.

Foster, I. and Kesselman, C. (1997). Globus: A metacomputing intrastructure toolkit. *Int. J. Supercomput. Appl.* **11**(2), 115–128.

Gannon, D., Plale, B., Christie, M., Fang, L., Huang, Y., Jensen, S., Kandaswamy, G., Marru, S., Lee Pallickara, S., Shirasuna, S., Simmhan, Y., Slominski, A., and Sun, Y. (2005). Service oriented architectures for science gateways on grid systems. In *International Conference on Service Oriented Computing.* B. Benatallah, F. Casati, and P. Traverso (Eds.). LNCS 3826, Springer-Verlag Berlin Heidelberg, pp. 21–32.

Graves, S. J., Hinke, T., and Kansal, S. (1996). Metadata: The golden nuggets of data mining. *First IEEE Metadata Conference,* Bethesda, MD, April 16–18, http://www.cs.uah.edu/~thinke/Mining/metapaper.html

Griffin, S. M. (1998). *NSF/DARPA/NASA Digital Libraries Initiative: A Program Manager's Perspectiv.* D-Lib Magazine, July/August. Retrieved from http://www.dlib.org/dlib/july98/07griffin.html

Hey, A. J. G. and Trefethen, A. (2003). The data deluge: An e-Science perspective. *Grid Computing, Making the Global Infrastructure a Reality.* F. Berman, G. Fox, and A. J. G. Hey (Eds.), John Wiley and Sons, p. 1060.

Hood, C. A. (March 9, 2005). *Implementation of GEOSS: A Review of All-hazards Warning System and its Benefits to Public Health, Energy, and the Environment.* Written testimony. US House Committee on Energy and Commerce.

IWGEO (2005). Strategic Plan for the U.S. Integrated Earth Observation System. *National Science and Technology Council Committee on Earth and Natural Resources.* Washington, DC, Retrieved from http://iwgeo. ssc.nasa.gov/

Lawrence, B. (June 24, 2003). Experiences with archiving databases in British Atmospheric Data Center. *6th DPC Forum focused on Open Source Software and Dynamic Databases.* London, UK.

Lemke, J. L. (1990). *Talking Science: Language, Learning, and Values*, Ablex Publishing, Norwood, NJ.

Manduca, C. (2002). *Using Data in the Classroom, Workshop sponsored by NSF.* Carleton College, April 2002. Retrieved from http://

dlesecommunity.carleton.edu/research education/usingdata/index.html

Michalakes, J., Chen, S., Dudhia, J., Hart, L., Klemp, J., Middlecoff, J., and Skamarock, W. (2001). Development of a Next Generation Regional Weather Research and Forecast Model, Developments in Teracomputing. *Proceedings of the Ninth ECMWF Workshop on the Use of High Performance Computing in Meteorology.* W. Zwieflhofer and N. Kreitz (Eds.). World Scientific, Singapore, pp. 269–276.

Middleton, D. (2001). *The Community Data Portal: Sustainable Strategies for Enabling Both Providers and Consumers of Earth System Data.* Retrieved from http://www.ncar.ucar.edu/stratplan/communitydata.html

Morss, R. (2002). Working Group Report. *Workshop on GIS in Weather, Climate, and Impacts Workshop*, Boulder. Retrieved from http://www.esig.ucar.edu/gis/02workshop/working groups/hazardsB.pdf

NRC (2001). *Grand Challenges in Environmental Sciences.* National Research Council, National Academy Press, p. 96. Retrieved from http://www.nap.edu/books/0309072549/html/

NSF (1997). *Geoscience Education: A Recommended Strategy.* Arlington, Virginia. Retrieved from http://www.geo.nsf.gov/adgeo/geoedu/97171.htm

NSF (2000). *NSF Geosciences Beyond 2000: Understanding and Predicting Earth's Environment and Habitability.* NSF Report No. 0027, p. 54. Retrieved from http://www.geo.nsf.gov/adgeo/geo2000/geo 2000 summary report.htm

NSF (2003). *Complex Environmental Systems: Synthesis for Earth, Life, and Society in the 21st Century.* NSF Report No. 03–27, p. 80. Retrieved from http://www.nsf.gov/geo/ere/ereweb/ac-ere/acere synthesis rpt full.pdf

OPeNDAP. Open-source Project for a Network Data Access Protocol. Retrieved from http://www.opendap.org/

Ramachandran, R., Conover, H., Graves, S. J., Keiser, K., and Rushing, J. (1999). *The Role of Data Mining in Earth Science Data Interoperability.* ASPRS Annual Conference, Conference on Remote Sensing Education (CORSE), Portland, OR, May 17–22.

Raymond, E. S. (1999). *The Cathedral & the Bazaar: Musings on Linux and Open Source by an Accidental Revolutionary.* O'Reilly and Associates, p. 268.

Rutledge, G. K., Alpert, J., Stouffer, R. J., and Lawrence, B. (2002). The NOAA Operational Archive and Distribution System (NOMADS). *Proceedings of the Tenth ECMWF Workshop on the use of High Performance Computing in Meteorology.* W. Zwiefhoferand and N. Kreitz (Eds.). World Scientific, pp. 106–129.

Suber, P. (2003). *Removing Barriers to Research: An Introduction to Open Access for Librarian.* College and Research Libraries News, 64 (February 2003), pp. 92–94.

Yoksas, T., Almeida, W., Garrana, D., Castro, V., and Spangler, T. (2005). Internet data distributionExtending real-time data sharing throughout the Americas. *2005 European Geosciences Union General Assembly.* Education Symposium, Session on Earth System Science Data Access, Distribution and User for Education and Research, Vienna, Austria, pp. 24–29.

Wikipedia (2006). *The Free Encyclopedia.* Retrieved from http://en.wikipedia.org/wiki/Grid_computing

2

Abdel-Fattah, A. N. (2004). Dead Sea black mud: Medical geochemistry of a traditional therapeutic agent. *Geological Society of America 2004 Annual Meeting, Abstracts with Programs.* **36**(5), Abstract 7–7, p. 25.

Abrahams, P W. (2005). The involuntary and deliberate (geophagy) ingestion of soil by humans and other members of the animal kingdom. In *Essentials of Medical Geology.* Selinus, et al. (Eds.). Elsevier, NY. p. 435–458.

Alam, H., Zheng, C., Jaskille, A., Querol, R., Koustova, K., Inocencio, R., Conran, R., Seufert, A., Ariaban, N., Toruno, K., and Rhee, P. (2004). Application of a zeolite hemostatic agent achieves 100% survival in a lethal model of complex groin injury in swine. *J Trauma Injury Critical Care* **56**, 974–983.

Aschoff, J. and Tashingang, T. Y. (2001). *Tibetan Jewel Pills: The Rinchen Medicine.* Fabri Verlag, Ulm, Germany, p. 139.

Berger, A., Selinus, O., and Skinner, C. (2001). Medical geology—An emerging discipline. *Episodes* **24**(1).

Bowman, C., Bobrowski, P. T., and Selinus, O. (2003). Medical geology: New relevance in the earth sciences. *Episodes* **26**(4), 270–278.

Bunnell, J. E. (2004). Medical geology: Emerging discipline on the ecosystem-human health interface. *EcoHealth* **1**, 15–18.

Ceruti, P., Davies, T., and Selinus, O. (2001). GEOMED—Medical geology, the African perspective. *Episodes* **24**(4).

De Courssou, L. B. (2002). Preliminary studies of debridement using clay materials on Mycobacterium ulcerans infections and ramifications: Geneva, Switzerland. Presented at the *5th WHO Advisory Group Meeting on Buruli Ulcers*, burulibusters.com

Dissanayake, C. (2005). Of stones and health: Medical geology in Sri Lanka. *Science* **309**, 883–885.

Dissanayake, C. B. and Chandrajith, R. (1999). Medical geochemistry of tropical environments. *Earth Sci. Rev*. **47**, 219–258.

Earthwise (2001). *Geology and Health*. British Geological Survey, Issue 17.

Finkelman, R. B., Orem, W., Castranova, V., Tatu, C. A., Belkin, H. E., Zheng, B., Lerch, H. E., Maharaj, S. V., and Bates, A. L. (2002). Health impacts of coal and coal use: Possible solutions. *Int. J. Coal Geol.* **50**, 425–443.

Finkelman, R. B., Skinner, C. W., Plumlee, G. S., and Bunnell, J. E. (2001). Medical geology. *Geotimes* **46**(11), 20–23.

Gienger, M. (2003). *Crystal Power, Crystal Healing: The Complete Handbook*. Translated from German. Sterling Publishing Co., NY, p. 416.

Hurlbut, Jr., C. S. (1961). *Dana's Manual of Mineralogy,* 17th edition. J. Wiley & Sons, NY, p. 609.

Knishinsky, R. (1998). *The Clay Cure*. Healing Arts Press, Rochester, VT. p. 104.

Komatina, M. M. (2004). Medical Geology: Effects of Geological Environments on Human Health. *Developments in Earth & Environmental Sciences*, Vol. 2. Elsevier, Amsterdam, p. 488.

Limpitlaw, U. (2005). *The Palliative and Curative uses of Minerals, Fossils, and Rocks.* University of Northern Colorado, Greeley, CO, Unpublished data.

Limpitlaw, U. G. (2004). The medical uses of minerals, rocks, and fossils. *Geological Society of America 2004 Annual Meeting,* Abstracts with Programs, Abstract 48–5, **36**(5), p. 130.

Min, S. J. (Nov/Dec 2004). *Lost in Translation: Journal of Science and Spirit.* Heldref Publication, Washington, DC, p. 36–41.

Natural Science and Public Health (2003). *Prescription for a Better Environment.* Abstracts from a conference on April 1–3, Reston Virginia. Convened by the US Geological Survey (USGS). Open File Report 03–097, US Department of the Interior, USGS.

Nunn, J. F. (1996). *Ancient Egyptian Medicine.* University of Oklahoma Press, Norman, OK.

Opalitho (2005). *The Journal of crystal healing EV*. Printing Speh GmBH, Fronreute, USA.

Pavetic, K., Hadzija, M., Pavelic, J., Katiz, M., Kralj, M., Bosnar, M. H., Kapitanovic, S., Poljak-Blazi, M., Stojkovic, R., and Subotic, B. (October 2000). Natural zeolite clinoptilolite: New adjuvant in anticancer therapy. *J. Mol. Med.* (published online).

Price, W. A. (2000). *Nutrition and Physical Degradation*, 6th edition. The Price-Pottenger Nutrition Foundation, Inc., La Mesa, CA. p. 524.

Reinbacher, R. (2003). *Healing Earths: The Third Leg of Medicine*. York University, Toronto, Canada, p. 244.

Selinus, O., Alloway, B., Centeno, J. A., Finkelman, R. B., Fuge, R., Lindh, U., and Smedley, P. (in press). *Essentials of Medical Geology.* Elsevier, Amsterdam.

Selinus, O. S. (2004). Medical Geology: An emerging speciality. *Terrae*, **1**(1), A1–8.

Skinner, H. C. W. and Berger, A. R. (Eds.) (2003). *Geology and Health: Closing the Gap.* Oxford Press, Oxford, p. 179.

Wang, B., Finkelman, R. B., Belkin, H. E., Palmer, C. A. (2004). A possible health benefit of 21st coal combustion. *Abstracts of the Annual Meeting of the Society for Organic Petrology* **21**, 196–198.

3

Abshire, J. B., Sun, X., Riris, H., Sirota, J. M., McGarry, J. F., Palm, S., Yi, D., and Liiva, P. (2005). Geoscience Laser Altimeter System (GLAS) on the ICESat Mission: On-orbit measurement performance, Geophys. *Res. Lett.* **32**, L21S02, doi:10.1029/2005GL024028.

Baum, B. A., Frey, R. A., Mace, G. G., Harkey, M. K., and Yang, P. (2003). Nighttime multilayered cloud detection using MODIS and ARM data. *J. Appl. Meteor.* **42**, 905–919.

Baum, B. A., Uttal, T., Poellot, M., Ackerman, T. P., Alvarez, J. M., Intrieri, J., Starr, D. O'C., Titlow, J., Tovinkere, V., and Clothiaux, E. (1995). Satellite remote sensing of multiple cloud layers. *J. Atmos. Sci.* **52**(23), 4210–4230.

Burkert, P., Fergg, F., and Fischer, H. (1983). A compact high resolution Michelson interferometer for passive atmospheric sounding (MIPAS). *IEEE Trans. Geosci. Remote Sensing GE*-21, 345.

Chang, F. L. and Li., Z. (2005). A new method for detection of cirrus overlapping water clouds and determination of their optical properties. *J.Atmos. Sci.* **62**, 3993–4009.

Chazette P., Pelon, J., and Mégie, G. (2001). Determination by spaceborne backscatter lidar of the structural parameters of atmospheric scattering layer. *App. Opt.* **40**, 3428–3440.

David, C., Bekki, S., Godin, S., Megie, G., and Chipperfield, M. P. (1998). Polar stratospheric clouds climatology over Dumont d'Urville between 1989 and 1993 and the influence of volcanic aerosols on their formation. *J. Geophys. Res.* **103**, 22163–22180.

Evan, A. T., Heidinger, A. K., and Vimont, D. J. (2007). Arguments against a physical long-term trend in global ISCCP cloud amounts. *Geo. Res. Lett.* **34**, L04701, doi:10.1029/2006GL028083.

Flatau, P. J., Cotton, W. R., Stephens, G. L., Dalu, G. A., and Heymsfield, A. J. (1989). Mixed layer model of cirrus clouds-Growth and dissipation mechanisms.*Symposium on the Role of Clouds in Atmospheric Chemistry and Global Climate*. Anaheim, CA, US, January 30–February 3, pp. 151–156.

Forster, P., Ramaswamy, V., Artaxo, P., Berntsen, T., Betts, R., Fa-hey, D. W., Haywood, J., Lean, J., Lowe, D. C., Myhre, G., Nganga, J., Prinn, R., Raga, G., Schulz, M., and Van Dorland, R. (2007). Changes in atmospheric constituents and in radiative forcing. In *Climate Change 2007: The Physical Science Basis*. Contribution of Working Group I to the Fourth Assessment Report of the Intergovernmental Panel on Climate Change. S. Solomon, D. Qin, M. Manning, Z. Chen, M. Marquis, K. B. Averyt, M. Tignor, and H. L. Miller (Eds.). Cambridge University Press, Cambridge, UK and New York, NY, USA.

Grant, W. B., Browel, E. B., Butler, C. F., and Nowicki, G. D. (1997). LITE measurements of biomass burning aerosols and comparisons with correlative airborne lidar measurements of multiple scattering in the planetary boundary layer. *Advances in Atmospheric Remote Sensing with Lidar*. A. Ansmann, R. Neuber, P. Rairoux, and U. Wandinger (Eds.). Springer-Verlag, Berlin, pp. 153–156.

Heidinger, A. K. and Pavolonis, M. J. (2005): Global daytime distribution of overlapping cirrus cloud from NOAA's Advanced Very High Resolution Radiometer. *J. Climate* **18**, 4772–4784.

Höpfner, M., Grabowski, U., Stiller, G. P., and Von Clarmann, T. (2007). *Climatology of Artic and Antarctic polar stratospheric clouds (PSCs) from 2002–2007 as observed by MIPAS*. POSTER EGU2007-A-08879 AS3.13-1FR1P-0120.

Hu, Y., Li, D., and Liu, J. (2007). Abrupt seasonal variation of the ITCZ and the Hadley circulation. *Geophys. Res. Lett.* **34**, L18814, doi:10.1029/2007GL030950.

Hugues, N. A. and Henderson-Sellers, A. (1985). Global 3-Dnephanalysis of total cloud amount: Climatology for 1979. *J. Clim. and Appl. Meteor.* **24**, 669–686, doi:10.1175/15200450(1985)024<0669:GNOTCA>2.0.CO;2.

Ingmann, P., Straume-Lindner, A. G., and Werh T. (2008). ESA spaceborne lidars in preparation or planned, In *Proceedings, 24th ILRC*. M. Hardesty and S. Mayor (Eds.). NCAR, Boulder, CO, pp. 1109–1110.

Jin, Y. and Rossow, W. B. (1997). Detection of cirrus overlapping low-level clouds. *J. Geophys. Res.* **102**, 1727–1737.

Kent, G. S., Winker, D. M., Osborn, M. T., and Skeens, K. M. (1993). A model for the separation of cloud and aerosol in SAGE II occultation data. *J. Geophys. Res.* **98**(D11), 20725–20735, doi:10.1029/93JD00340.

Li, T. (1997). Air-sea interactions of relevance to the ITCZ: Analysis of coupled instabilities and experiments in the hybrid coupled GCM. *J. Atmos. Sci.* **54**, 134–147.

Liu, Z., Vaughan, M. A., Winker, D. M., Hostetler, C. A., Poole, L. R., Hlavka, D., Hart, W., and McGill, M. (2004). Use of probability distribution functions for discriminating between cloud and aerosol in lidar backscatter data. *J. Geophys. Res.* **109**, D15202, doi:10.1029/2004JD004732.

Menzel, W. P., Baum, B. A., Strabala, K. I., and Frey, R. A. (2002). Cloud top properties and cloud phase-Algorithm. *Theorical Basis Document. ATBD-MOD-04*, p. 61. Retrieved from http//modis-atmos.gsfc.nasa.gov/

Nasiri, S. L. and Baum, B. A. (2004). Daytime multilayered cloud detection using multispectral imager data. *J. Atmos. Ocean. Technol.* **21**, 1145–1155.

Palm, S. P., Hart, W., Hlavka, D., et al. (2002). GLAS atmospheric data products. *Algorithm Theor. Basis. Doc. ATBD-GLAS-01*, version 4.2. Earth Obs. Syst. Proj. Off., Greenbelt, MD. Retrieved from http://www.csr.utexas.edu/glas/pdf/glasatmos.atbdv4.2.pdf

Palm, S. P. and Spinhirne, J. (1998). The detection of clouds, aerosol and marine atmospheric boundary layer characteristics from simulated GLAS data. *The 19th International Laser Radar Conference*. Annapolis, MD, July 6–10.

Pavolonis, M. J. and Heidinger, A. K. (2004). Daytime cloud overlap detection from AVHRR and VIIRS. *J. Appl. Meteorol.* **43**, 762–778.

Philander, S. G. H., Gu, D., Halpern, D., Lambert, G., Lau, N.-C., Li, T., and Pacanowski, R. C. (1996). Why the ITCZ is mostly north of the equator. *J. Climate* **9**, 2958–2972.

Pitts, C., Thomason, L. W., Poole, L. R., and Winker, D. M. (2007). Characterization of Polar Stratospheric Clouds with SpaceBorne Lidar: CALIPSO and the 2006 Antarctic Season, Atmos. *Chem. Phys. Discuss.* **7**, 7933–7985, available at www. atmos-chem-phys-discuss. net/7/7933/2007/

Platnick, S., King, M. D., Ackerman, S. A., Menzel, W. P., Baum, B. A., Riedi, J. C., and Frey, R. A. (2003). The MODIS cloud products: Algorithms and example from Terra. *IEEE Trans. Geosci.* **41**, 459–473.

Platt, C. M. R., Young, S. A., Manson, P. J., Patterson, G. R., Mars-den, S. C., Austin, R. T., and Churnside, J. H. (1998). The optical properties of equatorial cirrus from observations in the ARM pilot radiation observation experiment. *J. Atmos. Sci* **55**, 1977–1996, doi:10.1175/1520-0469(1998)055<1977:TOPOEC>2.0.CO;2.

Rossow, W. B., Mosher, F., Kinsella, E., Arking, A., Desbois, M., Harrison, E., Minnis, P., Ruprecht, E., Seze, G., Simmer, C., and Smith, E. (1985). ISCCP cloud algorithm intercomparison. *J. Climate Appl. Meteor.* **24**, 877–903, doi:10.1 175/15200450(1985)024<0887:ICAI>2.0.CO;2.

Rossow, W. B. and Schiffer, E. A. (1991). ISCCP Cloud Data Products, Bull. *Amer. Meteor. Soc.* **72**, 2–20, doi:10.1175/15200477(1991)072<000 2:ICDP>2.0.CO;2.

Schiffer, R. A. and Rossow, W. B. (1983). The International Satellite Cloud Climatology Project (ISCCP): The first project of the world climate research program. *Bull. Am. Meteorol. Soc.* **64**, 779–784.

Solomon, S. (1999). Stratospheric ozone depletion: A review of concepts and history. *Rev. Geophys.* **37**, 275–316.

Stephens, G. L. (2005). Cloud feedbacks in the climate system: A critical Review. *J. Clim.* **18**, 237–273, doi:10.1175/JCLI-3243.1.

Stowe, L. L. (1984). Evaluation of NIMBUS 7 THIR/CLE and air force three-dimensional nephanalysis estimates of cloud amount. *J. Geophys. Res.* **89**, 5370–5380.

Stubenrauch, C. J., Armante, R., Abdelaziz, G., Crevoisier, C., Pierangelo, C., Scott, N. A., and Chedin, A. (2006a). *Cloud properties from AIRS, 15th International TOVS Study Conference Proceeding*, Matera, Italy, October 4–10.

Stubenrauch, C. J., Chédin, A., Radel, G., Scott, N. A., and Serrar, S. (2006b). Cloud properties and their seasonal and diurnal variability from TOVS Path-B, 2007. *J. Climate* **19**(21), 5531–5553, doi:10.1175/JCLI3929.1.

Susskind, J., Reuter, D., and Chahine, M. T. (1987). Cloud fields retrieved from HIRS2/MSU sounding data. *J. Geophys. Res.* **92D**, 4035–4050.

Thome, K., Reagan, J., Geis, J., Bolt, M., and Spinhirne, J. (2004). Validation of GLAS calibration using ground-and satellite-based data, *Geoscience and Remote Sensing Symposium, 2004.* IGARSS '04. *Proceedings, 2004 IEEE International* **4**, 2468–2471, September 20–24, 2004. Retrieved from http://ieeexplore.ieee.org/iel5/9436/29947/01369793.pdf

Vaughan, M. A., Young, S. A., Winker, D. M., Powell, K. A., Omar, A. H., Liu, Z., Hu, Y., and Hostetler, C. A. (2004). Fully automated analysis of space-based lidar data: An overview of the CALIPSO retrieval algorithms and data products. *Proc. Spie.* **5575**, 16–30, doi:10.1117/12.572024.

Waliser, D. E. and Gautier, C. (1993). A satellite-derived climatology of the ITCZ. *J. Climate* **6**, 2162–2174, doi:10.1175/15200442(1993)006<2162:ASDCOT>2.0.CO;2.

Warren, S. G., Hahn, C. J., and London, J. (1985). Simultaneous occurrence of different cloud types. *J. Clim. Appl. Meteorol.* **24**, 658–667, doi:10.1175/15200450(1985)024<0658:SOODCT>2.0.CO;2.

Winker, D. M. (1998). Cloud Distribution Statistics from LITE. In *19th International Laser Radar Conference*. U. N. Singh, S. Ismail, and G. K. Schwemmer (Eds.). Annapolis, Maryland, pp. 955–958.

Winker, D. M., Couch, R. H., and McCormick, M. P. (1996). An overview of LITE: NASA's Lidar in space technology experiment. *Proc. IEEE* **84**, 164–180.

Winker, D. M. and Hunt, B. (2007). First Results from CALIOP, Proc. *Third Symposium on LIDAR Atmospheric Applications*. 87th AMS Annual Meeting, San Antonio, Texas, January 15–18.

Winker, D. M., Pelon, J., and McCormick, M. P. (2002). The calipso mission: Aerosols and cloud observation from space. L. Bissonnette, G. Roy, and G. Vallee (Eds.), *Proc. ILRC* **21**. 735–738.

Wylie, D. P., Jackson, D. L., Menzel, W. P., and Bates, J. J. (2005). Trends in global cloud cover in two decades of HIRS observations. *J. Climate* **18**, 3021–3031.

Yeh, H. Y. and Liou, K. N. (1983). Remote sounding of cloud parameters from a combination of infrared and microwave channels: A parameterization approach. *J. Appl. Meteor.* **22**, 201–213, doi:10.1175/1520-0450(1983)022<0201:RSOCPF>2.0.CO;2..

4

Araiza-Quijano, M. R. and Hern´andez-del-Valle, G. (1996). Some observations of atmospheric luminosity as a possible earthquake precursor. *Geofisica Internacional* **35**, 403–408.

Bernabe, Y. (1998). Streaming potential in heterogenous networks. *J. Geophys. Res.* **103**, 20827–20841.

Bernard, P., Pinettes, P., Hatzidimitriou, P. M., et al. (1997). From precursors to prediction: A few recent cases from Greece. *Geophys. J. Internatl.* **131**, 467–477.

Brace, W. F. (1975). Dilatancy-related electrical resistivity change in rocks. *Pure Appl. Geophys.* **113**, 207–217.

Brace, W. F., Paulding, B. W., and Scholz, C. (1966). Dilatancy in the fracture of crystalline rocks. *J. Geophys. Res.* **71**, 3939–3953.

Davis, K. and Baker, D. M. (1965). Ionospheric effects observed around the time of the Alaskan earthquake of March 28, 1964. *J. Geophys. Res.* **70**, 2251–2253.

Derr, J. S. (1973). Earthquake lights: A review of observations and present theories. *Bull. Seismol. Soc. Am.* **63**, 2177–2187.

Fraser-Smith, A. C., Bernardi, A., McGill, P. R., et al. (1990). Low-frequency magnetic field measurements near the epicenter of the Ms=7.1 Loma Prieta earthquake. *Geophys. Res. Lett.* **17**, 1465–1468.

Freund, F. (2002). Charge generation and propagation in rocks. *J. Geodynamics* **33**, 545–572.

Freund, F. T. (2003). On the electrical conductivity structure of the stable continental crust. *J. Geodynamics* **35**, 353–388.

Freund, F. T., Takeuchi, A., and Lau, B. W. (2006). Electric currents streaming out of stressed igneous rocks -A step towards understanding pre-earthquake low frequency EM emissions. *Phys. Chem. Earth* **31**, 389–396.

Fujinawa, Y. and Takahashi, K. (1990). Emission of electromagnetic radiation preceding the Ito seismic swarm of 1989. *Nature* **347**, 376–378.

Galli, I. (1910). Raccolta e classifzione di fenomeni luminosi osservati nei terremoti. *Bolletino della Societa Sismological Italiana* **14**, 221–448.

Gershenzon, N. and Bambakidis, G. (2001). Modeling of seismoelectromagnetic phenomena. *Russian J. Earth Sci.* **3**, 247–275.

Glover, P. W. J. and Vine, F. J. (1992). Electrical conductivity of carbon-bearing granulite at raised temperatures and pressures. *Nature* **360**, 723–726.

Gokhberg, M. B., Morgounov, V. A., and Pokhotelov, O. A. (1995). *Earthquake Prediction.* Gordon and Breach.

Gokhberg, M. B., Morgounov, V. A., Yoshino, T., and Tomizawa, I. (1982). Experimental measurements of electromagnetic emissions possible related to earthquakes in Japan. *J. Geophys. Res.* **87**, 7824–7828.

Gornyi, V. I., Salman, A. G., Tronin, A. A., and Shilin, B. B. (1988). The Earth's outgoing IR radiation as an indicator of seismic activity. *Proc. Acad. Sci. USSR* **301**, 67–69.

Hayakawa, M. (1989). Satellite observation of low latitude VLF radio noise and their association with thunderstorms. *J. Geomagn. Geoelectr.* **41**, 573–595.

Hedervari, P. and Noszticzius, Z. (1985). Recent results concerning earthquake lights. *Annal. Geophys.* **3**, 705708.

Holliday, J. R., Nanjo, K. Z., Tiampo, K. F., et al. (2005). Earthquake forecasting and its verification. *Nonlin. Processes Geophy.* **12**, 965–977.

Hough, S. E. (2002). *Earthshaking Science: What We Know (and Don't Know) about Earthquakes*. Princeton University Press, Princeton, NJ, p. 272.

Jouniaux, L., Bernard, M. L., Zamora, M., and Pozzi, J. P. (2000). Streaming potential in volcanic rocks from Mount Pelée. *J. Geophys. Res.* **105**, 8391–8401.

Kagan, Y. Y. (1997). Are earthquakes predictable? *Geophys. J. Internatl.* **131**, 505–525.

Kanamori, H. (1996). A seismologist's view of VAN. In *Earthquake Prediction from Seismic Electric Signals*. J. Sir Lighthill (Ed.). World Science Publication, Singapore, pp. 339345.

Karakelian, D., Beroza, G. C., Klemperer, S. L., and Fraser-Smith, A. C. (2002). Analysis of ultralow-frequency electrommagnetic field measurements associated with the 1999 M 7.1 Hector Mine, California, earthquake sequence. *Bull. Seismol. Soc. Am.* **92**, 1513–2524.

Keilis-Borok, V. (2002). Earthquake Prediction: State-of-the-Art and Emerging Possibilities. *Ann. Rev. Earth Planet. Sci.* **30**, 1–33.

King, C.-Y. (1983). Electromagnetic emission before earthquakes. *Nature* **301**, 377.

Knopoff, L. (1996). Earthquake prediction: The scientific challenge. *Proc. Natl. Acad. Sci. USA* **93**, 3719–3720.

Kopytenko, Y. A., Matiashvili, T., Voronov, P. M., et al. (1993). Detection of ultralow frequency emissions connected with the Spitak earthquake and its aftershock activity, based on geomagnetic pulsation data at Susheti and Vardzia observatories. *Phys. Earth Planet. Inter.* **77**, 85–95.

Liperovsky, V. A., Pokhotelov, O. A., Liperovskaya, E. E., et al. (2000). Modification of sporadic E-layers caused by seismic activity. *Surveys Geophysics* **21**, 449–486.

Liu, J. Y., Chen, Y. I., Chuo, Y. J., and Tsai, H. F. (2001). Variations of ionospheric total electron content during the Chi-Chi earthquake. *Geophy. Res. Lett.* **28**, 1383–1386.

Liu, J. Y., Chen, Y. I., Pulinets, S. A., et al. (2000). Seismo-ionospheric signatures prior to M=6.0 Taiwan earthquakes. *Geophys. Res. Lett.* **27**, 3113–3116.

Lomnitz, C. (1994). *Fundamentals of Earthquake Prediction*. New York, NY, p. 326.

Ma, Q.-Z., Jing-Yuan, Y., and Gu, X.-Z. (2003). The electromagnetic anomalies observed at Chongming station and the Taiwan strong earthquake (M=7.5, March 31, 2002). *Earthquake* **23**, 49–56.

Mack, K. (1912). Das süddeutsche Erdbeben vom 16. November 1911, Abschnitt VII: Lichterscheinungen. In *Würtembergische Jahrbucher fur Statistik and Landeskunde.* (Ed.). Stuttgart, p. 131.

Martelli, G. and Smith, P. N. (1989). Light, radiofrequency emission and ionization effects associated with rock fracture. *Geophys, J. Internatl.* **98**, 397–401.

Merzer, M. and Klemperer, S. L. (1997). Modeling low-frequency magnetic-field precursors to the Loma Prieta earthquake with a precursory increase in fault–zone conductivity. *Pure Appl. Geophys.* **150**, 217–248.

Milne, J. (1899). *Earthquakes and Other Earth Movements*. D. Appleton Co., New York, NY, p. 376.

Molchanov, O. A. and Hayakawa, M. (1998). On the generation mechanism of ULF seimogenic electromagnetic emissions. *Phys. Earth Planet. Int.* **105**, 201220.

Morrison, F. D., Williams, E. R., and Madden, T. D. (1989). Streaming potentials of Westerly granite with applications. *J. Geophys. Res.* **94**, 12449–12461.

Musya, K. (1931). On the luminous phenomena that attended the Idu earthquake, November 6th, 1930. *Bull. Earthquake Res. Inst.* **9**, 214–215.

Naaman, S., Alperovich, L. S., Wdowinski, S., et al. (2001). Comparison of simultaneous variations of the ionospheric total electron content and geomagnetic field associated with strong earthquakes. *Natural Hazards Earth System Sci.* **1**, 53–59.

Nitsan, U. (1977). Electromagnetic emission accompanying fracture of quartz-bearing rocks. *Geophys. Res. Lett.* **4**, 333–336.

Ogawa, T. and Utada, H. (2000). Electromagnetic signals related to incidence of a teleseismic body wave into a subsurface piezoelectric body. *Earth Planets Space* **52**, 253–256.

Ouzounov, D., Liu, D., Kang, C., et al. (2007). Outgoing long wave radiation variability from IR satellite data prior to major earthquakes. *Tectonophysics* **431**, 211–220.

Park, S. K. (1997). Electromagnetic precursors to earthquakes: A search for predictors. *Sci. Progr.* **80**, 65–82.

Pulinets, S. and Boyarchuk, K. (2004). *Ionospheric Precursors of Earthquakes.* Springer Verlag, p. 350.

Qiang, Z.-J., Xu, X.-D., and Dian, C.-D. (1990). Abnormal infrared thermal of satellite-forewarning of earthquakes. *Chinese Sci. Bull.* **35**, 1324–1327.

Qiang, Z.-J., Xu, X.-D., and Dian, C.-D. (1991). Thermal infrared anomaly—precursor of impending earthquakes. *Chinese Sci. Bull.* **36**, 319323.

Quing, Z., Xiu-Deng, X., and Chang-Gong, D. (1991). Thermal infrared anomaly-precursor of impending earthquakes. *Chinese Sci. Bull.* **36**, 319–323.

Rundle, J. B., Turcotte, D. L., Shcherbakov, R., et al. (2003). Statistical physics approach to understanding the multiscale dynamics of earthquake fault systems. *Rev. Geophys.* 1019–1049.

Sasai, Y. (1991). Tectonomagnetic modeling on the basis of the linear piezomagnetic effect. *Bull. Earthq. Res. Inst., Univ. Tokyo* **66**, 585–722.

Srivastav, S. K., Dangwal, M., Bhattachary, A., and Reddy, P. R. (1997). Satellite data reveals pre-earthquake thermal anomalies in Killari area, Maharashtra. *Curr. Sci.* **72**, 880–884.

St-Laurent, F. (2000). The Saguenay, Québec, earthquake lights of November 1988-January 1989. *Seismolog. Res. Lett.* **71**, 160–174.

Sykes, L., Shaw, B., and Scholz, C. (1999). Rethinking earthquake prediction. *Pure Appl. Geophys.* **155**, 207232.

Terada, T. (1931). On luminous phenomena accompanying earthquakes. *Bull. Earthquake Res. Inst. Tokyo Univ.* **9**, 225–255.

Tramutoli, V. et al. (2005). Assessing the potential of thermal infrared satellite surveys for monitoring seismically active areas: The case of Kocaeli (Izmit) earthquake, August 17, 1999. *Remote Sens. Environ.* **96**, 40426.

Tributsch, H. (1984). *When the Snakes Awake: Animals and Earthquake Prediction.* MIT Press, Cambridge, Mass, p. 264.

Tronin, A. A. (Ed.) (1999). *Satellite Thermal Survey Application for Earthquake Prediction.* Terra Sci. Publication, Tokyo, Japan, pp. 717–746.

Tronin, A. A. (2002). Atmosphere-lithosphere coupling: Thermal anomalies on the Earth surface in seismic process. In *Seismo-Electromagnetics: Lithosphere-Atmosphere-Ionosphere Coupling.* M. Hayakawa and O. A. Molchanov (Eds.). Terra Scientific Publication, Tokyo, pp. 173–176.

Tronin, A. A., Molchanov, O. A., and Biagi, P. F. (2004). Thermal anomalies and well observations in Kamchatka. *Int. J. Remote Sensing.* **25**, 2649–2655.

Tsukuda, T. (1997). Sizes and some features of luminous sources associated with the 1995 Hyogo-ken Nanbu earthquake. *J. Phys. Earth.* **45**, 73–82.

Turcotte, D. L. (1991). Earthquake prediction. *Ann. Rev. Earth Planet. Sci.* **19**, 263–281.

Uyeda, S. (1998). VAN method of short-term earthquake prediction shows promise. *EOS Trans. AGU.* **79**, 573–580.

Varotsos, P., Alexopoulos, K., Lazaridou, M., and Nagao, T. (1993). Earthquake predictions issued in Greece by seismic electric signals since February 6, 1990. *Tectonophys.* **224**, 269–288.

Varotsos, P., Alexopoulos, K., Nomicos, K., and Lazaridou, M. (1986). Earthquake prediction and electric signals. *Nature* **322**, 120.

Varotsos, P. A. (2005). *The Physics of Seismic Electric Signals.* Terra Scientific Publ. Co., Tokyo, p. 388.

Vershinin, E. F., Buzevich, A. V., Yumoto, K., et al. (1999). Correlation of seismic activity with electromagnetic emissions and variations in Kamchatka region. In *Atmospheric and Ionospheric Electromagnetic Phenomena Associated with Earthquakes.* M. Hayakawa (Ed.). Terra Sci. Publication, Tokyo, Japan, pp. 513–517.

Warwick, J. W., Stoker, C., and Meyer, T. R. (1982). Radio emission associated with rock fracture: Possible application to the great Chilean earthquake of May 22, 1960. *J. Geophys. Res.* **87**, 2851–2859.

Wyss, M. and Dmowska, R. (Eds.) (1997). *Earthquake Prediction—State of the Art.* Birkhäuser Verlag, p. 272.

Yasui, Y. (1973). A Summary of Studies on Luminous Phenomena Accompanied with Earthquakes. *Memoirs Kakioka Magnetic Observatory* **15**, 127–138.

Yen, H.-Y., Cheng, C-H., Yeh, Y-H., et al. (2004). Geomagnetic fluctuations during the 1999 Chi-Chi earthquake in Taiwan. *Earth Planets Space* **56**, 3945.

Yoshida, S., Manjgaladze, P., Zilpimiani, D., et al. (1994). Electromagnetic emissions associated with frictional sliding of rock. In *Electromagnetic Phenomena Related to Earthquake Prediction.* M. Hayakawa and Y. Fujinawa (Eds.). Terra Scientific, Tokyo, pp. 307322.

Yoshino, T. and Tomizawa, I. (1989). Observation of low-frequency electromagnetic emissions as precursors to the volcanic eruption at Mt. Mihara during November, 1986. *Phys. Earth Planet. Inter.* **57**, 32–39.

Zlotnicki, J. and Cornet, F. H. (1986). A numerical model of earthquake-induced piezomagnetic anomalies, *J. Geophys. Res.* **B91**, 709–718.

5

Bono, A. and Badiali, L. (2005). PWL Personal WaveLab 1.0, an Object-Oriented workbench for seismogram analysis on Windows Systems. *Comput.Geosci.* **31**, 1.

Burrato, P., Casale, P., Cultrera, G., Landi, P., Nappi, R., Nostro, C., Scarlato, P., Scotto, C., Stramondo, S., Tertulliani, A., Winkler, A., and Bonifaci, U. (2003). Geophysics for kids: The experience of the Istituto Nazionale di Geofisica e Vulcanologia. *Seismolog. Res. Lett.* **74**(5), 529–535.

CPTI Working Group (1999). *Catalogo parametrico dei terremoti italiani.* GNDT-ING-SGA-SSN (Eds.). Tipografia Compositori (publication), Bologna, p. 88.

De Dolomieu, D. (1785). Memoria sopra i tremuoti della Calabria dell'anno 1783. *Neaples.*

DiDa (2002). Working group INGV: Educational geophysics at INGV–Rome. *Italy–EOS. Trans. AGU* **83**(47), Fall Meeting Suppl., Abstract ED52A-0004.

Hamburger, M., Pavlis, G. L., Phinney, R. A., Steinberg, D., Owens, T. J., and Hall-Wallace, M. (2001). New science education initiative brings seismology into the classroom. *E–Trans. AGU* **82**, 266–267.

Johnson, C. (1999). Seismology and education: The IRIS Education and Outreach Program. *GSA Today* **9**, 8–10.

Jones, A. L., Braile, L. W., and Braile, S. J. (2003). A suite of educational computer programs for seismology. *Seismolog. Res. Lett.* **74**(5), 605–617.

Mallet, R. (1862). *Great Neapolitan earthquake of 1857.* Chapman and Hall, London.

Nostro C. and E&O-INGV Group (2004). Using earthquakes to uncover the Earth's inner secrets. *Interactive Exhibits for Geophysical Education, Es2 Geo-And Space-Physical Sciences* (Education, Outreach and Defining Users), Abstract-No. EGU04-A-06276, European Geosciences Union 1st General Assembly, Nice, April 25–30.

Plinius Caecilius Secundus (minor) (1968). *Epistolarum libri decem.* R. A. B. Mynors (Ed.). Oxford.

Tertulliani, A. and Donati S. (2000). A macroseismic network of schools for the collection of earthquake effects in a large city. *Seismol. Res. Lett.* **71**, 536–543.

Tertulliani, A., Nostro, C., Macr`ı, P., Winkler, A., Castellano, C., Burrato, P., Casale, P., Cultrera, G., Scarlato, P. G., Doumaz, F., Piscini, A., Di Felice, F., Vallocchia, M., Stramondo, S., Badiali, L., Alfonsi, L., Baroux, E., Ciaccio, M. G., Frepoli, A., and Marsili, A. (2004). Interactive exhibit for geophysical education: earthquakes! *SS-3 Special Session: Education and Outreach*, Abstract-No. SS-3 14, European Seismological Commission, XXIX General Assembly, Potsdam, September 2004.

Virieux, J. (2000). Educational Seismological Project: EDUSEIS. *Seismol. Res. Lett.* **71**, 530–535.

Winkler, A. and the E&O Group INGV-Roma (2004). Fatal attraction—A discovery of the Earth magnetic field. *Eos Trans. AGU* **85**(47), Fall Meet. Suppl., Abstract ED31C-0752.

6

Bodine, M. W., Jr. and Jones, B. F. (1986). The Salt Norm: A Quantitative Chemical-Mineralogical Characterization of Natural Water. *Geological Survey Water-Resources Investigations Report*, 86−4086, US, p. 130.

Feth, J. H., et al. (1965). *Preliminary Map of The Conterminous United States Showing Depth to and Quality of Shallowest Ground Water Containing More Than 1,000 Parts Per Million Dissolved Solids.* Geological Survey Hydrologic Investigations Atlas HA, US, p. 199.

Galloway, D. L., Alley, W. M., Barlow, P. M., Reilly, T. E., and Tucci, P. (2003). Evolving Issues and Practices. In *Managing Ground-Water Resources—Case Studies on The Role of Science.* Geological Survey Circular, US, **1247**, p. 73.

Hood, J. W. (1963). *Saline Ground Water in the Roswell Basin, Chaves and Eddy Counties.* US Geological Survey Water-Supply Paper, 1539−M, 1958–59, New Mexico, p. 46.

Kharaka, Y. K., Ambats, G., Thordsen, J. J., and Evans, W. C. (1997). Deep-Well Injections of Brine From Paradox Valley, Colorado: Potential Major Precipitation Problems Remediated By Nanofiltration. *Water Resour. Res.* **33**, 1013−1020.

Kharaka, Y. K., Leong, L. Y. C., Doran, G., and Breit, G. N. (1999). Can produced water be reclaimed? Experience with the Placerita oil field. In *Environmental Issues in Petroleum Exploration, Production and Refining: Proceedings 5th Ipec Meeting.* K. L. Sublette (Ed.). California, October 1998, CD−ROM format.

Kister, L. R. (1973). *Quality of Ground Water in the Lower Colorado River Region.* Arizona, Nevada, New Mexico, and Utah. US Geological Survey Hydrologic Investigations, Atlas HA−478, 2 sheets.

Krieger, R. A., Hatchett, J. L., and Poole, J. L. (1957). *Preliminary Survey of the Saline-Resources of the United States.* US Geological Survey Water-Supply Paper 1374, p. 172.

Lacombe, P. J. and Carleton, G. B. (2002). *Hydrogeologic Framework, Availability of Water Supplies, and Saltwater Intrusion Cape May County.* US Geological Survey Water-Resources Investigations Report 01−4246, New Jersey, p. 151.

Miller, J. A. (2000). *Ground-Water Atlas of the United States.* US Geological Survey.

Reynolds, S. E. (1962). *Twenty-Fifth Biennial Report of the State Engineer of New Mexico.* The Valliant Company, Albuquerque, New Mexico, p. 193.

Rice, C. A. and Nuccio, V. (2000). *Water Produced With Coal-Bed Methane.* US Geological Survey Fact Sheet FS–156–00, p. 2.

Robson, S. G. and Banta, E. R. (1995). *Ground-Water Atlas of the United States, Segment 2.* US Geological Survey Hydrologic Investigations Atlas HA–730C, Arizona, Colorado, New Mexico, and Utah, p. 32.

Sandia National Laboratories (2002). *Tularosa Basin National Desalination Research Facility Study.* US Bureau of Reclamation Report to Congress, p. 34.

US Bureau of Reclamation and Sandia National Laboratories (2003). *Desalination and Water Purification Technology Roadmap.* US Bureau of Reclamation, Desalination and Water Purification Research and Development Report #95, Denver, Colorado, p. 61.

Winslow, A. G. and Kister, L. R. (1956). *Saline-Water Resources of Texas.* US Geological Survey Water-Supply Paper 1365, p. 105.

Wood, W. W. and Jones, B. F. (1990) Origin of solutes in saline lakes and springs on the southern High Plains of Texas and New Mexico. In *Geologic Framework and Regional Hydrology. Upper Cenozoic Blackwater Draw and Ogallala Formations, Great Plains.* T. C. Gustavson (Ed.). The University of Texas at Austin, Bureau of Economic Geology Report, pp. 193–208.

7

Alexandrov, G. A., Yamagata, Y., Oikawa, T. (1999). Towards a Model for Projecting Net Ecosystem Production of the World Forests. *Ecolog. Model.* **123**, 183191.

Benítez, P. C. and Obersteiner, M. (2006). Site identification for carbon sequestration. In *Latin America: A Grid–based Economic Approach.* Forest Policy and Economics **8**, 636651.

Benítez, P., McCallum, I., Obersteiner, M., and Yamagata, Y. (2004). *Global Supply for Carbon Sequestration: Identifying Least-Cost Afforestation Sites Under Country Risk Consideration.* Tech. rep., International Institute for Applied System Analysis.

CIESIN (2005). Center for International Earth Science Information Network (CIESIN), Columbia University; and Centro Internacional de Agricultura Tropical (CIAT). Gridded Population of the World Version 3 (GPWv3): National Boundaries. Palisades, NY: Socioeconomic Data and Applications Center (SEDAC), Columbia University. Retrieved from http://sedac.ciesin.columbia.edu/gpw.

CIESIN (2005). Center for International Earth Science Information Network, Columbia University; and Centro Internacional de Agricultura Tropical (CIAT). Gridded Population of the World Version 3 (GPWv3): Population Density Grids. Palisades, NY: Socioeconomic Data and Applications Center (SEDAC), Columbia University. Retrieved from http://sedac.ciesin.columbia.edu/gpw.

FAO (2005). *Global Forest Resources Assessment 2005, Progress towards sustainable forest management. Volume 147.* FAO Forestry Paper. Rome: ood and Agriculture Organization of the United Nations.

Grubler, A., Nakicenovic, N., Riahi, K., Wagner, F., Fischer, G., Keppo, I., Obersteiner, M., O'Neill, B., Rao, S., and Tubiello, F. (in press). Integrated assessment of uncertainties in greenhouse gas emissions and their mitigation: Introduction and overview. *Technological Forecasting and Social Change.*

IPCC International Panel on Climate Change (2001). R. T Watson and Core Writing (Eds.). Team: [http://www.ipcc.ch/pub/syreng.htm] *Climate Change 2001: Synthesis Report. A Contribution of Working Groups I, II, and III to the Third Assessment Report of the Integovernmental Panel on Climate Change.* Cambridge University Press, Cambridge, UK, and New York, NY, USA.

JRC (2003). Retrieved from http://www-gvm.jrc.it/glc2000. *The Global Land Cover Map for the Year 2000. GLC2000 database.* European Commision Joint Research Centre.

Kaufmann, D., Kraay, A., and Mastruzzi, M. (2005). Retrieved from http://ssrn.com/abstract=718081. *Governance Matters IV: Governance Indicators for 1996–2004, World Bank*

Policy Research Working Paper Series No. 3630. World Bank.

Marland, G., Boden, T., and Andres, R. (2006). Global, Regional, and National CO_2 Emissions. In *Trends: A Compendium of Data on Global Change. Carbon Dioxide Information Analysis Center, Oak Ridge National Laboratory.* US Department of Energy, Oak Ridge, Tenn., USA.

Obersteiner, M., Alexandrov, G., Benítez, P. C., McCallum, I., Kraxner, F., Riahi, K., Rokityanskiy, D., and Yamagata, Y. (2006). Global Supply of Biomass for Energy and Carbon Sequestration from Afforestation/Reforestation Activities. *Mitigation and Adaptation Strategies for Global Change* 13812386.

R Development Core Team (2005). Retrieved from http://www.R-project.org. R: *A language and environment for statistical computing.* R Foundation for Statistical Computing. Vienna, Austria. [ISBN 3-900051-07-0].

Ramankutty, N., Foley, J. A., Norman, J., and McSweeney, K. (2002). The global distribution of cultivable lands: Current patterns and sensitivity to possible climate change. *Global Ecology & Biogeography* **11**(5):377392.

Tubiello, F. N. and Fischer, G. (in press). Reducing climate change impacts on agriculture: Global and regional effects of mitigation, 2000–2080. *Technological Forecasting and Social Change.*

WDPA Consortium (2004). *World Database on Protected Areas.* Copyright World Conservation Union (IUCN) and UNEP-World Conservation Monitoring Centre (UNEP-WCMC).

World Bank (2005). *World Development Indicators.* World Bank.

8

Almannavarnir: Skyrsla Bergrisinn 2006, Almannavarnadeild Ríkislögreglustjórans. p. 24.

Bird, D. and Dominey-Howes, D. (2006). Tsunami risk mitigation and the issue of public awareness. *Aust. J. Emerg. Mgmt.* **21**, 29–35.

Bird, D. and Dominey-Howes, D. (2008). Testing the use of a t't'questionnaire survey instrument" to investigate public perceptions of tsunami hazard and risk in Sydney, Australia. *Nat. Hazards* **45**, 99–122.

Bird, D., Roberts, M. J., and Dominey-Howes, D. (2008). Usage of an early warning and information system Web-site for real-time seismicity in Iceland. *Nat. Hazards* **47**, 75–94.

Björnsson, H. (2002). Subglacial lakes and jökulhlaups in Iceland, Global Planet. *Change* **35**, 255–271.

Björnsson, H., Pálsson, F., and Guomundsson, M. T. (2000). Surface and bedrock topography of the Myrdalsjökull ice cap, Iceland: The Katla caldera, eruption sites and routes of jökulhlaups, *Jökull*, **49**, 29–46.

Chester, D. K., Dibben, C. J. L., and Duncan, A. M. (2002). Volcanic hazard assessment in western Europe. *J. Volcanol. Geoth. Res.* **115**, 411–435.

Dominey-Howes, D. and Minos-Minopoulos, D. (2004). Perceptions of hazard and risk on Santorini. *J. Volcanol. Geoth. Res.* **137**, 285–310.

Dominey-Howes, D., Papathoma-Köhle, M., Bird, D., Mamo, B., and Anning, D. (2007). Letter to the Editor: The Australian Tsunami Warning System and lessons from the 2 April 2007 Solomon Islands tsunami alert in Australia. *Nat. Hazards Earth Syst. Sci.* **7**, 571–572. Retrieved from http://www.nat-hazards-earth-syst-sci.net/7/571/2007/.

Gregg, C. E., Houghton, B. F., Paton, D., Swanson, D. A., and Johnston, D. M. (2004). Community preparedness for lava flows from Mauna Loa and Hualalai volcanoes, Kona, Hawai'i. *B. Volcanol.* **66**, 531–540.

Guomundsson, M. T. (2005). Subglacial volcanic activity in Iceland. In *Iceland: Modern processes.* C. J. Caseldine, A. Russell, J. Hardardóttir, and Ó. Knudsen (Eds.). Past Environments, Elsevier, pp. 127–151.

Guomundsson, M. T. and Gylfason, Á. G. (Eds.) (2005). Hættumat vegna eldgosa og hlaupa frá vestanverdum Myrdalsjökli og Eyjafjallajökli, Ríkislögreglustjórinn and Háskólaútgáfan, Reykjavík, p. 210.

Guomundsson, M. T., Larsen, G., Höskuldsson, Á., and Gylfason, Á. G. (2008). Volcanic hazards in Iceland. *Jökull*, **58**, 251–268.

Haynes, K., Barclay, J., and Pidgeon, N. (2007). Volcanic hazard communication using maps: An evaluation of their effectiveness. *B. Volcanol.* **70**, 123–138.

Haynes, K., Barclay, J., and Pidgeon, N. (2008). Whose reality counts? Factors affecting the perception of volcanic risk. *J. Volcanol. Geoth. Res.* **172**, 259–272.

Hólm, S. L. and Kjaran, S. P. (2005). Reiknilíkan fyrir útbreidslu hlaupa úr Entujökli. In *Hættumat Vegan Eldgosa Og Hlaupa Frá Vestanverdum Myrdalsjökli Og Eyjafjallajökli.* M. T. Guomundsson and Á. G. Gylfason (Eds.). Ríkislögreglustjórinn and Háskólaútgáfan, Reykjavík, pp. 197–210.

Horlick-Jones, T., Sime, J., and Pidgeon, N. (2003). The social dynamics of environmental risk perception: Implications for risk communication research and practice. In *The Social Amplification of Risk.* N. Pidgeon, R. E. Kasperson, and P. Slovic (Eds.). Cambridge University Press, pp. 262–285.

Hughes, W. P. and White, P. B. (2006). The media, bushfires and community resilience. In *Disaster Resilience: An Integrated Approach.* D. Paton, D. Johnston, and C. Charles (Eds.). Thomas Publisher Ltd, Springfield, Illinois, pp. 213–225.

Johnston, D. M., Bebbington, M. S., Lai, C.-D., Houghton, B. F., and Paton, D. (1999). Volcanic hazard perceptions: comparative shifts in knowledge and risk. *Disast. Preven. Mgmt*, **8**, 118–126.

Kjartansson, G. (1967). The Steinsholtshlaup, Central-South Iceland on 15 January 1967. *Jökull* **17**, 249–262.

Larsen, G. (2000). Holocene eruptions within the Katla volcanic system, south Iceland: Characteristics and environmental impact. *Jökull*, **49**, 1–28.

Larsen, G., Smith, K., Newton, A., and Knudsen, Ó. (2005). Jökulhlaup til vesturs frá Myrdalsjökli: Ummerki um forsöguleg hlaup nidur Markarfljót. In *Hættumat Vegna Eldgosa Og Hlaupa Frá Vestanverdum Myrdalsjökli Og Eyjafjallajökl.* M. T. Guomundsson and Á. G. Gylfason (Eds.). Ríkislögreglustjórinn and Háskólaútgáfan, Reykjavík, pp. 75–98.

McGuirk, P. M. and O'Neill, P. (2005). Using Questionnaires in Qualitative Human Geography. In *Qualitative Research Methods in Human Geography.* I. Hay (Ed.).Oxford University Press, Australia, pp. 147–162.

Murdock, G., Petts, J., and Horlick-Jones, T. (2003). After amplification: rethinking the role of the media in risk communication. In *The Social Amplification of Risk.* N. Pidgeon, R. E. Kasperson, and P. Slovic (Eds.). Cambridge University Press, pp. 156–178.

Parfitt, J. (2005). Questionnaire design and sampling. In *Methods in Human Geography.* R. Flowerdew and D. Martin (Eds.). Pearson Education Limited, England, pp. 78–109.

Paton, D. and Johnston, D. (2001). Disasters and communities: vulnerability, resilience and preparedness, *Disast. Preven. Mgmt*, **10**, 270–277.

Paton, D., Smith, L., Daly, M., and Johnston, D. (2008). Risk perception and volcanic hazard mitigation: Individual and social perspectives. *J. Volcanol. Geoth. Res.* **172**, 170–178.

Patton, M. Q. (1990). *Qualitative Evaluation and Research Methods*. Sage Publications, Newbury Park, p. 532.

Ronan, K. R., Paton, D., Johnston, D. M., and Houghton, B. F. (2000). Managing societal uncertainty in volcanic hazards: a multidisciplinary approach, Disaster Prevention and Management, **9**, 339–349.

Russell, A. J., Tweed, F. S., and Knudsen, Ó. (2000). *Flash flood at Sólheimajökull heralds the reawakening of an Icelandic subglacial.*

Sarantakos, S. (1998). *Social Research.* Macmillan Publishers Australia Pty Ltd, South Yarra, p. 488.

Smith, K. T. (2004). Holocene jökulhlaups, glacier fluctuations and palaeoenvironment, Myrdalsjökull, south Iceland, Institute of Geography, School of Geosciences, University of Edinburgh, Edinburgh, pp. 139.

Smith, K. T. and Haraldsson, H. (2005). A late Holocene jökulhlaup, Markarfljót, Iceland: nature and impacts, *Jökull*, **55**, 75–86.

Sturkell, E., Einarsson, P., Roberts, M. J., Geirsson, H., Guomundsson, M. T., Sigmundsson, F., Pinel, V., Guomundsson,G. B., Ólafsson, H., and Stefánsson, R. (2008). Seismic and geodetic insights into magma accumulation at Katla subglacial volcano, Iceland:

1999 to 2005. *J. Geophys. Res.* **113**, B03212, doi:10.1029/2006JB004851.

Thordarson, T. and Larsen, G. (2007). Volcanism in Iceland in historical time: Volcano types, eruption styles and eruptive history. *J. Geodyn.* **43**, 118–152.

9

Bendimerad, F. (2001). Loss estimation: A powerful tool for risk assessment and mitigation. *Soil. Dynam. Earthquake Eng.* **21**, 467–472.

Erdas Field Guide (2005). *Geospatial Imaging.* LLC Norcross, Georgia.

Gonzalez, R.C. and Wintz, P. (1977). *Reading, Massachusetts.* Digital Image Processing, Addison-Wesley Publishing Company.

Kaya, S., Curran, P. J., and Llewellyn, G. (2005). Post-earthquake building collapse: A comparison of government statistics and estimates derived from SPOT HRVIR data. *Int. J. Remote Sens.* **26**, 2731–2740.

Lillo-Saavedra, M., Gonzalo, C., Arquero, A., and Martinez, E. (2005). Fusion of multispectral and panchromatic satellite sensor imagery based on tailored filtering in the Fourier domain. *Int. J. Remote Sens.* **26**, 1263–1268.

Pal, S. K., Majumdar, T. J., and Bhattacharya, A. K. (2006). Extraction of linear and anomalous features using ERS SAR data over Singhbhum Shear Zone, Jharkhand using fast Fourier transform. *Int. J. Remote Sens.* **27**, 4513–4528.

Sertel, E., Kaya, S., and Curran, P. J. (2007). Use of semi-variograms to identify earthquake damage in an urban area. *IEEE Trans. Geosci. Remot. Sen.* **45**, 1590–1594.

Stramondo, S., Bignami, C., Chini, M., and Pierdicca, N. (2006). Tertulliani, A. Satellite radar and optical remote sensing for earthquake damage detection: Results from different case studies. *Int. J. Remote Sens.* **27**, 4433–4447.

Tan, O., Tapirdamaz, M. C., and Yoruk, A. (2008). The earthquake catalogues for Turkey. Turkish. *J. Earth Sci.* **17**, 405–418.

TSI (Turkish Statistical Institute) (2008). Available at www.tuik.gov.tr. (accessed on October 10).

Turker, M. and San, B. T. (2003). The SPOT HRV data analysis for detecting earthquake-induced changes in Izmit, Turkey. *Int. J. Remote Sens.* **24**, 2439–2450.

USGS. (2008). US Geological Survey. *Earthquake Hazards Program 2008.* Retrieved from http://earthquake.usgs.gov. (accessed on October 10).

Westra, T. and De Wulf, R. R. (2007). Monitoring Sahelian floodplains using Fourier analysis of MODIS time-series data and artificial neural networks. *Int. J. Remote Sens.* **28**, 1595–1610.

10

Alperovich, L. and Fedorov, E. (1999). Perturbation of atmospheric conductivity as a cause of the lithosphere-ionosphere interaction. In *Atmospheric and Ionospheric Electromagnetic Phenomena Associated with Earthquakes.* M. Hayakawa (Ed.). TERRAPUB, Tokyo, pp. 591–596.

Chmyrev, V. M., Isaev, N. V., Bilichenko, S. V., and Stanev, G. A. (1989). Observation by spaceborn detectors of electric field and hydromagnetic waves in the ionosphere over an earthquake centre. *Phys. Earth Plan In.* **57**(1–2), 110–114.

Earthquake Summary Poster. (2008). Retrieved from http://earthquake.usgs.gov/eqcenter/eqarchives/poster/2006/20060820.php, last access April 15.

Gagnon, K., Chadwell, C., and Norabuena, E. (2005). Measuring the onset of locking in the Peru-Chile trench with GPS and acoustic measurements. *Nature* **434**, 205–208.

Goff, J., Walters, R., and Callaghan, F. (2006). *Tsunami Source Study.* National Institute of Water and Atmospheric Research, New Zealand, Environment Waikato Technical Report 2006/49, pp. 1–26.

Gousheva, M., Glavcheva, R., Danov, D., Angelov, P., and Hristov, P. (2005a). Influence of earthquakes on the electric field disturbances in the ionosphere on board of the Intercosmos-Bulgaria-1300 satellite. *Compt. Rend. Acad. Bulg. Sci.* **58**(8), 911–916.

Gousheva, M., Glavcheva, R., Danov, D., Angelov, P., Hristov, P., Kirov, B., and Georgieva, K. (2005b). Observation from the Intercosmos-

Bulgaria-1300 Satellite of Anomalies. In *The Ionosphere Associated With Seismic Activity.* Poster Proceedings of 2nd International Conference on Recent Advances in Space Technologies: Space in the Service of Society, RAST'2005, June 9–11, Istanbul, Turkey, pp. 119–123.

Gousheva, M., Glavcheva, R., Danov, D., Angelov, P., Hristov, P., Kirov, B, and Georgieva, K. (2006b). Satellite monitoring of anomalous effects in the ionosphere probably related to strong earthquakes. *Adv. Space Res.* **37**(4), 660–665.

Gousheva, M., Glavcheva, R., Danov, D., and Boshnakov, I. (2006a). Satellite observations of ionospheric disturbances associated with seismic activity. *Compt. Rend. Acad. Bulg. Sci.* **59**(8), 821–826.

Gousheva, M., Glavcheva, R., Danov, D., Hristov, P., Kirov, B., and Georgieva, K. (2007). Possible pre- and post-earthquake effects in the ionosphere. *IEEE Proceedings of 3rd International Conference on Recent Advances in Space Technologies*, June 14–16 Istanbul, Turkey, 754–759.

Gousheva, M., Glavcheva, R., Danov, D., Hristov, P., Kirov, B, and Georgieva, K. (2008a). Electric field and ion density anomalies in the mid latitude ionosphere: Possible connection with earthquakes? *Adv. Space Res.* **42**(1), 206–212.

Gousheva, M., Danov, D., Hristov, P., and Matova, M. (2008b). Quasi-static electric fields phenomena in the ionosphere associated with pre- and post earthquake effects. *Nat. Hazards Earth Syst. Sci.* **8**, 101–107, Retrieved from http://www.nat-hazards-earth-syst.sci.net/8/101/2008/

Hepper, J. P. (1977). Empirical models of high-latitude electric field. *J. Geophys. Res.* **82**, 1115–1125.

Heppner, J. P. and Maynard N. C. (1987). Empirical high-latitude electric field models. *J. Geophys. Res.* **92**, 4467–4489.

Heelis, R. A., Lowell J. K., and Spiro R. W. (1982). A model of the high-latitude ionospheric convection. *J. Geophys. Res.* **87**, 6339–6345.

Indonesia (2008). Retrieved from http://earthquake.usgs.gov/regional/world/indonesia/seismotectonics.php, last access: June 10.

Kim, V. P and Hegai, V. V. (1999). A possible presage of strong earthquake in the night-time mid-latitude F2 region ionosphere, In *Atmospheric and Ionospheric Electromagnetic Phenomena Associated with Earthquakes.* M. Hayakawa (Ed.). TERRAPUB, Tokyo, pp. 619–628.

Kolobov, V. Y., Tikunov, Y. V., Travin, A. V., and Akimtsev, V. A. (2006). Geochemistry of volcanic rocks of the new hebrides trench: Evidence of changes in geodynamic regime. *Geochem. Inter.* **43**, 29–38.

Nanang, T. P. and Gunawan, I. (2005). Tsunami sources in the Sumatra region, Indonesia and simulation of the December 26, 2004 Aceh tsunami. *ISET J. Earth. Tech.* **42**, 111–125.

Pedersen, A., Cattell, C. A., Falthammar, C.-G., Formisano, V., Lindqvist, P.-A., Mozer, F., and Torbert, R. (1984). Quasi-static electric field measurements with spherical double probes on the GEOS and ISEE satellites. *Space Sci. Rev.* **37**, 269–312.

Pedersen, A., Mozer, F., and Gustafsson, G. (1998). Electric field measurements in a tenuous plasma with spherical double probes. In *Measurement Techniques in Space Plasmas: Fields (AGU Geophysical Monograph, 103)*. J. Borovsky, R. Pfaff, and D. Young (Eds.). American Geophysical Union, pp. 1–12.

Sorokin, V. M. and Chmyrev, V. M. (2002). Electrodynamic model of ionospheric precursors to earthquakes and certain types of disasters. *Geomagn. Aeron.* **42**, 784–792.

Sorokin, V. M., Chmyrev, V. M., and Yaschenko, A. K. (2001a). Electrodynamic model of the lower atmosphere and ionosphere coupling. *J. Atmos. Solar-Terr. Phys.* **63**, 1681–1691.

Sorokin, V. M., Chmyrev, V. M., and Yaschenko, A. K. (2001b). Perturbation of the electric field in the Earth-ionospherhe layer at the charged aerosols injection. *Geomagn. Aeron.* **41**(2), 187–191.

Sorokin, V. M., Chmyrev, V. M., and Yaschenko, A. K. (2006). Possible DC electric field in the ionosphere related to seismicity. *Adv. Space Res.* **37**(4), 666–670.

Sorokin, V. M. and Yaschenko, A. K. (2000). Electric field disturbances in the Earth-ionosphere layer. *Adv. Space Res.* **26**(8), 1219–1223.

Sorokin, V. M., Yaschenko A. K., Chmyrev, V. M., and Hayakawa, M. (2005). DC electric field amplification in the mid-latitude ionosphere over seismically active faults. *Nat. Hazards Earth Syst. Sci.* **5**, 661–666, Retrieved from http://www.nat-hazards-earth-systsci.net/5/661/2005/

The Mid-Atlantic Ridge (2008). 2nd International Mid Atlantic Ridge Expert Workshop, March 27–28, 2008, Azores, Portugal, Retrieved from http://whc.unesco.org/en/activities/504/, last access: June 14.

Walters, R., Goff, J., and Wang, K. (2006). Tsunamigenic sources in the bay of plenty, New Zealand. *Sci. Tsunami Haz.* **24**, 339–359.

11

Cai, Y., Ning, X., and Liu, C. (2002a). Distribution characteristics of size-fractionated chlorophyll a and productivity of phytoplankton in the northern South China Sea and Beibu Gulf during August 1999. *Studia Marina Sinica* **44**, 11–21 (in Chinese).

Cai, Y., Ning, X., Liu, C., and Hao, Q. (2007). Distribution pattern of photosynthetic picoplankton and heterotrophic bacteria in the northern South China Sea. *J. Integrative Plant Biol.* **49**, 282–298.

Cai, Y., Ning, X., and Liu, Z. (2002b). Studies on primary production and new production of the Zhujiang Estuary, China. *Acta Oceanol. Sinica* **24**, 101–111 (in Chinese).

Chai, F., Xue, H., and Shi, M. (2001). Formation and distribution of upwelling and downwelling in the South China Sea. *Oceanogr. in China* **13**, 105–116 (in Chinese).

Chen, L., Zhu, W., Wang, W., Zhou, X., and Li, W. (1998). Studies on climate change in China in recent 45 years. *Acta Meteorol. Sinica* **56**, 257–271 (in Chinese).

Chen, Q. (1985). *Report on the Multidisciplinary Investigation of the South China Sea (Ii)*. China Science Press, Beijing, pp. 432 (in Chinese).

Chen, Q. (1992). *Compiling of the Information on Resources of the Seas and Islands in Guangdong Province*. Guangdong People Press, Guangzhou, pp. 622 (in Chinese).

Chen, S. and Ma, J.: *The Analytical Methods and Its Application of the Disposal of Marine Data.* China Ocean Press, Beijing, pp. 660, 1991 (in Chinese).

Chen, T., Qian, G., and Zeng, X. (1999). Trends of the changes in air temperature in Hong Kong, Macao and Guangzhou during 20th century. *Studies Develop.South China Sea* **1**, 12–17.

Chen, X., Qi, Y., and Wang, W. (2005). Seasonal and annual changes in the characteristics of mesoscale eddies in the South China Sea. *J. Tropical Oceanogr.* **24**, 51–57 (in Chinese).

, S. P. (1949). Experimental studies on the environmental factors influencing the growth of phytoplankton. *Con. Fish. Res. Inst. Dept. Fish Nat. Univ. Shantung*, **1**, 37–52.

Fan, J. (1985). *Report on the Multidisciplinary Investigation of the South China Sea (Ii)*. China Sci. Press, Beijing, pp. 432 (in Chinese).

Fang, G., Fang, W., Fang, Y., and Wang, K. (1998). A survey of studies on the South China Sea upper ocean circulation. *Acta Oceanogr. Taiwanica* **37**, 1–16.

Fu, Q., Celeste, M., Johanson, J. M. W., and Thomas, R. (2006). Enhanced mid-latitude tropospheric warming in satellite measurements. *Science* **312**, 1179.

Gong, G. C., Liu, K. K., Liu, C. T., and Pai, S. C. (1992). Chemical hydrography of the South China Sea and a comparison with the West Philippine Sea. *Terr. Atmos. Oceanic Sci.* **3**, 587–602.

Guan, W., Wong, L., and Xu, D. (2003). Modeling nitrogen and phosphorus cycles and dissolved oxygen in the Zhujiang River Estuary, Part II. Modeling results. *Acta Oceanol. Sinica* **25**, 61–68 (in Chinese).

Guo, J. and Chen, P. (2000). Study on exploitation of cephalopod stock in the South China Sea. *J. Tropic Oceanogr.* **19**, 51–58 (in Chinese).

Han, W. (1998). *Chemistry Oceanography of the South China Sea*. China Science Press, pp. 289 (in Chinese).

Han, W., Wu, L., and Lin, H. (1990). Preliminary research on cultivating hydrochemistry of waters off Hong Kong. *Mar. Sci. Bull.* **9**, 37–44 (in Chinese).

Hao, Q., Ning, X., Liu, Y., Cai, Y., and Le, F. (2007). Satellite and situ observations of primary production in the northern South China. *Acta Oceanol. Sinica* **29**, 58–68 (in Chinese).

Harvey, H. W. (1957). *Chemistry and Fertility of Seawater*. Cambridge Univ. Press, Cambridge, pp. 234.

He, G., Yuan, G., and Li, F. (2004). Effect of economic development on water quality in the Pearl Estuary. *Mar. Environ. Sci.* **23**, 50–52 (in Chinese).

He, Y. Z., Cheng, C., and Guan, C. (2003). Relationship between long-term changes of summer rainfall in the northern China and sea surface temperature over the South China Sea. *J. Tropical Oceanogr.* **22**, 1–8 (in Chinese).

Holm-Hansen, O., Lorenzen, C. J., Holmes, R. W., and Strickland, J. D. H. (1965). Fluorometric determination of chlorophyll. *J. Cons. Perm. Int. Explor. Mer* **30**, 3–15.

Hong, W., Hu, Q., Wu, Y., and Zhang, S. (1983). *Marine Phytoplankton*. China Agriculture Press, Beijing, pp. 299 (in Chinese).

Houghton, J. T., Filho, L. G. M., Callander, B. A., Harris, N., Kattenberg, A., and Maskell, K. (1996). Climate change in 1995: *The Science of Climate Change, Contribution of Working Group 1 To the Second Assessment Report of the Intergovernmental Panel on Climate Change*. Cambridge University Press, Cambridge, pp. 572.

Hu, M., Yang, Y., Xu, C., and Halisen, J. P. (1989). Limitation of phosphate on phytoplankton growth in the Changjiang River estuary. *Acta Oceanol. Sinica* **11**, 439–443 (in Chinese).

Huang, L. (1992). Vertical variations of chlorophyll a and fluorescence values in different areas of the South China Sea. *J. Tropical Oceanogr.* **11**, 89–95 (in Chinese).

Huang, Y., Qian, Q., Ou, Q., Huang, Y., and Liu, J. (1990). Investigation of zooplankton in the around area of Hainan-Island of the South China Sea. *Acta Ji-Nan University* **6**, 74–80 (in Chinese).

Hutchins, D. A. and Bruland, K. W. (1998). Iron-limited diatom growth and Si: N uptake in a coastal upwelling regime. *Nature* **393**, 561–564.

Jia, X., Du, F., Lin, Q., Li, C., and Cai, W. (2003). A study on comprehensive assessment method of ecological environment quality of marine fishing ground. *J. Fish. Sci. China* **10**, 160–164 (in Chinese).

Jia, X., Li, C., Gan, J., Lin, Q., Cai, W., and Wang, Z. (2005). Diagnosis and assessment on the health status and quality of the fisheries eco-environment of the northern South China Sea. *J. Fish. Sci. China* **12**, 757–765 (in Chinese)

Jiang, J. and Yao, R. (1999). Effects of nitrogen to phosphorus ratio on growth of Dunaliella salina and accumulation of glycerol and pigments. *J. Tropical Oceanogr.* **18**, 68–72 (in Chinese).

Justic, D., Rabalais, N. N., and Turner, R. E. (1995). Changes in nutrient structure of river-dominated coastal waters: Stoichiometric nutrient balance and its consequences, Estuarine. *Coast. Shelf Sci.* **40**, 339–356.

Lei, Y., Yang, X., Hu, Q., Luo, Z., and Yang, G. (2003). Hydraulic geometry features and channel evolution of the Dong-Ping waterway in last decades. *Tropical Geography* **23**, 204–208 (in Chinese).

Levimson, D. H. (2005). State of the Climate in 2004. *B. Am. Meteorol. Soc.* **6**, 17–18.

Li, Y., Cai, W., Li, L., and Xu, D. (2003). Seasonal and interannual variabilities of mesoscale eddies in the northeastern South China Sea. *J. Tropical Oceanogr.* **22**, 1–8 (in Chinese).

Li, Z. and Chen, K. (1998). Seawater quality and its change trends along the coast of the northern South China Sea. *Res. Develop. South China Sea* **1**, 15–23 (in Chinese).

Liao, G., Yuan, C., and Wang, Z. (2006). Three dimensional structure of the circulation in the South China Sea during of 1998. *Acta Oceanol. Sinica* **28**, 15–25 (in Chinese).

Lin, C., Ning, X., Su, J., Lin, Y., and Xu, B. (2005). Environmental changes and the responses of the ecosystems of the Yellow Sea during 1976–2000. *J. Mar. Syst.* **55**, 223–234.

Lin, C., Su, J., Xu, B., and Tang, Q. (2001). Long-term variations of temperature and salinity of the Bohai Sea and their influence on its ecosystem. *Prog. Oceanogr.* **49**, 9–17.

Lin, H. and Han, W. (1998). Study on the dissolved oxygen flux in the South China Sea. *Chin. J. Oceanol. Limnol.* **29**, 61–66 (in Chinese).

Lin, Y. (1985). *Report of the Synthetic Investigation of the South China Sea (Ii)*. China Sci. Press, Beijing, pp. 432 (in Chinese).

Lin, Y., Su, J., Hu, C., Zhang, M., Li, Y., Guan, W., and Chen, J. (2004). N and P in water of the Zhujiang River estuary in summer. *Acta Oceanol. Sinica* **26**, 63–73 (in Chinese).

Liu, K. K., Chao, S. Y., Shaw, P. T., Gong, G. C., Chen, C. C., and Tang, T. Y. (2002). Monsoon-forced chlorophyll distribution and primary production in the South China Sea: observations and a numerical study. *Deep-Sea Res. I* **49**, 1387–1412.

Liu, K. K., Kao, S. J., Hu, H. C., Chou, W. C., Hung, G. W., and Tseng, C. M. (2007). Carbon isotopic composition of suspended and sinking particulate organic matter in the northern South China Sea – from production to deposition. *Deep Sea Res. Part II* **54**, 1504–1527.

Martin, H. and Arun, K. (2003). Perfect ocean for drought. *Science* **299**, 691–696.

Mcphaden, M. J. (2004). Evolution of the 2002–2003 E1 Nino. *B. Am. Meteorol. Soc.* **5**, 677–695.

NBTS (National Bureau for Technique Supervision) (1991). Marine Biology Survey. In *Specifications for Oceanographic Survey, GB 12763.6-91*, pp. 104 (in Chinese).

Ning, X., Chai, F., Xue, H., Cai, Y., Liu, C., and Shi, J. (2004). Physical-biological oceanographic coupling influencing phytoplankton and primary production in the South China Sea *J. Geophys. Res.* **109**(C10), C10005, doi:10.1029/2004JC002365.

Nitani, H. (1972). Beginning of the Kuroshio. In *Kuroshio*. H. Stommel and K. Yoshida (Eds.). University of Washington Press, Seattle, pp. 517.

Peng, X., Ning, X., Sun, J., and Le, F. (2006). Responses of phytoplankton growth on nutrient enrichments in the northern South China Sea. *Acta Ecol. Sinica* **22**, 3959–3968 (in Chinese).

Peng, Y., Chen, L., and Chen, H. (1994). Relationship between dissolved oxygen and nutrients in the Pearl River estuary. *J. Tropical Oceanogr.* **13**, 96–100 (in Chinese).

Qian, H., Huang, Y., Ou, J., Huang, Y. L., Liu, J., and Huang, Y. (1990a). Investigation of zooplankton in the around area of the Huang-Island in the South China Sea. *Acta Ji-Nan. Univ.* **6**, 81–84 (in Chinese).

Qian, H., Huang, Y., Ou, J., Huang, Y. L., Liu, J., and Yu, W. (1990b). Investigation of zooplankton in the northeastern area of the northern South China Sea. *Acta Ji-Nan. Univ.* **6**, 70–73 (in Chinese).

Qin, D. (2003). *El Nino Phenomena*. Meteorology Press, Beijing, pp. 180 (in Chinese).

Qiu, Y., Wang, S., and Zhu, L. (2005). Variation trend of nutrient and chlorophyll a contents and their effects on eco-environment in Daya Bay. *J. Oceanogr. Taiwan Strait* **24**, 131–139 (in Chinese).

Richardson, K. (1997). Harmful or exceptional phytoplankton blooms in the marine ecosystem. *Adv. Mar. Biol.* **31**, 301–385.

Sadler, J., Lander, C. M. A., and Oda, L. K. (1985). *Tropical Marine Climate Atlas, I.* Indian Ocean and Atlantic Ocean. University of Hawaii, Honolulu, pp. 100.

Seitzinger, S. P., Kroeze, C., Bouwman, A. F., Caraco, N., Dentener, F., and Styles, R. V. (2002). Global patterns of dissolved inorganic and particulate nitrogen inputs to coastal systems: recent conditions and future projections. *Estuaries* **25**(4), 640–655.

Shaw, P. T. (1991). Seasonal variation of the intrusion of the Philippine Sea water into the South China Sea. *J. Geophys. Res.* **96**, 821–827.

Shen, S. (1985). *Report of the Multidisciplinary Investigation of the South China Sea (Ii)*. China Science Press, Beijing, 403–424 (in Chinese).

Shi, N., Chen, J., and Tu, Q. (1995). 4-phase climate change feature in the last 100 years over China. *Acta Meteorol. Sinica* **53**, 431–439 (in Chinese).

Shi, P., Du, Y., Wang, D., and Gan, Z. (2001). Annual cycle of mixed layer in the South China Sea. *J. Tropical Oceanogr.* **20**, 10–17 (in Chinese).

SOAC (State Oceanic Administration of China) (1975). *Specification for Oceanogr*. Survey, Beijing, pp. 550 (in Chinese).

SOAC (2000). *Seawater Quality Results of Survey*, SOAC, pp. 15 (in Chinese).

SOAC (2001a). *Status of the Environ. Quality of the China Sea in the End of 20th Century.* SOAC, pp. 167 (in Chinese).

SOAC (2001b). *Seawater Quality Results of Survey*, SOAC, pp. 30 (in Chinese).

SOAC (2002). *Seawater Quality Results of Survey*, SOAC, pp. 20 (in Chinese).

SOAC (2003). *Seawater Quality Results of Survey*, SOAC, pp. 22, (in Chinese).

SOAC (2004). *Seawater Quality Results of Survey*, SOAC, pp. 23 (in Chinese).

Sournia, A. (1978). *Phytoplankton Manual.* UNESCO, Paris, pp. 337.

Steemann-Nielson, E. (1952). The use of radioactive carbon (14C) for measuring organic production in the sea. *J. Cons. Int. Expor. Mer.* **18**, 117–140.

Strickland, J. D. H. and Parsons, T. R. (1972). A practical handbook of seawater analysis. *Bull. Fish Res. Bd. Can.* **167**, pp. 310.

Su, J. (1998). Circulation dynamics of the China Sea: North of 18°N, in: The Sea, 11. *The Global Coastal Ocean: Regional Studies and Syntheses.* A. R. Robinson and K. Brink (Eds.). John Wiley Press, pp. 506.

Takano, K., Akira, H., and Takaya, N. (1998). A numerical simulation of the circulation in the South China Sea preliminary results. *Acta Oceanogr. Taiwanica* **37**, 165–186.

Takeuchi, N. (1987). Auspice of ENSO. *Physical geography research report of Hokkaido University* **492**, 381–386 (in Japanese).

Tang, D., Wang, S., and Di, B. (2006). *Harmful Algal Blooms in the Coastal Waters of the South China Sea.* Marine Science and Technology in Asia, June 11–15, PACON.

Wang, D. and Peng, Y. (1996). Nutrient elements of the Zhujiang (Pearl River) estuary waters, China. In *Biogeochemistry Study of China Major River Estuaries*. J. Zhang (Ed.). China Ocean Press, Beijing, pp. 241 (in Chinese).

Wang, S. and Gong, D. (1999). ENSO events and their intensity during the past century. *Meteorology* **25**(1), 9–14 (in Chinese).

Wang, W., Wang, D., and Shi, P. (2001). Annual and interannual variations of large-scale dynamic field in South China Sea. *J. Trop. Oceanogr.* **20**(1), 61–68 (in Chinese).

Wei, M., Lai, T., and He, B. (2002). Changing trends of the chemical items in Qinzhou Bay during the last twenty years: Nutrient condition. *Marine Environ. Sci.* **21**, 49–52 (in Chinese).

Wei, M, Lai, T, and He, B. (2003). Change trend of nutrient conditions and influencing factors during flood and dry seasons in Qinzhou Bay. *J. Tropical Oceanogr.* **22**, 16–21 (in Chinese).

White, B. B., Meyes, G. A., and Songu, Y. J. R. (1985). Short-term climatic variability in the thermal structure of the Pacific Ocean during 1979–1982. *J. Phys. Oceanogr.* **15**, 917–935.

Wolfe, D. A. and Schelske, C. L. (1967). Liquid scintillation and geiger counting efficiencies for carbon-14 incorporated by marine phytoplankton in productivity measurements. *ICES J. Mar. Sci.* **31**, 31–37.

Wyrtki, K. (, 1961). Physical oceanography of the southeast Asia waters, in Scientific Results of Marine Investigations of the South China Sea and the Gulf of Thailand. *NAGA Rep. Scripps Inst Oceanogr.* **2**, 1–195. Lajolla, Calif.

Xie, Z. and Zhang, Y. (2003). Review of the Pearl River flood calamity. *The Pearl River* **4**, 13–16 (in Chinese).

Xu, J., Xue, H., Shi, M., and Liu, Z. (2001). Observation and study of the circulation and mesoscale eddies in the upper layer of the South China Sea in summer of 1998. In *Oceanography in China*. H. Xue, F. Chai, and J. Xu (Eds.). China Ocean Press, Beijing, **13**, 178–187.

Xu, J., Yin, K., He, L., Yuan, X., Ho, A. Y. T., and Harrison, P. J. (2008). Phosphorus limitation in the northern South China Sea during late summer: Influence of the Pearl River. *Deep-Sea Res. I* **55**(10), 1330–1342.

Xue, H., Chai, F., Pettigrew, N., Xu, D., Shi, M., and Xu, J. (2004). Kuroshio intrusion and the circulation in the South China Sea. *J. Geophys. Res.* **109**(C2), C02217, doi:10.1029/2002JC001724.

Xue, H., Chai, F., Xu, D., and Shi, M. (2001a). Characteristics and seasonal variation of the coastal currents in the South China Sea. *Oceanogr. China* **13**, 64–73 (in Chinese).

Xue, H., Chai, F., Wong, L., and Chen, J. (2001b). Zhujiang River estuarine circulation model. *Oceanogr. China* **13**, 138–151 (in Chinese).

Yang, H. and Liu, Q. (1998). Seasonal feature of temperature distribution in the upper layer of the South China Sea. *Chin. J. Oceanol. Limnol.* **29**, 501–507 (in Chinese).

Yin, K. D., Qian, P. Y., Wu, M. C. S., Chen, J. C., Huang, L., Song, X., and Jian, W. (2001). Shift from P to N limitation of phytoplankton growth across the Pearl River estuarine plume during summer. *Mar. Ecol. Prog. Ser.* **221**, 17–28.

Yuan, S. and Deng, J. (1997a). Thermal-salinity structure in the northern South China Sea, I. *Study and Development of the South China Sea* **2**, 23–27 (in Chinese).

Yuan, S. and Deng, J. (1997b). Thermal-salinity structure in the northern South China Sea II. *Study and Development of the South China Sea* **3**, 19–24 (in Chinese).

Yuan, S. and Deng, J. (1998). Thermal-salinity structure in the northern South China Sea, III. *Study and Development of the South China Sea* **2**, 28–36 (in Chinese).

Zeng, Q., Li, R., and Ji, Z. (1989). Computation of monthly average current. *Atmosp. Sci.* **13**, 127–138.

Zhai, P. and Ren, F. (1997). On changes of maximum and minimum temperature in China over the recent 40 years. *Acta Meteorol. Sinica* **55**, 418–429 (in Chinese).

Zhao, H., Tang, D., and Wang, S. (2005). Spatial distribution of chlorophyll a concentration in summer in western South China Sea and its response to oceanographic environmental factors. *J. Tropical Oceanogr.* **24**, 31–37 (in Chinese).

Zhang, J. L., Yu, Z., and Zhang, J. (1999). Wet and dry deposition and its influence on marine ecosystem. *Mar. Environ. Sci.* **18**, 70–76 (in Chinese).

Zhang, Q. and Huang, R. H. (1993). The space-time character of the development of ENSO events. *Chin. J. Atmos. Sci.* **17**, 395–402 (in Chinese).

Zhang, Q., Tao, S., and Chen, L. (2003). The interannual variability of East Asian summer indices and its association with the pattern of general circulation over East Asia *Acta Meteor. Sinica* **61**, 559–567 (in Chinese).

Zhang, S. (1984). Ecological study of the planktonic copepoda of the South China Sea. *J. Tropical Oceanogr.* **3**, 46–55 (in Chinese).

Zhao, W., Jiao, N., and Zhao, Z. (2000). Distribution and variation of nutrients in aquaculture waters of the Yantai Sishili Bay. *Mar. Sci.* **4**, 31–34 (in Chinese).

Zhu, C., He, J., and Wu, G. (2000). East Asian monsoon index and its inter-annual relationship with large scale thermal dynamic circulation. *Acta Meteor. Sinica* **58**, 391–401 (in Chinese).

Zou, J., Dong, L., and Qin, B. (1983). Preliminary analysis of the eutrophication and red tide in the Bohai Sea. *Mar. Environ. Sci.* **2**, 42–54 (in Chinese).

12

Bhattacharya, J. P. and Walker, R. G. (1992). Deltas. In *Facies Models: Response To Sea Level Change*. R. G. Walker and N. P. James (Eds.). Geological Association of Canada. St. John's Newfoundland, Canada, pp. 157–177.

Blacut, G. and Kleinpell, R. M. (1969). A stratigraphic sequence of benthonic smaller foraminifera from the La Boca Formation, Panama Canal Zone. *Contributions from the Cushman Foundation for Foraminiferal Research* **20**, 1–22.

Burton, K. W., Ling, H. F., and O'Nions, R. K. (1997). Closure of the Central American isthmus and its effect on deep-water formation in the North Atlantic. *Nature* **386**, 382–385.

Case, J. E. (1974). Oceanic crust forms basement of eastern Panama. *Geological Society of America Bulletin* **85**, 645–652.

Case, J. E. and Holcombe, T. L. (1980). Geologic-tectonic map of the Caribbean region. *US Geological Survey Miscellaneous Investigations Series Map I-1100*, scale 1:2,500,000, 1 sheet.

Coates, A. G. (1999). Lithostratigraphy of the Neogene strata of the Caribbean coast from Limon, Costa Rica, to Colon, Panama. In A paleobiotic survey of Caribbean faunas from the Neogene of the Isthmus of Panama. *Bulletins of*

American Paleontology. L. S. Collins and A. G. Coates (Eds.), pp. **357**, 17–38.

Coates, A. G., Aubry, M. P., Berggren, W. A., and Collins, L. S. (2003). Early Neogene history of the Central American arc from Bocas del Toro, western Panama. *Geological Society of America Bulletin* **115**, 271–287.

Coates, A. G., Collins, L. S., Aubry, M. P., and Berggren, W. A. (2004). The geology of the Darien, Panama, and the late Miocene-Pliocene collision of the Panama arc with northwestern South America. *Geological Society of America Bulletin* **116**, 1327–1344.

Coates, A. G., Jackson, J. B. C., Collins, L. S., Cronin, T. M., Dowsett, H. J., et al. (1992). Closure of the Isthmus of Panama: The near-shore marine record of Costa Rica and western Panama. *Geological Society of America Bulletin* **104**, 814–828.

Coates, A. G. and Obando, J. A. (1996). The geologic evolution of the Central American isthmus. In *Evolution and Environment in Tropical America*. J. B. C. Jackson, A. F. Budd, and A. G. Coates (Eds.). University of Chicago Press. Chicago pp. 21–56.

Collins, L. S. and Coates, A. G. (1999). Introduction. In A paleobiotic survey of Caribbean faunas from the Neogene of the Isthmus of Panama. *Bulletins of American Paleontology*. L. S. Collins and A. G. Coates (Eds.) pp. **357**, 5–13.

Collins, L. S., Coates, A. G., Berggren, W. A., Aubry, M. P., and Zhang, J. (1996). The late Miocene Panama isthmian strait. *Geology* **24**, 687–690.

Collins, L. S., Coates, A. G., Jackson, J. B. C., and Obando, J. A. (1995). Timing and rates of emergence of the Limon and Bocas del Toro basins: Caribbean effects of Cocos Ridge subduction? In *Geologic and Tectonic Development of the Caribbean Plate Boundary in Southern Central America*. P. Mann (Ed.). Geological Society of America Special Paper. pp. **295**, 263–289.

Compton, R. R. (1985). *Geology in the Field*. John Wiley and Sons. New York.

Duque-Caro, H. (1990). Neogene stratigraphy, paleoceanography and paleobiogeography in northwest South America and the evolution of the Panama seaway. *Palaeogeography, Palaeoclimatology, Palaeoecology* **77**, 203–234.

Emiliani, C., Gartner, S., and Lidz, B. (1972). Neogene sedimentation on the Blake Plateau and the emergence of the Central American isthmus. *Palaeogeography, Palaeoclimatology, Palaeoecology* **11**, 1–10.

Escalante, G. (1990). The geology of southern Central America and western Colombia. In *The Caribbean Region*. G. Dengo and J. E. Case (Eds.). The Geological Society of America. Boulder, pp. 201–230.

Gradstein, F., Ogg, J., and Smith, A. (2004). *A Geologic Time Scale 2004*. Cambridge University Press, Cambridge.

Graham, A., Stewart, R. H., and Stewart, J. L. (1985). Studies in neotropical paleobotany. III. The Tertiary communities of Panama–Geology of the pollen-bearing deposits. *Annals of the Missouri Botanical Garden* **72**, 485–503.

Haq, B. U., Hardenbol, J., and Vail, P. R. (1987). Chronology of fluctuating sea levels since the Triassic. *Science* **235**, 1156–1167.

Haug, G. H. and Tiedemann, R. (1998) Effect of the formation of the Isthmus of Panama on Atlantic Ocean thermohaline circulation. *Nature* **393**, 673–676.

Haug, G. H., Tiedemann, R., Zahn, R., and Ravelo, A. C. (2001). Role of Panama uplift on oceanic freshwater balance. *Geology* **29**, 207–210.

Hill, R. T. (1898). The geological history of the Isthmus of Panama and portions of Costa Rica. *Bulletin of the Museum of Comparative Zoology* **28**, 151–285.

Hodell, D. A., Mueller, P. A., and Garrido, J. R. (1991). Variations in the strontium isotopic composition of sea water during the neogene. *Geology* **19**, 24–27.

Hodell, D. A. and Woodruff, F. (1994). Variations in the strontium isotopic ratio of seawater during the Miocene: Stratigraphic and geochemical implications. *Paleoceanography* **9**, 405–426.

Jackson, J. B. C., Jung, P., Coates, A. G., and Collins, L. S. (1993). Diversity and extinction of tropical American mollusks and emergence of the Isthmus of Panama. *Science* **260**, 1624–1626.

Jackson, J. B. C., Jung, P., and Fortunato, H. (1996). Paciphilia revisited: Transisthmian evolution of the Strombina group (Gastropoda: Columbellidae). In *Evolution and Environment in Tropical America*. J. B. C. Jackson, A. F. Budd, and A. G. Coates (Eds.). University of Chicago Press, Chicago, pp. 234–270.

Johnson, K. G. and Kirby, M. X. (2006). The Emperador Limestone rediscovered: Early Miocene corals from the Culebra Formation, Panama. *J Pal.* **80**, 283–293.

Jones, D. S., Mueller, P. A., Hodell, D. A., and Stanley, L. A. (1993). 87Sr/86Sr geochemistry of Oligocene and Miocene marine strata in Florida. In *The Neogene of Florida and adjacent regions, Proceedings of the 3rd Bald Head Island Conference on Coastal Plain Geology*. V. A. Zullo, W. B. Harris, T. M. Scott, and R. W. Portell (Eds.). Florida Geological Survey Special Publication, pp. **37**, 15–26.

Jones, S. M. (1950). Geology of Gatun Lake and vicinity. *Geol. Soc. Am. Bull.* **61**, 893–920.

Keigwin, L. D. (1978). Pliocene closing of the Isthmus of Panama, based on biostratigraphic evidence from nearby Pacific Ocean and Caribbean cores. *Geology* **6**, 630–634.

Keigwin, L. D. (1982). Isotopic paleoceanography of the Caribbean and east Pacific: Role of Panama uplift in Late Neogene time. *Science* **217**, 350–352.

Kirby, M. X. and MacFadden, B. J. (2005). Was southern Central America an archipelago or a peninsula in the middle Miocene? A test using land-mammal body size. *Palaeogeography, Palaeoclimatology, Palaecology* **228**, 193–202.

Lowrie, A., Stewart, J., Stewart, R. H., Van Andel, T. J., and McRaney, L. (1982). Location of the eastern boundary of the Cocos plate during the Miocene. *Marine Geology* **45**, 261–279.

MacDonald D. F. (1913) Isthmian geology. *Annual Report of Isthmian Canal Commission*, pp. 564–582.

MacDonald, D. F. (1919). The sedimentary formations of the Panama Canal Zone, with special reference to the stratigraphic relations of the fossiliferous beds. *US National Museum Bulletin* **103**, 525–545.

MacFadden, B. J. (2006). North American Miocene land mammals from Panama. *J. Verteb. Paleont.* **26**, 720–734.

MacFadden, B. J. and Higgins, P. (2004) Ancient ecology of 15-million-year-old browsing mammals within C3 plant communities from Panama. *Oecologia* **140**, 169–182.

Mallinson, D. J., Compton, J. S., Snyder, S. W., and Hodell, D. A. (1994) Strontium isotopes and Miocene sequence stratigraphy across the northeast Florida Platform. *J.Sedimen. Res.* **B64**, 392–407.

Mann, P. (1995). Preface. In *Geologic and tectonic development of the Caribbean plate boundary in southern Central America*. P. Mann (Ed.). Geological Society of America Special Paper, pp. **295**, xi–xxxii.

Martin, E. E., Shackleton, N. J., Zachos, J. C., and Flower, B. P. (1999). Orbitally-tuned Sr isotope chemostratigraphy for the late middle to late Miocene. *Paleoceanography* **14**, 74–83.

McArthur, J. M., Howarth, R. J., and Bailey, T. R. (2001). Strontium isotope stratigraphy: LOWESS Version 3: Best fit to the marine Sr-isotope curve for 0-509 Ma and accompanying look-up table for deriving numerical age. *J. Geol.* **109**, 155–170.

Miall, A. D. (1978). *Fluvial sedimentology*. Canadian Society of Petroleum Geologists. Calgary.

Miall, A. D. (1992). Alluvial deposits. In *Facies Models: Response to Sea Level Change*. R. G. Walker and N. P James (Eds.). Geological Association of Canada. St. John's Newfoundland, Canada, pp. 119–142.

Miller, K. G., Feigenson, M. D., Wright, J. D., and Clement, B. M. (1991). Miocene isotope reference section, Deep Sea Drilling Project site 608: An evaluation of isotope and biostratigraphic resolution. *Paleoceanography* **6**, 33–52.

Miller, K. G., Kominz, M. A., Browning, J. V., Wright, J. D., Mountain, G. S., et al. (2005). The Phanerozoic record of global sea-level change. *Science* **310**, 1293–1298.

Miller, K. G. and Sugarman, P. J. (1995). Correlating Miocene sequences in onshore New Jersey boreholes (ODP Leg 150 X) with global δ18O and Maryland outcrops. *Geology* **23**, 747–750.

Oslick, J. S., Miller, K. G., and Feigenson, M. D. (1994). Oligocene-Miocene strontium isotopes: Stratigraphic revisions and correlations to an inferred glacioeustatic record. *Paleoceanography* **9**, 427–443.

Pettijohn F. J., Potter P.E., and Siever R. (1987) *Sand and Sandstone.* Springer-Verlag, New York.

Pin, C. and Bassin, C. (1992). Evaluation of a Sr-specific extraction chromatographic method for isotopic analyses in geological materials. *Analytica. Chimica. Acta* **269**, 249–255.

Pratt, T. L., Holmes, M., Schweig, E. S., Gomberg, J., and Cowan, H. A. (2003). High resolution seismic imaging of faults beneath Limón Bay, northern Panama Canal, Republic of Panama. *Tectonophysics* **368**, 211–227.

Reineck, H. E. and Singh, I. B. (1975). *Depositional Sedimentary Environments, With Reference To Terrigenous Clastics*. Springer-Verlag. New York.

Retallack, G. J. (2001). *Soils of the Past: An Introduction to Paleopedology*. Blackwell Science, Oxford.

Retallack, G. J. and Kirby, M. X. (2007). Middle Miocene global change and paleogeography of Panama. *Palaios* **22**, 667–679.

Savin, S. M. and Douglas, R. G. (1985). Sea level, climate, and the Central American land bridge. In *The Great American Biotic Interchange*. F. G. Stehli and S. D. Webb (Eds.). Plenum Press, New York, pp. 303–324.

Slaughter, B. H. (1981). A new genus of geomyoid rodent from the Miocene of Texas and Panama. *J. Verteb. Paleont.* **1**, 111–115.

Stewart, R. H., Stewart, J. L., and Woodring, W. P. (1980). Geologic map of the Panama Canal and vicinity, Republic of Panama. *US Geological Survey Miscellaneous Investigations Series Map I-1232*, scale 1:100,000, 1 sheet.

Tedford, R. H., Albright, L. B., Barnosky, A. D., Ferrusquia-Villafranca, I., Hunt, R. M., et al. (2004). Mammalian biochronology of the Arikareean through Hemphillian interval (late Oligocene through early Pliocene epochs). In *Late Cretaceous and Cenozoic Mammals of North America*. M. O. Woodburne (Ed.). Columbia University Press, New York, pp. 169–231.

Terry, R. A. (1956). *A Geological Reconnaissance of Panama*. Occasional Papers of the California Academy of Sciences **23**, 1–91.

Van den Bold, W. A. (1972). Ostracoda of the La Boca Formation, Panama Canal Zone. *Micropaleontology* **18**, 410–442.

Vermeij, G. J. (1978). *Biogeography and adaptation: Patterns of marine life.* Harvard University Press, Cambridge.

Vermeij, G. J. and Petuch, E. J. (1986). Differential extinction in tropical American molluscs: Endemism, architecture, and the Panama land bridge. *Malacologia* **27**, 29–41.

Webb, S. D. (1985). Late Cenozoic mammal dispersals between the Americas. In. *The Great American Biotic Interchange*. F. G. Stehli and S. D. Webb (Eds.). Plenum Press, New York, pp. 357–386.

Weyl, P. K. (1968). The role of the oceans in climate change; a theory of the ice ages. *Meteorological Monographs* **8**, 37–62.

Whitmore, F. C. and Stewart, R. H. (1965). Miocene mammals and Central American seaways. *Science* **148**, 180–185.

Woodring, W. P. (1957–1982). *Geology and paleontology of Canal Zone and adjoining parts of Panama*. US Geological Survey Professional Paper **306**, 1–759.

Woodring, W. P. (1964). *Geology and Paleontology of Canal Zone and Adjoining Parts of Panama*. US Geological Survey Professional Paper 306-C, 241–297.

Woodring, W. P. (1965). Endemism in middle Miocene Caribbean molluscan faunas. *Science* **148**, 961–963.

Woodring, W. P. (1966). The Panama land bridge as a sea barrier. *Proceedings of the American Philosophical Society* **110**, 425–433.

Woodring, W. P. and Thompson, T. F. (1949). Tertiary formations of Panama Canal Zone and adjoining parts of Panama. *Bull. Am. Assoc. Pet. Geol.* **33**, 223–247.

Wright, D. B. (1998). Tayassuidae. In *Evolution of Tertiary Mammals of North America*. C. M. Janis, K. M. Scott, and L. L. Jacobs (Eds.). Cambridge University Press, Cambridge, pp. 389–401.

13

Akiwumi, F. A. and Butler, D. R. (2008). Mining and environmental change in Sierra Leone, West Africa: A remote sensing and hydrogeomorphological study. *Environ. Monit. Assess.* **142**, 309–318.

Central Weather Bureau (2000). *Report on Typhoons in 2000.* Ministry of Transportation and Communications, Taiwan, pp. 130–162.

Central Weather Bureau (2001). *Report on typhoons in 2001.* Ministry of Transportation and Communications, Taiwan, pp. 84–111.

Chang, K. T., Chiang, S. H., and Hsu, M. L. (2007). Modeling typhoon- and earthquake- induced landslides in a mountains watershed using logistic regression. *Geomorphology* **89**, 335–347.

Cohen, W. B. and Goward, S. N. (2004). Landsat's role in ecological applications of remote sensing. *Bioscience* **54**, 535–545.

Curran, P. J. and Atkinson, P. M. (1998). Geostatistics and remote sensing. *Prog. Phys. Geog.* 22, 61–78.

CWB (Central Weather Bureau) (1999). Retrieved from http://scman.cwb.gov.tw/eqv3/eq_report/special/ 19990921/921, Isomap, GIF.

CWB (Central Weather Bureau) (2008). *Typhoon Database*. Retrieved from http://rdc28.cwb.gov.tw/

DeMets, C., Gordon, R. G., Argus, D. F., and Stein, S. (1990). Current plate motions. *Geophys. J. Int.* **104**, 425–478.

Deutsch, C. V. and Journel, A. G. (1992). *GSLIB: Geostatistical Software Library and User's Guid.*, Oxford University Press, New York.

Edwards, G. and Fortin, M. J. (2001). Delineation and analysis of vegetation boundaries. *Spatial Uncertainty in Ecology: Implications for Remote Sensing and Gis Application.* C. T. Hunsaker, M. A. Goodchild, A. Friedl, and T. J. Case (Eds.). Springer, New York, pp. 158–174.

Fortin, M. J. and Edwards G. (2001). *A Cognitive View of Spatial Uncertainty.* C. T. Hunsaker, M. F. Goodchild, M. A. Friedl, and T. J. Case (Eds.), Springer, New York, pp. 133–157.

Foster, D. R., Knight, D. H., and Franklin, J. F. (1998). Landscape patterns and legacies resulting from large, infrequent forest disturbances. *Ecosystems*, **1**, 497–510.

Fox, D. M., Maselli, F., and Carrega, P. (2008). Using SPOT images and field sampling to map burn severity and vegetation factors affecting post forest fire erosion risk. *Catena* **75**, 326–335.

Fredericks A. K. and Newman, K. B. (1998). A comparison of the sequential Gaussian and Markov-Bayes simulation methods for small samples. *Math. Geol.* **30**, 1011–1032.

Garrigues, S., Allard, D., and Baret, F. (2006). Influence of the spatial heterogeneity on the non linear estimation of Leaf Area Index from moderate resolution remote sensing data. *Remote Sens. Environ.* **106**, 286–298.

Garrigues, S., Allard, D., and Baret, F. (2008a). Modeling temporal changes in surface spatial heterogeneity over an agricultural site. *Remote Sens. Environ.* **112**, 58–602.

Garrigues, S., Allard, D., Baret, F., and Morisette, J. (2008b). Multivariate quantification of landscape spatial heterogeneity using variogram models. *Remote Sens. Environ.* **112**, 216–230.

Giriraj, A, Irfan-Ullah, M., Murthy, M. S. R., et al., (2008). Modelling spatial and temporal forest cover change patterns (1973–2020): A case study from South Western Ghats (India). *Sensors* **8**, 6132–6153.

Goovaters, P. (1997). *Geostatistics for Natural Resources Evaluation.* Oxford University Press, New York.

Hayes, D. J. and Cohen, W. B. (2007). Spatial, spectral and temporal patterns of tropical forest cover change as observed with multiple scales of optical satellite data. *Remote Sens. Environ.* **106**, 1–16.

Iman, R. L. and Conover, W. J. (1980). Small sample sensitivity analysis techniques for computer models, with an application to risk assessment. *Commun. Statist. Theory Meth.* **A9**, 1749–1874.

Jensen, T. R. (1996a). *Introductory Digital Image Processing: A Remote Sensing Perspective.* Prentice Hall, New York, pp. 116–121, 179-186.

Jensen, T. R. (1996b). *Introductory Digital Image Processing: A Remote Sensing Perspective*, Prentice Hall, New York, pp. 116–121, 179–186,.

Keefer, D. K. (1984). Landslides caused by earthquakes. *Geol. Soc. Am. Bul.* **95**, 406–421.

Keefer, D. K. (1994). The importance of earthquake-induced landslides to long term slope erosion and slope-failure hazards in seismically active regions. *Geomorphology* **10**, 265–284.

Kyriakidis, P. C. (2001). Geostatistical models of uncertainty for spatial data. *Spatial Uncertainty in Ecology: Implications for Remote Sensing and Gis Application*. C. T. Hunsaker, M. F. Goodchild, M. A. Friedl, and T. J. Case (Eds.). Springer, New York, pp. 175–213.

Lee, M. F., Lin, T. C., Vadeboncoeur, M. A., and Hwong, J. L. (2008). Remote sensing asseeement of forest damage in relation to the 1996 strong typhoon Herb at Lienhuachi Experimental Forest, Taiwan. *Forest Ecol. Manag.* **255**, 3297–3306.

Lin, C. W., Liu, S. H., Lee, S. Y., and Liu, C. C. (2006a). Impacts of the Chi-Chi earthquake on subsequent rainfall-induced landslides in central Taiwan. *Eng. Geol.* 87–101.

Lin, C. W., Shieh, C. L., Yuan, B. D., Shieh, Y. C., Liu, S. H., and Lee, S. Y. (2003). Impact of Chi-Chi earthquake on the occurrence of landslides and debris flows: Example from the Chenyulan River watershed, Nantou, Taiwan. *Eng. Geol* **71**, 49–61.

Lin, C. Y., Wu, J. P., and Lin, W. T. (2001). The priority of revegetation for the landslides caused by the catastrophic Chi-Chi earthquake at ninety-nine Peaks in Nantoun area. *J. Chinese Soil Water Cons.* **32**, 59–66.

Lin, G. W., Chen, H., Chen, Y. H., and Horng, M. J. (2008b). Influence of typhoons and earthquakes on rainfall-induced landslides and suspended sediments discharge. *Eng. Geol.* **97**, 32–41.

Lin, W. T., Chou, W. C., and Lin, C. Y. (2008c). Earthquake-induced landslide hazard and vegetation recovery assessment using remotely sensed data and a neural network-based classifier: a case study in central Taiwan. *Nat. Hazards* **47**, 331–347.

Lin, W. T., Lin, C. Y., and Chou, W. C. (2006c). Assessment of vegetation recovery and soil erosion at landslides caused by a catastrophic earthquake: A case study in Central Taiwan. *Ecol. Eng.* **28**, 79–89.

Lin, Y. B., Lin, Y. P, and Deng, D. P. (2008a). Integrating remote sensing data with directional two-dimension wavelet analysis and open geospatial techniques for effective disaster monitoring and management. *Sensors* **8**, 1070–1089.

Lin, Y. P., Chang, T. K., Wu, C. F., Chiang, T. C., and Lin, S. H. (2006b). Assessing impacts of typhoons and the ChiChi earthquake on Chenyuland watershed landscape patterns in Central Taiwan. *Environ. Manage.* **38**, 108–125.

Lin, Y. P., Yen, M. H., Deng, D. P., and Wang, Y. C. (2008d). Geostatistical approaches and optimal Additional sampling schemes for spatial patterns and future samplings of bird diversity. *Global Ecol. Biogeogr.* **17**, 175–188.

Lobo, A., Moloney, K., Chic, O., and Chiariello, N. (1998). Analysis of fine-scale spatial pattern of a grassland from remotely-sensed imagery and field collected data. *Landscape Ecol.* **13**, 111–131.

McKay, M. D., Beckman, R. J., and Conover, W. J. A. (1979). comparison of three methods for selecting values of input variables in the analysis of output from a computer code. *Technometrics* **21**, 239–245.

Millward, A. A. and Kraft, C. E. (2004). Physical influences of landscape on a large-extent ecological disturbance: The northeastern North American ice storm of 1998. *Landscape Ecol.* **19**, 99–111.

Minasny, B. and McBratney, A. B. (2006). A conditioned Latin hypercube method for sampling in the presence of ancillary information. *Comput. Geosci.* **32**, 1378–1388.

Minasny, B. and McBratney, A. B. (2007). The variance quadtree algorithm: Use for spatial sampling design. *Comput. Geosci.* **33**, 383–392.

Pebesma, E. J. and Heuvelink, G. B. M. (1999). Latin hypercube sampling of Gaussian random fields. *Technometrics* **41**, 303–312.

Press, W. H., Flannery, B. T., Teukolsky, S. A., and Vetterling, W. T. (1992). *Numerical Recipes in Fortran: The Art of Scientific Computing, 2nd edition.* Cambridge University Press, Cambridge, UK.

Roger, B. and Yu, T. T. (2000). The morphology of thrust faulting in the 21 September 1999, Chi-Chi, Taiwan earthquake. *J. Asian Earth Sci.* **18**, 351–367.

Sellers, P. J. (1985). Canopy reflectance, photosynthesis and transpiration. *Int. J. Remote Sens.* **6**, 1335–1372.

Sellers, P. J. (1997). Modeling the exchange of energy, water, and Carbon between continents and atmosphere. *Science* **275**, 602–609.

Swanson, F. J., Johnson, S. L., Gregory, S. V., and Acker, S. A. (1998). Flood disturbance in a forested mountain landscape. *Bioscience* **48**, 681–689.

Tarnavsky, E., Garrigues, S., and Brown, M. E. (2008). Multiscale geostatistical analysis of AVHRR, SPOTVGT, and MODIS global NDVI products. *Remote Sens. Environ.* **112**, 535–549.

Thompson, S. K. (1992). *Sampling.* Wiley Interscience, New York.

Turner, M. G. and Dale, V. H. (1998). Comparing large, infrequent disturbances: What have we learned? *Ecosystems* **1**, 493–496.

Urban, D. L., O'Niell, R. V., and Shugart, H. H. (1987). Landscape ecology. *Bioscience* **37**, 119–127.

Ward, D., Phinn, S. R., and Murray, A. T. (2000). Monitoring growth in rapidly urbanizing areas using remotely sensed data. *Prof. Geogr.* **52**, 371–386.

Wiens, J. A. (1989). Spatial scaling in ecology. *Funct. Ecol.* **3**, 385–397.

Xu, C., He, H. S., Hu, Y., Chang, Y., Li, X., and Bu, R. (2005). Latin hypercube sampling and geostatistical modeling of spatial uncertainty in a spatially explicit forest landscape model simulation. *Ecol. Model.* **185**, 255–269.

Zhang, Y. and Pinder, G. F. (2004). Latin-hypercube sample-selection strategies for correlated random hydraulic-conductivity fields. *Water Resour. Res.* **39**, 1226.

Zomeni, M., Tzanopoulos, J., and Pantis, J. D. (2008). Historical analysis of landscape change using remote sensing techniques: An explanatory tool for agricultural transformation in Greek rural areas. *Landscape Urban Plan.* **86**, 38–46.

15

Aktug, B. and Kilicoglu, A. (2006). Recent crustal deformation of Izmir, Western Anatolia and surrounding regions as deduced from repeated GPS measurements and strain field. *J. Geodyn.* **41**(5), 471484.

Akyol, N., Zhu, L., Mitchell, B. J., Sozbilir, H., and Kekovali, K. (2006). Crustal structure and local seismicity in western Anatolia. *Geophys. J. Int.* **166**, 1259–1269.

Ayan, T. (1981). Optimization of geodetic networks. *Associate Professorship Thesis*. Istanbul Technical University, Istanbul, Turkey.

Bagci, G. (2000). Seismic risk of Izmir and its surrounding region. *Proceedings of International Symposia on Seismicity of Western Anatolia.* Izmir, pp. 2427.

Blewitt, G. (2000). Geodetic Network Optimization for Geophysical Parameters. *Geophys. Res. Lett.* **27**(22), 36153618.

Emre, O. and Barka, A. (May 2427, 2000). Active faults between Gediz Graben and Aegean Sea (Izmir Region). *Proceedings of International Symposia on Seismicity of Western Anatolia.*

Emre, O., Ozalp, S., Dogaz, A., Ozaksoy, V., Yildirim, C., and Goktas, F. (2005). *The Report on Faults of Izmir and its Vicinity and their Earthquake Potentials*. General Directorate of Mineral Research and Exploration Report No. 10754.

Genc, C., Altunkaynak, S., Karacik, Z., Yazman, M., and Yilmaz, Y. (2001). The Cubuklu Graben, South of Izmir: Tectonic significance in the Neogene Geological evolution of the Western Anatolia. *Geodinamica Acta* **14**(1/3), 4555.

Gerasimenko, M. D., Shestakov, N. V., and Kato, T. (2000). On optimal geodetic network design for fault-mechanics studies. *Earth Planets Space* **52**, 985987.

Halicioglu, K. (2007). Network design and optimization for deformation monitoring on Tuzla Fault-Izmir and its vicinity. *MSc Thesis.* Bogazici University.

Jackson, J., Haines, J., and Holt, W. (1994). A comparison of satellite laser ranging and Seismicity Data in the Aegean Region. *Geophys. Res. Lett.* **21**, 28492852.

Kaymakci, N. (2006). Kinematic development and paleostress analysis of the Denizli Basin (Western Turkey): Implications of spatial variation of relative paleostress magnitudes and orientations. *J. Asian Earth Sci.* **27**, 207–222.

KOERI Earthquake Catalog (2008). Bogazici University, Kandilli Observatory and Earthquake Research Institute, National Earthquake Monitoring Center, Waveform Data Request System, May 2006. Retrieved from http://www.koeri.boun.edu.tr/sismo/mudim/katalog.asp

Kreemer, C. and Chamot-Rooke, N. (2004). Contemporary Kinematics of the Southern Aegean and the Mediterranean Ridge. *Geophys. J. Int.* **157**, 13771392.

McClusky, S., Balassanian, S., Barka, A., Demir, C., Ergintav, S., Georgiev I., Gurkan, O., Hamburger, M., Hurst, K., Kahle, H., kastens, K., Kekelidze, G., King, B., Kotzev, V., Lenk, O., Mahmoud, S., Mishin, A., Nadaria, M., Ouzouns, A., Paradissis, D., Peter, Y., Prilepin, M., Reilinger, R., Sanli, I., Seeger, H., Tealeb, A., Toksoz, M. N., and Veis, G. (2000). Global positioning system constraints on plate kinematics and dynamics in the Eastern Mediterranean and Caucasus. *J. Geophys. Res.* **105**(B3), 56955719.

McKenzie, D. (1978). Active Tectonics of Alphine-Himalayan Belt: The Aegean Region and Surrounding Regions. *Geophysical J. R. Ast. Soc.* **55**, 217254.

Mercier, J., Sorel, D., Vergely, P., and Simeakis, K. (1989). Extensional tectonic regimes in the Aegean basins during the Cenozoic. *Basin Res.***2**, 4971.

Müller, S., Kahle, H. G., and Barka, A. (1997). Plate Tectonics situation in the Anatolian-Aegean Region, Active Tectonics of Northwestern Anatolia- the Marmara Poly-Project.

Nur, A. and Cline, E. H. (2000). Poseidon's Horses: Plate Tectonics and Earthquake Storms in the Late Bronze Age Aegean and Eastern Mediterranean. *J. Archaeol. Sci.* **27**, 43–63.

Ocakoglu, N., Demirbas, E., and Kuscu, I. (2005). Neotectonic structures in Izmir Gulf and Surrounding Regions: Evidences of Strike-Slip Faulting with Compression in the Aegean Extensional Regime. *Marine Geol.* **219**, 155–171.

Ozcep, F. (2000). Has Central Anatolia A Micro-Plate Behavior Within Turkish Plate? *A Paleomagnetic Discussion. International Conference on Earth Sciences and Electronics (ICESE 2002).*

Ozmen, B., Nurlu, M., and Güler, H. (1997). *Investigation of Earthquake Zones with Geographic Information Systems.* The Ministry of Public Works and Settlement General Directorate of Disaster Affairs.

Papazachos, C. B. (1999). Seismological and GPS evidence for the Aegean Anatolia Interaction. *Geophys. Res. Lett.* **17**, 26532656.

Piper, J., Gursoy, H., and Tatar, O. (May 8, 2001). Palaemagnetic analysis of Neotectonic Crustal Deformation in Turkey. *Proceeding of Symposia on Seismotectonics of the North-Western Anatolia-Aegean and Recent Turkish Earthquakes.*

Reilinger, R., McClusky, S., Vernant, P., Lawrence, S., Ergintav, S., Cakmak, R., Ozener, H., Kadirov, F., Guliev, I., Stepanyan, R., Nadariya, M., Hahubia, G., Mahmoud, S., Sakr, K., ArRajehi, A., Paradissis, D., Al-Aydrus, A., Prilepin, M., Guseva, T., Evren, E., Dmitrotsa, A., Filikov, S. V., Gomez, F., Al-Ghazzi, R., and Karam, G. (2006). GPS Constraints on Continental Deformation in the Africa-Arabia-Eurasia Continental Collision Zone and implications for the Dynamics of Plate Interactions. *J. Geophys. Res.* **111**, B05411.

Reilinger, R. E., Ergintav, S., Burgmann, R., McClusky, S., Lenk, O., Barka, A., Gurkan, O., Hearn, L., Feigl, K. L.; Cakmak, R., Aktug, B., Ozener, H., and Toksoz, M. N. (2000). Coseismic and Postseismic fault slip for the 17 August 1999 M=7.5 Izmit, Turkey Earthquake. *Science* **289**(5484), 15191524.

Saroglu, F., Emre, O., and Kuscu, I. (1992). Turkish Active Faults Map, General Directorate of Mineral Research and Exploration, Ankara.

Schmitt, G. (1985). Review of network design: Criteria, risk functions, design ordering. In *Optimization and Design of Geodetic Network.* E. Grafarend and F. Sanco (Eds.). Springer, Berlin, pp. 6–10.

Segall, P. and Davis, J. L. (1997). GPS Applications for Geodynamics and Earthquake Studies. *Ann. Rev. Earth Planet Sci.* **25**, 30136.

Taymaz, T. (May 8, 2001). Active Tectonics of the North and Central Aegean Sea, Proceeding of Symposia on Seismotectonics of the North-Western Anatolia-Aegean and Recent Turkish Earthquakes.

Türkelli, N., Kalafat, D., and Gündodu, O. (1995). November, 6, 1992 Izmir (Doganbeyli) Earthquake, Field Observations and Focal Mechanism Solutions. *Geophysics* (in Turkish).

Türkelli, N., Kalafat, D., and Ince, S. (1992). After shocks of November, 6, 1992 Izmir (Doganbeyli) Earthquake. Bulletin of Earthquake Investigations.

Wu, J., Tang, C., and Chen, Y. Q. (November 37, 2003). First-order Optimization for GPS Crustal Deformation Monitoring, *Proceedings of the 7th South East Asian Surveying Congress*, Hong Kong, China.

Yilmaz, Y. (2000). Active tectonics of Aegean Region. *Proceedings of International Symposia on Seismicity of Western Anatolia.* pp. 2427.

Zhu, L., Akyol, N., Mitchell, B. J., and Sozbilir, H. (2006). Seismotectonics of Western Turkey from high resolution earthquake relocations and moment tensor determinations. *Geophys. Res. Lett.* **33**, L07316, doi:10.1029/2006GL025842.

Index

D

F

H

I

J

K

L

M

N

W

Y

Z